About this booĸ

- This is an almost 98% true story of how mineral exploration was done in the 2nd half pf the 20th century, dating back to 1960.
- It is partly autobiographal, but I am a bit player and use my experiences as a post to anchor other sub-stories. The main characters are the bush, the equipment used and the people I have associated with – most of them <u>real</u> characters, if you catch my drift.
- Although mineral exploration still continues into the 21st century, it is not quite the same these days.
- Sit back, dust off your own memories and enjoy.

I Call Myself a Prospector
Book One

By
Bob Durnin
With
Frank Durnin

Prospectors' Burro - Boreal Forest Sub-species

Acknowledgements

AKA
Thank you list

To Carl Haglund and all those who read the first pitiful efforts, offering encouragement and useful input.

To Rick and Nicole who supplied me with a spiffy core shack to work in. - They kept the woodpile topped up and never once let me down.

To Ed Tetu, who believed in us.

To Bob Marvin, who had faith in a member of the Over-The-Hill Gang.

To Bell Canada & Virgin Mobile - 99% of the work was done with the two authors separated by 800 miles (1300k.)

To Bill Gates – couldn't have done it without ya.

To my co-writer – a published author – who gave me adjectives when I needed them.

To Janet, who stepped in when Little Brother got typist's cramp.

****

And to all those whose spirits still walk the Interlocking Olympic Rings

– our snowshoe trails will cross again someday.

****

Foreword

Post WWII, International Nickel, (Inco) already well-established with mines, mill and smelter in the Sudbury Basin, began to seek new projects. Using in-house expertise and knowledge gleaned from (for example) sub-hunting airborne magnetometer development, Inco began over-flying known mineral occurrences in greenstone belts in the Canadian Shield.

One promising target in Manitoba was at Moak Lake. A shaft had been sunk and a company compound had been built, when, in 1956 a diamond drill hole on another showing 20 miles south at Cook Lake pulled a bonanza sulphide intersection with high nickel assays.

With this new zone rapidly proving to be the centre of attention, and with a massive claim staking, line-cutting, and geophysical program underway, the Moak Lake compound was converted to an exploration centre, housing Canadian Nickel staff. (Canico – Inco's Manitoba exploration arm.)

By 1960 the claim group extended from Moak Lake 90 miles south to Setting Lake, and was up to 20 miles wide.

With so much ground to cover, jobs were available for northern Manitobans and farm boys from Saskatchewan to Northern Ontario.

This was and is the Thompson Nickel Belt.

Note: This book has many terms and names unfamiliar to the average reader. Rather than slow down the flow of the story, they have been explained in the glossary.

Prologue

How does one describe one's profession in a few words? Some jobs are self-explanatory, others not so much.

At Christmas in 1960 I came home from Thompson on a three-week break, my first significant holiday after six months in a tent. Now I re-entered civilization, as it were, running into old acquaintances and schoolmates at a party or two. Of course the conversation often came around to "What are you doing now?

It was an easy question for everyone else to answer. One might be upwardly mobile in a bank, some might be pursuing higher education, and others would already be well into their livese at the local paper mill – stuff like that.

Me, I was a geophysical operator - and what does a geophysical operator do?

I would try to explain, eyes would glaze over and other conversations would start up before I was finished. Worst of all was that a good-looking chick worthy of being impressed, would just turn her attention to a much more interesting articling accountant.

Something had to be done and when I returned to Thompson for the winter season I studied on the problem. Over spring break-up, once again back in the land of flush toilets, I played my Ace. I told a girl I was a prospector – and just like that, I was a star!

So there you have it. Blame it on hormones if you will, but the die was cast many years ago.

And I still call myself a prospector.

.
*****

Contents

Chapter I   Thompson 1960

It is early June and I am in a passenger car jiggling and swaying over the Churchill Line permafrost on my way to a new job. I have little idea what lies ahead, only that it sounds romantic. I know three things for sure. One is that I will be training as a geophysical operator (whatever that is.) Two is that I will be living in a tent. Item three, once in the bush I will be expected to stay until Christmas. This might be daunting to a more timid soul but I welcomed the challenge, and I had packed accordingly.

I took every stitch of work clothing I owned and topped up with some of my older siblings' stuff. My duffel bag was full of pants, shirts, socks and what-have-you, including a pair of leather work boots and a pair of rubber boots. (These barely survived that first summer and had to be replaced by more bush-friendly footwear in September.)

I travelled in jeans and a casual shirt with an extra set in my dad's old Gladstone suitcase. Should I be invited to a ball I included my penny loafers. It's a good thing my mom helped me pack. She insisted on long-johns despite my pleas that it was summertime – moms know a thing or two.

Mom also gave me a "soldiers' friend," which she made herself. It was fabricated out of recycled denim with little pockets containing needles of all sizes, scissors, tweezers, spools of thread, and even a stash of yarn to darn holes in wool socks. Also – being Mom – she made sure I had a pillow, a blanket, and writing paper and envelopes for those letters home with instructions to write often – which I promised to do.

I bussed to Winnipeg and I had just enough time to hop aboard the westbound for the 24-hour trip to Thompson.

In 1960 there were only two ways to get to Thompson. One could charter a LambAir floatplane at the Pas, or ride the CNR all the way. I was paying the shot, so I rode the rails.

When construction kicked off post-1956 a spur line had been built running northwest from Sipewisk to Thompson. An airport was planned, as was a highway, and both had been cleared of trees, but Inco was in no rush to finish them just yet. They had guaranteed X ton/miles of freight, and X passengers, and why would they compete with themselves? The airport would be finished in 1962 and the highway a few years later.

I fall asleep in my coach class seat and when the train is split in the wee hours at Hudson Bay Junction my dreams are interrupted as more people enter my car. The trans-continental will continue on to Vancouver – four or five cars head up the Churchill line. When we leave the Pas, just after sunup, the cars are packed. (I was later told that Thompson had three shifts; one coming in, one working and one going out.)

I detrain at the Thompson station and am met by Wray Dayson, Canico's Thompson expediter. I think I am getting the VIP treatment until I find out that Wray meets all the trains. There is always freight of some sort destined for Moak Lake and I am freight, and have been upgraded to fast track. Wray whisks me past houses and businesses under construction, through a security gate to the mine site, past more construction and on down to the Cook Lake dock where a Cessna 180 waits.

I meet Keith Olson, the pilot, and without further ado I am on the plane and we are taxiing down to the end of the lake.

He tells me to buckle up and asks me if I have ever flown before. I tell him I haven't, and he points out the location of the barf-bag, should I need it. Keith turns the LambAir 180 into the wind, pours on the power, we lift off and bank north towards Moak Lake. It is my first flight of many and I can still recall every second.

8

It is only a 15-minute flight to Moak Lake. It is perhaps 3 pm on a warm June afternoon, with only a few scattered, puffy clouds. Below me is an expanse of different shades of green, interspersed with lakes in shades of blue. A little creek meanders through a green swamp, and we cross the Mystery River. Keith is a good tour guide. He tells me Mystery Lake empties into the river, which joins the Burntwood down towards Thompson – what magical names! He also points out parts of a bush road not used in the summer months. It is a winter road link to Moak from Thompson.

The old water tower at Moak. Note the all-wood frame construction. And how in the world did they get that tank up there without a huge crane?

One thing that surprises me is that although it is a relatively calm day, the ride is bumpy. I will be clued in on thermals tomorrow.

Moak Lake comes into focus through the spinning prop. The lake is quite large and the Canico compound looks like a postcard tourist camp. We tie up at the dock and I carry my stuff to the office. Keith stays behind to fuel up – another routine I will learn. Gas up after landing – not just before you leave.

Canico headquarters at Moak looks like a village to me and I am told that it was the beginnings of a town before the hit at Thompson. From the beach a broad gravel road leads upgrade and I can see the upper half of the mothballed headframe partially hidden behind the hill. To the right on top of the hill sits a water tower. Also lower down on the right is a large shop with other smaller buildings nearby. One is a light plant – Moak has 24/7 power. Others, I will learn, are semi-permanent tents over frames and serve as core shacks and storage areas.

Just to the left of the dock is a large, neat, frame building. The front half is divided into office space – one room holds the two-way radio telephone, the sole communications system in Moak at that time. Other spaces have desks, and one larger area contains map tables and drafting boards. The rear half of the building is used for single men's accommodation with semi-private double-cot rooms. There is a large common room with a card table and chairs. There is no TV of course, but they have a shower and flush toilets.

Not far from the office door is a kitchen/dining room building. This is where we single men eat our meals and have milk and cookies before bedtime.

West of the office/ bunkhouse are three or four houses set back a bit from the lake. These are for married upper-echelon personnel – my supervisor, two geologists, and the road and maintenance boss. The whole deal on the north shore of Moak Lake is on a gentle sandy slope peppered with spruce, jackpine, and some broadleafs. There is no underbrush and little grass. The ground is covered with short Labrador tea and everything is as neat as a pin.

I will not spend much time here after today but I have described Moak in detail for two reasons, one being that it was such a marvellous place to hang one's hat.

The other reason is that in the early '60s Moak was on its own with no connection to Thompson other than Wray Dayson and the Cook Lake dock. Sure – Midwest Diamond Drilling had their own tent camp near Cook Lake, but it only served to shuttle their own men in and out, and as a parts/repair depot. Moak directed all the drilling and I was told that for every man in Moak or at Cook there were ten in the bush. Thompson was mining – Moak was exploration – and in 1960 Moak Lake ruled the bush.

The two hours following my arrival in Moak are hectic (at least for me.) I meet my supervisor and he tells me that I will be going to Setting Lake tomorrow to join Charlie McLeod Jr's crew. He also tells me I will be paid $260 per month including free room and board. I like the 260 part and resist the impulse to tell him he had me at free grub.

Then he passes me on to Ralph Johnson, who is from my hometown and the only familiar face in the bunch, and Ralph explains the drill.

My once-a-month paycheque is issued from Moak and Ralph has authority to open an account for me at the Royal Bank in Thompson. My cheque after deductions will be deposited and I will get my pay stub from Moak and monthly statements from the Royal. So far so good – but how do I withdraw cash? Ralph says that if I need cash he will send me some and take it off my cheque, and the Royal will send me cheques to pay other bills by mail. I can also buy stuff from Eaton's Mail Order – they have a depot in Thompson and the order comes in C.O.D. Wray pays for it out of petty cash and Moak takes it off my cheque – it's a well-oiled machine. Now I need some other stuff.

The land survey crew often used a chopper. Note the Bombardier fleet on wheels awaiting summer tune-ups. The one on the left would not look out of place on the streets of New York - Auto-Neige must have run out of blue paint.

Stubby Bjornson has an eiderdown sleeping bag he doesn't need. I get a half-price steal – a hundred bucks – this will come off my first cheque.

10

Now I am issued a Silva Prospector fluid-filled compass, also a payroll deduction (ten bucks? Can't remember.) I hang the compass around my neck and try it out. When I slide the needle thingy out of its case a little mirror drops down. Ralph tells me that is so I can look at who's lost (haha.)

I get my choice of axes, and I pick a two-tone with automatic transmission. The axe is free – right – I am still paying for that axe!

I am shown my room and I just have time to stow my gear when the steel triangle beside the cookhouse door clangs. It's suppertime and I meet the best cook ever – bar none.

The dining room is actually two rows of tables and benches. It is spartan, but hospital clean, and though there is room for more than twenty only a dozen or so are here tonight. Two tables are loaded with mashed potatoes, gravy and a choice of veggies. Places have been set, and I wait until others are seated before choosing a spot. I catch on fast – this is pork chop night and the chops are on warming trays on the huge oil-burning cook stove. Tom (the cook) is standing behind the stove. He points out that I have the choice of medium-rare to well-done. Tom is a friendly sort.

After supper we retire to the common room where I continue to listen and learn, and over the years there have been many listening-and-learning common rooms. Whether at a single-crew cookhouse table, a cup of tea on a stump at a noon campfire, or an impromptu meeting in a bar, they all have one thing in common: they are places to tell tales of success and derring-do. Stories become legends, heroes become superheroes, and villains are villianized. Most of the stories are uplifting, some are downright funny and the downers, while sad, have their own heroes. So that first night at Moak I listen and file, and eventually in the coming years I will repeat these tales, adding some embellishments and a few new stories of my own.

At 7pm I am invited to listen in on the "Sked" in the radio room. This is a regularly scheduled contact with outpost camps. Daily progress reports are noted and additions to weekly grocery orders might be necessary. One camp needs an outboard rewind rope – another needs some canoe repairs. If a man is sick with the flu it is also noted as a reason for lack of production. It's like a big party line – every camp has a radio and everyone gathers around to get the news. Camps are contacted in rotation by their call sign. For instance: "GYP-49, are you by?" "49 here – go ahead, Moak." At the end of the conversation they sign off – "That's QRU from 49." I am told that QRU is ham radio talk for "No further transmission."

Setting Lake is contacted. Charlie McLeod Jr's progress is not as impressive as the rest. When he is told I will join them tomorrow he sounds relieved.

With the geophysical camps all entered in the log the radio is passed on to the Moak Lake geologist. He will now get his own progress reports from the Midwest diamond drill camps. It is pretty boring stuff and I don't need to clutter up my already overloaded brain cells. I go back to the common room feeling better about one nagging worry: I may be in semi-isolation this summer, but the radio binds us all together.

It's already evening coffee-and-tea time so I follow the guys back to the cookhouse. Again I am faced with a plethora of food – leftover chops are back on the warming

11

trays, bracketed by apple, raisin, and lemon meringue pies that have survived supper and need to be eaten. A big pot of coffee and a pot of hot tea water are simmering and though I try, I can't handle more than two pieces of those perfect pies. We don't have to wash our own dishes – a cookee will do the job. These Moak Lakers are spoiled, I'm thinkin', and we trail back to the common room for another BS session.

It's a history lesson of sorts. I am fresh ears, and everyone has a story to tell. It's not a hazing ritual by any means – just a "welcome to the club" sort of thing. (And, by gosh, they did have an ersatz club. Next winter I will purchase a club jacket – heavy black corduroy with white leather sleeves. On the back it says "Canico Geophysics," on the front "Moak Lake." One shoulder patch holds my first name – the other a crest – "Club 55 ½." This indicates that Thompson is halfway between the 55th and the 56th parallel. Cool, huh?)

Common Room - Moak Lake.
Irwin Wilson rakes in the pot while Lyle Fennell studies his dwindling reserves. Maybe Neil McAskill shouldn't be in the game - he needs both hands to hold four cards.

So the boys make sure I know about important things in the North.

I am told about Vic, a man who was as strong as he was lazy. I hear about how a young, freshly graduated geophysicist, proud of his new degree and unearned authority, caught Vic napping in his tent one afternoon. He read Vic a few chapters until Vic picked the young man up by his collar and hip pockets, used him as a door opener and tossed him halfway to the lake before resuming his nap. (I worked for that geophysicist the following year, and as far as I was concerned a daily toss should have been mandatory.)

I learn about Ed G, famous for his go-for-broke work and play ethic. Ed had transferred back to the Sudbury Basin before I arrived at Moak but I will work and play with Ed in the future, and I liked him. Heck – I liked him before I knew him.

Ditto for ultra-famous Henry Levac, whom I will also meet within a couple of years. Added to his supreme work ethic was a hair-trigger temper. The year before, Henry, at Setting Lake at the time, was also caught napping one afternoon. The thing was that he had a good reason for doing so – he was as sick as a dog. When the Beaver tied up at the floating dock Henry hauled himself down to meet it, only to be harangued by H, the super at that time. Henry didn't hear much of the sermon before he performed his own body toss – this time directly into the lake. Man! What great stories!

The last tale of the evening is a sad one. A young bush pilot, socked in by bad weather, was determined to get his load of Christmas presents into a reserve. When the sky cleared after sunset he took a shot at it, and never arrived. The reserve built a monument honouring that young man and as far as I know it still stands – a sad but true story.

It is bedtime. As I crawl into my first comfy eiderdown and my first bed in 48 hours I couldn't be happier. My first day of my dream job has been a real-life dream and I sleep a dreamless sleep.

When my travel alarm goes off, I am already awake. Breakfast is at seven sharp, and today's weather is a repeat performance. Summer solstice is only ten days away and at 55 ½ degrees the sun has already been shining since 4:30. I go to the cookhouse expecting a re-run of yesterday's supper drill.

There are no pancakes or slices of toast on the warming oven, and I see no eggs or such, but Tom is standing behind the stove again and other men are already digging in. I pour myself a coffee and Tom asks me for my breakfast order. This startles me – I had not expected personal service, and I am still the slightly shy greenhorn. I mumble that whatever he has is OK with me. Tom is impatient, and pretending to be stern, he tells me he hasn't got all day – spit it out, lad!

Well, I figure never say "Whoa" in a mud hole, so I jump in with both feet. I tell him I want two slices of toast, with bacon on the toast and two sunny-side eggs on top of the bacon.

Tom doesn't bat an eye. He tells me to sit down and mere minutes later my perfect dream breakfast sits in front of me. (I returned to Moak in late November to join an anomaly-chasing crew. At supper I said "Hi" to Tom, and the next morning I was greeted with exactly the same breakfast he had cooked for me 5 ½ months ago – can you beat that!?!)

(Tom, like many camp cooks, was a loner and a part-time alcoholic. He preferred to stay in the bush and support his family from a distance, but when the vanilla extract order exceeded potato poundage, Tom would be sent to dry out. I was told that he was on a twelve-and-two cycle – twelve months in – two months out.)

Shortly after 8 am I leave Moak Lake in the LambAir 180 with Keith for my one-hour flight to Setting Lake. On board along with me, my suitcase, and brand new Woods Tent and Awning #9 packsack, is a grocery order for the camp. Usually this stuff is flown in on a weekly grocery run with Norm Kearn's Beaver servicing more than one camp, but today I have my own personal aircraft and chauffeur. I get to ride shotgun – enjoy it Pal, it won't happen too often in the next few years.

The packsack is also mine, and will also be deducted from my first cheque. There are no flies on Canico, they know that if I own things such as compass, eiderdown and personal packsack that I will take care of them. All other stuff related to work is supplied by the company.

It's a nice, clear, hot day and the view is superb. Once again I am surprised at how rough the flight is – not really rough, but from time to time we bounce around. The sun has already been up for hours and Keith explains the concept of thermal air currents to me. It seems that as we pass over the lakes, the air, heated by the sun reflecting off the water, rises, and as we move over the cooler forested areas, the air cools and drops. I

13

am probably not telling you anything you don't already know, but this is new to me. Those airplanes flying overhead had always seemed to look like a smooth ride from my point of view on the ground. I will later learn that the smoothest flights are on overcast, quiet, even drizzly days.

Keith tells me it will get even bumpier as the day warms up and again mentions that

there is a barf-bag in the pocket on the back of the seat if I need one. No problem - I never will have tummy troubles while flying, although things can get a little queasy if someone else in the plane is selling Buicks.

Keith does not fly in a straight line to Setting Lake. He does not stray far from the path, but Keith is a tour guide. He shows me the twenty miles of #6 Hwy under construction south of Thompson. By this fall it will be up to grade and gravel will follow. On southward the right-of-way has been cleared to Sasagiu Rapids at the north end of Setting Lake.

View from the right-hand seat on any given Manitoba summer day.

We catch glimpses of Canico's muskeg tractor/winter Bombardier road network and fly over a spot where bush has been cleared beside the road. It is a service station – rows of oil drums stacked three high. Keith tells me they had been brought in by cat swing a few years before. There are diesel fuel, stove oil and gas barrels – I estimate more than 200. He says the empties are returned to Thompson in the winter months. As we fly overhead a tiny man is loading tiny barrels onto a tiny muskeg tractor. He gives us a tiny wave and gets a wing-waggle in return.

At Setting Lake, Charlie, Jacob and Willie are on the dock, and I get a warm welcome. We off-load the 180, and watch Keith taxi out and take off. I will learn that this is proper camp protocol – you always watch the plane leave. I will also learn that this is usually a seasonal protocol only. In the wintertime you get the heck out of there. It's cold enough already, and you don't hang around to get blasted by snow and ice pellets from the prop-wash.

We carry the grub order up to the kitchen and leave it for Willie to unpack. I go back for my stuff and Charlie shows me what will be my home for the next three months. It looks pretty hospitable to me and it really is a good camp. In fact, as I look back over the years, that Setting Lake camp was the best of the bunch, rivalling even the best of hotel/motel rooms, and lacking only a three-piece bath. When the Moak Lake camp construction crew built it, they did themselves proud.

14

WILLIE AND THE KITCHEN AT SETTING THE TALL POLE AT THE FRONT CORNER HOLDS WILLIE'S RADIO AERIAL. BUNK TENT OFFSTAGE LEFT

I set my alarm clock, put it on my orange crate, and it turns out to be totally useless what with Willie, his dishpan and wooden spoon only a few feet away, and after the first morning I never set that alarm again. The clock only tells me when to go to bed. Willie tells me when to get up.

I stash my toothbrush, shaving kit and writing materials in the orange crate and slide my suitcase under the bed, where it will sit, seldom opened for the rest of the summer. I take my trusty old army blanket out of my packsack, neatly tuck it in around the mattress, pull my eiderdown out of its carry case and roll it up at the head of the cot. Fifteen minutes and I am moved in: will life ever be so simple again?

The camp sits on a point on the west side of Setting Lake about five miles south of its northern end. The lake is quite large, running north-south perhaps 25 miles. At camp the east shoreline (only four miles away) is relatively straight, with the town of Wabowden on the Churchill Line near the south end. The west side of the lake is more irregular, and two miles south of our camp it curves west into a broad bay. The south boundary of the Nickel Belt claim group is just north of this bay.

Two tents on wood frames had been erected at camp just after spring break-up. I learn that the Moak Lake construction crew had built them as well as a couple of other similar camps. These semi-permanent tents are almost house-like, with plywood floors on sills and with six-foot walls. Four by eight plywood sheathes the first four feet of wall and a Woods Tent and Awning 14'x16' tent is held in place by the walls and 2x4 rafters. The canvas walls extend halfway down the 4x8 plywood and are secured by wood strips. Above the tent a canvas fly is draped over a second ridgepole and tied off at the sides on "squaw poles," leaving a six-inch air space between itself and the tent roof. With two separate layers of canvas even the hardest downpour will not get through to us.

Inside are four steel cots with mattresses – one in each corner. There is a door in the centre of the south wall. Inside the door to the right is a 2'x2' washstand which holds a water pail and a tin washbasin. Just above the washstand is a small window, and beneath the stand sits a clean, empty five-gallon oil pail. This is for our used wash water and is carried back to the trees when necessary – no tossing water out of the door.

15

In the centre is a 24" "Airtight" tin stove for the cool nights. In the middle of the back wall is a map table which looks a bit like Bob Cratchit's office desk. It is four feet wide and built on a 2x4 frame with a shelf below. The top is ¾" good-one-side plywood and is covered with heavy vellum to render a smooth surface. A narrow 1" x2" strip is nailed along the lower edge to keep pens and pencils from rolling off the slanted table. The mapmaker sits on a nifty hand-built stool and the shelf beneath holds rolls of maps. This deal, along with the kitchen table are the two most expensive and important pieces of wood furniture in camp.

The kitchen is an identical 12x16 tent. A 4x8 table with two benches sits to the right of the door, a work counter and propane range to the left. The cook sleeps in the north end and the rest of the space is used for shelf storage. There is a 45-gallon water barrel, which we keep topped up for the cook. The two tents sit about 100 feet back from the point and perhaps fifteen feet above the lake. From the tents the land slopes gently downward and the point has been cleared of trees and brush. This allows the prevailing summer breezes to keep the camp bug-free. Behind the tents to the north is thick forest, which serves two seasonal purposes. This summer an 8x10 tent in the shade of these trees holds our perishables – there is no fridge, freezer or light plant in camp. In the coming winter the forest will protect the camp from cold northern winds.

THE 18'FREIGHTER TIED TO THE FLOATING DOCK. A LITTLE BAY ON THE WEST SIDE OF OUR POINT OFFERED SOME PROTECTION IF SETTING DECIDED TO ROLL.

On the west side of the point, downslope from the bunk tent, is a 15'x15' floating dock. Nearby is a barrel of mixed gas. (Since our five-horse Evinrude short shaft and our EM motor are both two-cycle we have no need for unmixed gas.)

Another small tent behind us serves as miscellaneous storage. In here we keep a ten-gallon drum of white gas (naphtha) for our gas lamps.

My bed is in the northwest corner and I even have a night table. It is an empty orange crate and has a middle shelf – cool. Above my bed, five-inch nails in the top wall plate serve as a clothes closet. I hang my packsack on one end and the others will hold my work clothes and evening clothes. I will learn that we always change before supper. Our pants are often wet below the knees, sometimes our socks are wet also, and of course, our work shirts get sweat-drenched. Suppertime in camp is always black tie. (The five-inch nail is far superior to a clothes closet. In coming years I will often work from a hotel/motel room and I will always have the urge to bang a few spikes into the drywall.)

16

Jacob's bed is in the southeast corner and he has an orange crate also. His area is uncluttered – Jacob is a simple man with few wants or needs.

Charlie's bed is in the northeast corner and his night table is the two-way radio, so Charlie has built a shelf to hold his stuff.

The fourth bed is not in use and must be for visitors.

So that was my first camp and it set the bar for camps to come. I will live and work out of many camps in the years ahead, but none will equal Setting Lake for comfort, surroundings and fellow workers. How lucky can a young man be!

I have already been introduced to the three men I will spend the summer with and after I stow my gear it is time for an early lunch. This is an opportune time to find out more about each other.

Our party leader is Charlie McLeod Jr. from Cross Lake. Charlie is a tall, Jimmy Stewart kinda guy – quiet, competent, and as I was told at Moak, is held in high regard by everyone. He is in his late 30s and has a family in Cross Lake. He is saving his money to build a restaurant/poolroom and will turn his dream into reality next year. (Charlie built it himself out of logs. He hauled in pool tables and such over the winter road from Wabowden and did the whole ball of wax with no government subsidy.) Charlie also has a well-developed, although subtle sense of humour. It will bite me more than once before this year is done.

Jacob Brightnose, the axe man, is from Nelson House. He is illiterate, but that is no big deal. Jacob knows his job and does it well, rarely waiting for orders. He laughs a lot and gets a little homesick at times, but I will never ever see him or Charlie get angry.

Willie Chartier, the cook, is from Churchill. He is also a happy guy, but can get irritated at times, as I will find out. Willie is a widower, sends his cheque to his daughter in Churchill, and is a northern historian, as befits a man nearing 65. One thing I notice after a week or two is that Charlie and Jacob help Willie quite often. We keep his water barrel full, and both my tent-mates will dry dishes for him. When I find out that Willie is a TB survivor and has only one lung I take my turn on the dishtowel. Also, in the evening when Charlie and I are doing map-work and such, Jacob will spend an hour or two visiting Willie in the kitchen. Of course, at 9pm it's common room time, with coffee, pie and conversation around the kitchen table. Pretty tough life, Huh?

(And get this – I am the only white man in camp. To illustrate how times have changed – not only did anyone not have a problem with that, but I never once thought it could have been a problem.)

After lunch Charlie shows me stuff while Jacob putters around down at the dock and helps Willie with a kitchen shelf renovation. Although I will not hit the bush until tomorrow morning, and although it is all Greek to me, I have to be at least a little familiar with the gear and our purpose for being here.

First Charlie shows me the EM (Electromagnetic) gear, which consists of:

A: A motor generator mounted on a Woods packboard. The motor, a small two-cycle single cylinder, is coupled to a small generator, which feeds through an equally small rectifier box and produces a signal of 1000 cycles per second. By using aluminium in certain parts, the weight of this unit is kept to 75 lbs.

B: A transmitter coil, made up of concentric rubber sheathed copper wire wrapped with rubberized tape. When attached to an aluminium pipe unipod, the triangular coil measures fifteen feet on each side. In transit it is carried in the ubiquitous #9 Woods packsack, which is also mounted on a pack board – weight – fifty pounds.

C: The unipod: Three sections of aluminium pipe fit together to be used as an upright mast. At the top three light ropes are attached. These are used to tie off on trees or bushes to hold the mast vertical. There is also a small pulley and rope at the top section to hoist the transmitter coil. Another three sections of smaller aluminium pipe serve as a spreader bar, forming the triangle. Two wires from the transmitter coil are plugged into the rectifier box. When dismantled, the unipod becomes a bundle of six poles wrapped with the ropes and is easy to carry on a shoulder or under an arm. Weight – insignificant.

D: Last but not least is the receiving coil. It is about the size of a large steering wheel and a crossbar holds an inclinometer. When held level the inclinometer needle is on zero. Tip it left or right and the pendulum needle indicates degrees from horizontal. A wire plugged into this coil leads to a small battery pack kept in one's hip pocket, and from the battery pack another wire leads to a set of earphones.

So there is quite a lot of stuff to carry for two men, and if I am on the crew the load can be distributed more equitably.

Note to readers:
The following three or six pages deal with many things I had to learn in a short time span and may be boring for some. Therefore, management has agreed to issue a free pass to those who may want to skim through this section.

Me reading the A-2 mag (winter '61.) We are checking the gear before starting the season. The motor/gen is on the left, the transmitter coil is laid out on the snow to the right. We are checking for signal, so we don't have to hang the coil. The bunk tent is behind me, the kitchen is offstage right. The camp fuel dump is in the background.

Now I meet the Sharps A2 magnetometer. (a.k.a. mag.) This will be my main instrument this summer, so once again I pay close attention. The A2 operates kind of like the scale used when your dad bought a pound of nails at the hardware store. Instead of a steel beam and steel balance point, the A2 has a quartz fulcrum with a quartz bar balanced on it. On the top of the bar is a tiny scale divided into values of 100 and with smaller marks representing tenths. Instead of a bucket to hold the nails,

18

the end of the quartz bar has a magnet which interacts with the natural magnetic pull of the rock type lying beneath the unit – the more magnetic the rock, the stronger the pull on the magnet. The whole deal is about the size of a Super-8 movie camera and mounts on a tripod with adjustable legs.

To take a reading you first plant the tripod firmly on the ground, levelling the mag with a tiny built-in bulls-eye bubble. Now you orient the head (which swivels on its base) to face north. To do this you unlock a little compass needle (also built in,) swivel the head to line up the compass needle and re-lock the compass to neutralize it. (Leaving it unlocked can affect your reading.) Next, using a larger knurled knob on the side of the head, you release a locking button with your thumb and carefully lower the quartz bar onto the quartz fulcrum, wait a second or two for it to balance, and read the scale on the crosshair. Now you lock the quartz bar back in travel mode, record the reading in your notebook, pick up the mag, fold the legs and chug on down the line to the next station to repeat the process. You are always aware of your surroundings and you also make notes regarding shorelines, topography and outcrop – all pertinent info to be included on the map. If you cross a claim line or spot a claim post, you are to make a note. (Leave out that pile of moose droppings, but keep your eyes peeled.)

You also do base station checks. You have a base station in camp to check every morning, and on the grid you check a base line station periodically to control diurnal drift caused by sunspot flare-ups and their effect on our atmosphere. (Never happens at night – ergo diurnal instead of nocturnal.)

Back at camp you have a couple of hours of homework to do in the evening. First you correct your readings as to drift, then you convert the scale to a gamma value. We have established a background in a relatively un-magnetic area (i.e. granite) of just over 1000 gammas, and anything more than 200G above background is considered anomalous. As you get more experienced on the survey you learn to recognize these areas by your readings –if they are higher than 200G above background you fill in 50-foot stations to nail things down a bit. You'd better learn to do this; otherwise Charlie will send you back the next day to fill in the detail.

So the whole thing sounds complicated but it comes together in my mind in a few days. I know where north is, which speeds up the little compass regimen, and I learn to plant the tripod more effectively to minimize levelling. I learn to carry readings in my noggin, stopping every three or four hundred feet to record them. I also learn my noggin limitations – trying to remember more than four readings plus topography can cause a short circuit and I have to walk back and re-do.

After a few evenings of turning my notes over to Charlie, he teaches me to do the map entries. Then a few days later I am introduced to contouring – essentially the same concept as contouring elevations and I contour the mag readings at 200 gamma intervals. Curses – the more I learn the more I have to do. I've always hated homework, but somehow it's easier to handle when you're getting paid to do it.

So, I will become a capable, polished mag man. I will expected to do a minimum of two line-miles a day, and I usually meet my target and more, but long walks will sometimes lead to lower production.

Before we hit the bush I am taken to the map table and given a tutorial on the grid system. A cut grid is basically like a highway with crossroads at regular intervals. The main highway is the baseline and the crossroads are the picket lines. The baseline direction (azimuth) is determined by the general strike of the geology in the area. Because the structure of the Thompson Nickel Belt is generally north/south, we will be working with north/south baselines. The baseline may not be true north – the Setting Lake baseline runs 15 degrees east of true north, but for our purposes it is N/S.

On a basic grid system the baseline is designated as zero (0) or 0BL. It is cut wide enough to be able to use a surveyor's transit to keep the line straight. Then the baseline is chained (measured) using a 100' steel chain (measuring tape) and pickets are planted every 100 feet, taking care to keep all pickets in a straight line.

Ok, let's assume we want to cover a claim group three miles long and two wide, we start our baseline at 0 and extend it three miles north. As we chain and plant the baseline pickets we mark the chainage on each picket face, dropping the zeroes. Therefore, the first picket (where we started) is 0BL/0 and the next is 0BL/1N, the next is 0BL/2N, and so on until we have 15,840 feet of baseline cut and chained. Our last picket, if the claims have been accurately staked, is 0BL/158+40N.

SHIELD GEOPHYSICS LIMITED					
VLF Survey					

AREA_____ PROPERTY_____

OPERATOR_____ DATE_____
LINE NO._____

STATION	TRANS STAT			TRANS STAT	
	DIP ∢	F.S.		DIP ∢	F.S.

SHIELD GEOPHYSICS LIMITED					
Magnetometer Survey					

PROJECT_____ DATE_____

OPER._____ INST._____ SENS._____ S/S.D.

STATION	TIME	READING	DRIFT	CORRECTED READING		TOPO. NOTES

Inco's grid pattern generally used cross-lines at 400-foot intervals, so using the transit, we turn 90-degree angles at each appropriate picket – 4N, 8N, 12N etc. We cut about 50 feet of picket line with at least two pickets in a straight line. Thus, along with the baseline picket, we have a good line established to keep our cross-line straight.

We don't cut our cross-lines as wide as the base line. We just want to clear enough brush, branches and trees to allow us to check our back pickets for line accuracy. We do brush out a path through willow or alder thickets – after all, we have to be able to walk

20

the line winter and summer, carrying bulky equipment. In open areas we blaze trees along the line. When working we will have to cross between the lines from time to time, and it is easy to miss a line if it is not well marked. We may have a claim line to tell us when our line is complete, but a good line cutter always knows roughly where he is, and will make sure he has cut far enough to complete the line. When chaining it is a pain if we have to cut a couple hundred feet at the end.

Two men chain each line, planting pickets at 100-foot intervals and marking the chainage on the picket faces as we proceed. Let's say we are chaining the next cross-line west – this is the 4N line, and the first picket will be marked 4N/1W – got it? – 400 feet north and 100 feet west of 0/0. This goes on thusly – 4N/2W, 4N/3W and so on to the claim boundary. We use the same pickets the line cutter used to keep his line straight. We toss any extra pickets aside – we don't want strays hanging around to confuse the issue. When we are done we will have 40 or 41 cross lines extending a mile on either side of our baseline.

If you understand the grid system you can't get lost, as long as you are inside the perimeter. If I am at 32N/15W, I know I am 3200 feet north of, and 1500 feet west of 00.

Of course the lines to the east are done the same way as we go along. When completed, the grid will total 80 – 82 line-miles, roughly equivalent to 80 claims in total area.

Note: in an actual, on-the-job situation, there are four to six men on the crew. As the baseline progresses and angles are turned, some of the crew will start cutting lines immediately. We only need three men to cut and chain the baseline. We don't want to be tripping over each other.

I have described a theoretical grid, but the reality on the Thompson Nickel Belt in the '50s/'60s was on a much larger scale. Over the years Canadian Nickel had solidified a claim block stretching from our Setting Lake camp to Moak Lake, a distance of 90 miles and up to 20 miles wide in some areas. A single base line system would be virtually unmanageable (can you imagine the size of a picket face required to write a co-ordinate of 11400N/725S?) so the belt was split into areas. Ours was the Setting Lake block. Little Joe Lake, Paint Lake, Thompson, Mystery Lake and Moak Lake All had their own grid system and I have probably missed some.

All these are tied together on master maps and of course rock formations can vary in strike. Thus the Grass River baseline azimuth may be closer to true north instead of our 15 degrees east of north declination in the Setting Lake grid.

It didn't take me long to figure out the grid system. I had spent a month two years before working on a survey crew just east of Fort Frances, Ontario, so I knew a bit about transits, chains and co-ordinates.

Over years to come I will work on many cut grids of many sizes, but none will equal the magnitude of the Thompson Nickel Belt. I now believe this was the largest continuous grid system ever used in Canada.

Can you imagine the man/days needed to install and maintain that grid? That it worked so well with little or no confusion is a testament to the talents of the people involved – from those in charge down to the hard-working axe-men.

21

There is a coffee break at 3pm after which it is time to go down to the dock to show me our summer's transportation. It is a 18' Peterborough freighter canoe, a strong, heavy workhorse. It is post-birchbark, pre-aluminium, and pre-fibreglass, of cedar strip construction covered with canvas. The canvas is waterproofed with a strong coat of red epoxy-based paint. On each side of the bow, lettering identifies it as Canico -17'. This is confusing and I silently wonder why an 18-foot canoe would be called a 17-footer.

Bogus lettering on our 18' freighter. Note injury to my index finger - someone must have hit me in the face while I was pickin' and singin'.

Sidebar: Later on, Norm Kearns will solve that one for me when I am on the dock helping unload a grocery order. Our freighter is on its cradle beside the dock, and I ask Norm "Why the 17 on an 18?"

Norm chuckles – he tells me it is illegal to move an 18' canoe with a floatplane smaller than an Otter or a Norseman. He is also pretty sure that Moak thinks that both he and/or the DOT are stupid.

The freighter is a square-stern with a 20-inch transom. It is a little more than three feet wide amidships, and although there are no seats, strong thwarts can be used as such. Because the floor is relatively flat it is very stable in the water and can easily haul the three of us and our work equipment. Charlie tells me it can handle up to 15 horses but we use an Evinrude five-horse with a separate Cruise-a-Day gas tank. The power is adequate for our purposes and the motor is easy to portage. We have paddles should the Evinrude let us down – it never will.

Last, but not least, is a small tool kit with a pair of pliers, a ten-inch crescent wrench, a screwdriver, spare shear pins, cotter pins, and a tube of Ambroid. Ambroid is an epoxy used to seal rips in the canvas. We will have to tie up on many places along the shorelines and it is easy to snag a sharp rock. The canoe weighs over 150 pounds, and if water seeps into the canvas it will add to the weight.

The shear pins are actually four-inch common nails. If we shear a pin, the strong pliers nip off a piece of nail for a replacement – simple yet effective.

And, of course, we have three life jackets. The first morning I don mine and feel rather foolish when Charlie and Jacob use theirs for seat cushions. We will wear them only once later on this summer when an unexpected windstorm drives us out of the bush. I swim like a rock, yet I will never feel unsafe in that canoe.

Now it's suppertime. After eating and a post-supper relax I get my first introduction to the bush end of the 7pm Sked. All camp radios are WWII surplus. I am told they were used specifically in Lancaster bombers. (See photograph – Chapter two)

Inco had bought a ton of them and modified them for camp use. The radio works are in a varnished plywood box about 20 inches to each side and 30 inches high. Inside the box is the mysterious tech stuff, the carrier and crystals, I am told. On the front near the top are the controls and a five-inch speaker. On one side of the speaker is an on-off switch and a channel selector. On the other side is a round gauge with a needle. The handset is like any telephone, but has a button to push to transmit. When this button is pushed it activates the carrier which sounds like an electric motor and the gauge needle shows the outgoing signal strength. The whole deal is powered by two twelve-volt car batteries hooked in series equalling 24 volts. These are stored under Charlie's bed – they will bite me next winter.

We always use channel one. Occasional atmospheric conditions will necessitate a switch to channel two, but if there is too much static on both channels the Sked will be cancelled until tomorrow at 7am. Channel 3 is a privacy deal – only Moak can receive a channel 3 transmission.

My first outpost Sked is so interesting. Camps are called in turn by their radio number – I already know that from last evening at Moak. Now Charlie fills me in on the camps. That one is Neil McAskill, another is Bob Tilden – both are camps like ours. Others are fly camps, for instance Sailboat McLeod (Charlie's cousin) and Henry Linklater. I can't remember all of them, but Charlie, Jacob and Willie know them well, and by mid-summer we know who is working for whom. We now have the worldwide internet, but in 1960 the Nickel Belt was tied together with the Lancaster Party Line.

When the Sked is finished I write a letter to go out next week. Charlie and Jacob each received a letter today. Charlie reads Jacob's letter for him. They talk in Cree and I don't understand a word, but it must be a fun letter, because they laugh a lot.

It is nearly 9pm and I am thinking about crawling into my sleeping bag, when Charlie turns the radio on again. Now I learn that this is chat time – time to find out how your buddy in another camp is making out, time to ask a cousin how aunt Leila is feeling, time ask a guy at another camp (who is a neighbour back home) how his wife and your kids are doing (haha from both ends) – I love this stuff! Although the Moak radio is on 24/7 the supervising personnel are in their own company houses by now and because the radio room and the common room are at opposite ends of the office/single men's building, it's more or less a camp-to-camp conflab.

I drift off to sleep with the flapping of the tent fly in the gentle breeze and the occasional call of a night bird to keep me company. Farther down the lake a pair of loons are singing a lullaby. The last thought before dropping off is that I am being paid for this!

Willie is up at 4am baking bread and starting breakfast. The muffled noises from the kitchen have already begun to filter through our tent wall when we hear the wooden spoon on the dishpan and a yodel at 6:30. No alarm clocks are needed at Setting Lake.

Breakfast is basic here – with only three men to feed, Willie can't offer a choice of meat, so it's a rotation of sausages, ham and bacon. There is always more than enough for us, and surplus breakfast meat will be added to the soup of the day. Eggs come in two styles – over easy or sunny side up. As at Moak, Willie serves them fresh off the griddle and there is toasted homemade bread every day. Because we have no light plant the bread is toasted on a pyramid shaped deal over a burner on the propane stove. So I can build my own Tom-style dream breakfast if I wish and I sometimes do so. What really gets a chuckle out of Willie is when I add some pancake syrup – he thinks it's hilarious.

While we eat breakfast, Willie lays out sandwich stuff on his work counter and after eating we build our own lunch, and again there is not much to choose from. Today it's bung bologna – not pre-sliced. The whole bung sits on the counter and I can cut off a full inch if I wish. The greenhorn takes his lead from Charlie and Jacob this morning. We each take two sandwiches, an apple, an orange and a can of preserved fruit – peaches, pears or such for noon dessert. We don't take a thermos with us due to weight considerations, nor do we build a tea fire in the hot, dry days of summer. We will sweat a lot, and we'll keep hydrated with lake or creek water and the juice from the canned fruit.

Food Protocol as Dictated by Mother Canico and Common Sense
A: Order what you want, keeping in mind that while hard-working men must eat well, you don't want to waste grub. Charlie always helps Willie with the grub order with some input from me and Jacob regarding our personal likes or dislikes. Food comes in by floatplane once a week and next week's order goes back to Moak on the return flight. Errors or omissions can be rectified on the evening or morning Sked.

B: There is no refrigeration in camp. Fresh meat is kept in a wooden butter box with a screen on one side to allow air circulation. The box sits on a small platform nailed to a large spruce tree behind camp. The platform is twelve feet above ground and some short pieces of 2x4 nailed to the tree form a rudimentary ladder. The theory is: the bear can't climb the ladder. (However, one bear, unclear on the concept, did climb it that summer.) But the screen is strong, and it does discourage flies, squirrels, and roving fisher, marten and mink. Willie can't climb the ladder either, so we fetch the meat of the day every morning before leaving camp.

C: Fresh meat must be eaten in order of perishability, so on grub day it's pork chops for supper, on day two we have pork sausage for breakfast, and pork roast for supper. On days three, four, five, six and seven, we alternate bacon and ham for breakfast – both keep well in hot weather. Hams are stored in the meat box and slabs of cured bacon are hung on a nearby branch. An agile squirrel will sample the bacon, and when the slab is brought into the kitchen Willie just trims the squirrel bites – we don't mind sharing. Flies, thank goodness, won't lay eggs on cured bacon.

On day three and four we eat steak – T-bone or ribeye. Day five is roast beef, and of course the roasts have enough left over to give us a break from bologna and summer sausage in our sandwiches.

Day six and seven can be anything still edible. There is kielbasa served boiled and hot – wieners and beans are a break, and sometimes Willie makes excellent bannock, also served with delicious beans (not from a can.)

Bannock (1 ½ lbs)
3 cups flour (sifted three times)
1 ½ tsp baking powder
½ tsp salt
¼ cup lard
1 to 1 ¼ cups warm water

Mix dry ingredients in bowl.
Make a well in centre and add fat.
Mix as for pastry.
Add warm water.
Mix with hands.
Mix thoroughly and knead well – at least five minutes.
Cut into two balls.
Allow to rest ten to 15 minutes.
Roll ½ inch thick flat cakes.
Prick well with fork on both sides.
Put into pans.
Bake at 450 F for 20 minutes or until hollow sound when tapped (the bannock, not your head.)
Optional: Add raisins, blueberries or saskatoons.

One fact of life in a bush camp is that the weather does not always cooperate, and the plane may be a day or two late. Then Willie dives into the emergency rations, and comes up with a canned, whole chicken. It's pretty darned good, and let's not forget that there are fish in the lake.

D: Naturally, there is no fresh milk – canned evaporated milk only. There is always an open can on the table to add to tea or coffee, and canned milk diluted with lake water can be used with cereal or porridge, but none of us are fussy about that. As a favour to me, Willie orders a Kellog's Variety Pack but I never do sample each variety. At home I liked puffed wheat, but in the bush large bags of puffed wheat tend to become bug infested.

E: Soft drinks (aka soda pop) are <u>not</u> on anyone's suggested menu. All pop comes in glass bottles and they are expensive and unhandy to ship by any form of transportation. We get canned juice, and Canico mandates that geophysical camps can only order large cans of tomato, apple, or orange juice which are too bulky and heavy to carry in our lunch bag. Also – always important to us conscientious non-wasters, is the fact that an open 26 ouncer would be tossed in the bush after lunch, still half full. (Yet – diamond drill camps get unlimited juice of any style and size.)

Sidebar: Part way through my first season the juice protocol is explained. Geophysical camps are always on fresh water lakes. Our gear is portable and we can and often do walk a few miles to work.

Diamond drills are not as portable. They can move to a nearby hole using their own winch and longer moves are done by muskeg tractor, thus drill camps can often be far from potable water. I also learn that a land-locked drill camp on a loonshit lake too small for Norm's Beaver, has to rendezvous via muskeg tractor to pick up the fly-in order. In the coming winter I will learn that drill camps are more often served by Bombardier on the Canico bush highway network.

F: Fresh fruit protocol is even weirder and is never explained to my satisfaction. We get crates of fresh apples and oranges. Drill camps get the same, but they also get bananas! Not us – we have no bananas today! I guess we are supposed to get our daily dose of potassium from the occasional blueberry bog (which we often do.)

But there are no flies on this boy, so I pull the odd scam. Sometimes the Beaver will have an undelivered drill order still on the plane when Norm comes to our dock. If I think he isn't looking I swap a couple of cans of the hated unsweetened orange juice for two cans of tropical fruit punch. One day there were <u>two</u> boxes of bananas for the drill. Let's be reasonable here folks – one for me – one for them. Norm knows what's going on but he doesn't rat me out – Norm likes me.

So rules and scams notwithstanding I eat good, tasty and healthy food in my first two bush seasons. It could not have come at a better time for my physical development. I had arrived at Moak tall and lanky, weighing 210 lbs. One year later I will leave Moak tipping the scale at 235 – the extra 25 pounds is all muscle – not an ounce of fat. And like the man said "If I knew I would live this long, I would have taken better care of myself." Life control panels should include a rewind button.

After breakfast we go back to our tent to change into our work footwear. Then we carry our EM gear and our lunch to the dock and load the canoe. The mag will remain in camp for now – I will work with Charlie and Jacob for the next four days. Charlie has to find out if the greenhorn can cut the mustard.

With the canoe loaded and ready to go I am all set to jump in, but what is this? Charlie and Jacob head back up to the kitchen! I tag along, wondering what they are up to now.

It's coffee time! I am a bit mystified, thinking we should already be Evinruding up the lakeshore. We go inside, and Willie has a fresh pot on the stove. (Now – you might think this is overkill to be describing the next fifteen minutes in detail, but it is not only an indication of how much I have to learn – it is also quite funny.)

First of all I have to tell you about our up-to-date kitchen table place settings. All the plates, soup/dessert bowls, cups, and serving bowls are enamelware – white with a narrow blue trim around the edges. They are solid, easy to clean, and they retain heat until the cows come home.

I drink my coffee black. I pour myself a full cup but by the time I set it in the table the cup handle is almost too hot to touch. Charlie and Jacob choose soup bowls, and fill theirs halfway to the brim. Then they fill the rest of the bowl with canned milk, and drink it soup-kitchen style, trying not to laugh while I blow across my cup and burn the bejeesus out of my lips and inner mouth. Ten minutes later they are out the door and I give up and follow, leaving Willie chuckling behind me. Tomorrow I will do the 50-50 deal and will not get back onto the black coffee wagon until next winter.

Our first job is near camp. We tie up the canoe a mile north of the dock and carry the gear a few hundred feet up the picket line. After we set up the transmitter coil I follow Charlie. He takes the receiving unit to the next line 400 feet away, stops at a station (picket), points the coil at the set-up station and hollers. Jacob aligns the transmitter coil with the sound of Charlie's voice, starts the generator unit, runs it 30 seconds and shuts it off. Then he waits for Charlie to move to the next picket, and they repeat the process.

Reading an inclinometer - cigarette optional. Note the blazed line station in the background.

To take the reading Charlie tilts the coil first one way and then the other. In a non-conductive area the horizontal component of the primary field likes a horizontal receiver coil, thus the signal passes through with little resistance. Tilt the coil one way and you get a strong signal. Slowly tilt the coil the other way – the signal will fade and then strengthen again. Rock the coil until you have the weakest signal. This is your reading. In a non-conductive zone it will be zero.

So let's say the motor/gen is set up at 12N–5W. Charlie goes to 8N-0BL and starts his loop. He will read each station to 10W, 500 feet on each side of the set-up. Then he will walk across to 4N/10W and read back to the baseline, thus covering 2000 feet of picket line on the two lines adjacent to the south of the line we are set up on.

Now he walks to 16N and repeats the deal, covering 16N and 20 N. No conductor is found, so no further work is required here. We tear down and move on to the next set-up. One man carries the motor/gen, one man carries the coil bag and the third brings the unipod bundle and the lunch pail (we all carry our axes.) It is a short move of 800 feet to 20N/5W.

We brush out a 14-foot circle and set up again. We may have to knock down a couple of trees, because we must plant the unipod as close as possible to the picket. As long as we can see the picket line, we have a pretty good idea where Charlie is when he hollers. Voice orientation in the bush can be tricky as wind and/or topography can make it difficult to pin Charlie down, but if we are on the ball we know if Mother Nature is trying to fool us.

So Charlie reads the back line (12N – the line we had set up on previously) and walks over to 20N to read the 20N – 24N loop. But hold on, Charlie is picking something up!

Charlie's readings tell him there is a conductive structure beneath the surface to his north. Our primary field has triggered a secondary field around a north/south trending mineralization occurrence. The secondary field combines with the primary, and Charlie's receiver coil is picking up the resultant effect, and if you draw a 90 degree line through the centre of Charlie's receiver coil it will point at an angle to his west. As he approaches the spot directly over the conductor his readings will flatten out, and as he crosses the apex of the conductor his readings will change to point south.

This time we do not necessarily set up on a station. The cross-over is at, let's say, 6 +50 W (650 west of 0BL) and it is important for the integrity of the survey that we transmit from a spot directly over the conductor.

Now with the conductor nailed down, we trace it out as per our previously described actions until we are satisfied that enough work has been done in this particular area. We then move on to another spot.

Within three or four days I have a good basic knowledge of what is required for an EM survey. Later on over the summer months I will soak up some of the nuances involved – out of phase readings, for instance (broad, hard to establish null points.) These can indicate a flat lying structure or perhaps a conductive clay bed. I learn that the dip angles can tell me the shape and attitude (dip) of the conductive body. I learn about magnetic correlation – high mag readings coinciding with a strong, sharp cross-over could be caused by an iron formation associated with graphite. Weaker conductors with flanking magnetics are much more interesting. Maybe ore-bearing sulphides hide below – a follow-up drill program will tell the story (perhaps.)

I also learn that a sharp, easily-defined crossover with fuzzy readings on either side with no magnetic association is probably a water-filled shear zone – and on and on.

The classroom is the bush and lectures take place over lunch breaks or at the map table in the evening. Charlie seldom offers info, but if I show interest he is glad to share his knowledge – and I ask a lot of questions.

It boils down to this: there are needles hidden in the haystack and Inco is counting on us to do our part in finding them and we do our best to help out.

For the next three days I alternate, between Charlie and Jacob, and by day four I've got a handle on the EM and some unwritten rules, one of which is that you always carry your axe with you no matter what you may be doing. The picket lines may have been cut last year or three or four years ago, and although underbrush grows more slowly north of 55, we may have to freshen up parts of the line. If a picket is hard to read a fresh-cut face and a black magic marker restores the chainage info thereon. A picket may be missing. Why? Charlie says maybe a big woodpecker took it home to feed the kids. He's joking (or is he?)

So we carry an axe in the bush and it becomes such a habit that when I return to civilization on my seasonal breaks I feel empty-handed on cement sidewalks.

I learn to pace. On the picket line you always count paces between stations so you know when to look for a fallen or missing stake. You count paces when crossing between the 400' spaced cross-lines to be aware of where you should pick up the next line – sometimes in open bush they're easy to miss.

I learn that while 40 paces equals 100 feet for average length legs, both Charlie and I need only to count up to 36. With that in my noggin I learn that is easier to remember smaller numbers, so I count every second step – 18 doubles = 100 feet.

I learn to pay attention – no daydreaming! If I stop to put in a fresh blaze on a tree I catch myself continuing to count every second axe swing – now, where the heck was I? Lack of concentration can cause one to return to the last picket to start over, and no one likes to make extra work for oneself.

Sidebar: Next summer while staking claims in Ontario I learn another trick. Claims in Ontario in unsurveyed territory are 1320' to a side and that is a lot of paces to remember. So I count up to 72 doubles, break off a twig and put it in my pocket. Three twigs make 1200' – another 120' and I cut a corner post, throw away the twigs and start over.

(Civilization will bite me when I detrain in Winnipeg on my Christmas break. I walk down Broadway Ave. counting paces between utility poles. Folks sharing the sidewalk must think I am certifiable. Here's this guy, smelling like tea-fire smoke, carrying an imaginary axe and mumbling numbers. The transition from real life in the bush to phoney city life can be difficult.)

I learn that while we don't punch a time clock, we always put in a full day, weather permitting. We leave the dock at 8am, take a short break at tennish, and a half-hour lunch break. We may skip the afternoon break and we always try to get back to the canoe around 4:30. At the dock we upend the canoe, putting the motor and life jackets underneath. We carry our survey equipment to the storage tent, change into camp clothing and relax for 15 or 20 minutes. Willie will holler at us at 5:30.

If we get an overnight rain we must be patient in the morning, waiting for the sun and wind to do their dry-the-bush job. The thing is that the EM receiver stuff can die if it gets wet, so after breakfast we veg out for an hour or two. When Charlie figures the bush is dried out enough, we leave late. The lower tree branches and Labrador tea may still hold water, but we are quite often wet from the knees down anyway.

An afternoon shower will send us home early. We always listen to the weather forecast at breakfast – Willie has a good battery-powered radio. Charlie has numerous targets to choose and if rain threatens we will hit an area not far from shore. We can trot back to the canoe lickety-split, and if the rain catches us before camp we throw a jacket over the receiving apparatus. Men can dry out – electronics tend to stay wet for a day or two.

I learn that we don't always carry everything out of the bush at day's end as more work may need to be done in a given area. We hang the pack boards on a tree, electronics are tucked into the coil packsack and we carry a small plastic sheet to cover the gen-set. The mag can also be left on site with a canvas sample bag over the mag head. On those days we only pack out our lunch bag and the one-gallon gas can.

29

I learn about banking work. This is tied to Canico's day off protocol – work thirteen days – take one day off. I was told about this at Moak but I had not read the fine print. Now I find out that a rain day is considered a day off, and the count is reset at zero. It's a bum deal. One can do many things on a sunny, warm day off – fish, swim, tackle an interesting non-work related project, or borrow the canoe to explore the unknown south of camp. Also, on nice days off, you can wash clothes and hang them outside on the clothesline. But on a rainy day you cannot do that stuff. You read a book, play cribbage with Willie, nap – repeat – and you do that most evenings anyway. So to get a sunny day off Charlie has opened a progress report savings account.

I already know that daily progress is reported on the 7pm Sked. This report is basically how many EM set-ups, how many mag readings and so on. Canico does not push us, but they know what a day's work is. For a camp such as ours they expect three set-ups from the EM crew, and 100 stations from the mag man. If I am working with Charlie and Jacob it's four set-ups – zero mag readings.

So we do extra work on good days and save some in the bank – thus on a rainy day we can still report a day's work. Keep in mind that we are 70 air miles from Thompson, 90 from Moak, and if Moak should query our report Charlie can say the front passed us to the north. Also keep in mind that we know that if it is socked in up north the planes are grounded and no one will rat us out. It sounds like a convoluted process to get a nice day away from the bush but we are actually successful once or twice.

I learn something else that first week – something slightly discouraging, personally. Up to now I had considered the accumulated seasonal time off with pay as fair and equitable. Maybe I should have given my brain a rest, but one day with nothing better to do I start counting.

We bank one day every two weeks for seasonal time off. We get one day off every two weeks after working 13 days. I was told that my pay was based on a five-day workweek. Add it up – ten days of work, one day off, and one day in the seasonal bank equals twelve. I ask Charlie what gives, and Charlie laughs. He says the two missing days pay for the free room and board. I am very, very disappointed that Canico would bait and switch. For the first time in my life I know that while figures don't lie, liars can figure.

And get this – I also find out that although my starting wage at Moak is $260/mo, greenhorns working out of Copper Cliff in Ontario start at $300! I file this in red magic marker in my noggin under "TRANSFER TO ONTARIO SOONEST!"

Our Setting Lake crew is doing review work this summer. We are, in fact, designated by Moak as the "Setting Lake Review Camp." Over the years, EM and mag surveys have covered the entire Thompson Nickel Belt grid system. Field maps are generated and sent to Moak, where a crew of draftsmen transfer them to master base maps, three feet long by two feet wide. At a scale of 1" = 400' each map covers a section of grid almost three miles by two. Bordering maps continue the grid in all four directions to the claim boundaries. Since we are on the southern-most grid system, our north boundary is at the south edge of the Grass River grid, our other three boundaries are the edge of the claim block. The eastern edge goes out into Setting Lake, which will be surveyed this coming winter on the ice.

Our review work consists of filling in the gaps on Moak's maps. Charlie has quite a few of these maps on hand with sections to be covered circled in red. Sometimes EM and mag work is missing, sometimes just one or the other. When Charlie is confident I have a chance of finding the canoe in the pm, he turns me loose to do my own work, but usually we are working in the same area. Occasionally we have to fill in or extend a picket line – not often, but when we do I practise my axe work. I find it all interesting. Same old same old makes Jack a dull boy.

There are various reasons that gaps need to be filled. A previous crew may have suffered a brain cramp and neglected to follow out a conductor, or simply ran out of time. Some areas may have been flooded by beaver dams. If they move on, the water eventually recedes and we can fill in the info. Areas that are still wet are noted and will be worked next winter. Some of our work is to check on the quality of a previous survey. The results may have looked a bit fishy to Moak and they want us to check things out. With so many men in the bush there is bound to be some who would rather sit on a stump to generate survey readings. Charlie has nothing but disdain for the practise, and his attitude rubs off on me.

We also have a second grid to review. The Soab Lake grid, relatively small in area, warrants a separate grid due to a change in baseline azimuth.

Soab Lake is a one-mile circle of undrinkable water surrounded by a two-mile circle of Labrador tea and short, scrawny black spruce. It is what we call a "loonshit lake." The water, although shallow, looks deep and dark. The lake bottom consists of loose, black semi-solids, and will not support a man. Loonshit is bottomless.

We spend a couple of days in the area, and it is tough, tough walking. We sink up to our hips in the tea and beneath our feet there is really no solid ground. It's like walking on a huge underinflated air mattress. Seventy pounds of gear on our backs exacerbates our difficulties and we stop often to catch our breath and spit out the black flies. I help Charlie and Jacob at Soab. With three men we can swap off the heavy loads and we put in two long days. The walk from Setting is a long one, five miles each way, and we want this job in our rear view mirror. Back at camp an evening dip in the lake is welcome, and when the job is done, Charlie has some work in the bank. We take a day off to wash clothes, repair Labrador tea damaged jeans and rest our weary bones. Thank you, Charlie.

One day while sharing our lunch with some Soab Lake black flies, I ask Charlie where the name came from - it seems such an insignificant body of water, hardly worthy of a name. Charlie tells me that it has been on everyone's hate list since the first claims were staked. The general consensus is that it is a son-of-a-bitch – hence Soab.

Table Scrap

While researching the books we were fortunately able to contact Kieth Olson, now retired and living near Winnipeg. He shared this tale: One day he flew a geologist into Soab, waited until the geologizing was done and flew him out. As he circled back towards the north he overflew the lake and he could see his pontoon tracks in the loonshit.!!

Table Scrap: Fun and games

It's the first day of our Soab Lake jaunt. We arise at the crack of dawn, have an early breakfast, leave camp at 6am and are nearing our target at 8am. I am already feeling a bit tuckered out when, as we trudge along, we start to get flak from a squirrel who is jumping from spruce to spruce saying nasty squirrel-talk. Suddenly Charlie and Jacob drop their packs and start to chase the chattering rodent. At first I am dumbfounded, but I soon join in. We grab the squirrel's tree and shake it – easy to do as the roots are poorly anchored in the loonshit below the tea. The squirrel hangs on for dear life until the whipping action dislodges him. He flies to the ground and scampers up another small tree. He is not nearly so mouthy now, and with each of us guarding a nearby tree the little bugger's only option is to escape to a larger grove of spruce. Eventually he does so and we collapse – out of adrenaline and laughing. We pick up our gear and continue on, leaving the squirrel mumbling to himself. Tomorrow on our second and last trek into Soab, the squirrel will be elsewhere gathering pine cones in a safer area.

That's the thing about working with Charlie. Work comes first, but one never knows when that ever-present playfulness will break up a difficult day.

Take chasing rabbits for instance. We don't really chase rabbits (or do we?)

We are carrying up to 75 lbs on our backs and our natural centre of gravity has changed – we are top-heavy. We are walking at a brisk pace through bush or on cut picket lines. Sooner or later we trip over a deadfall or snag a tag alder stump. To keep from falling flat on our faces we stagger ahead to get the load balanced again. We are not always successful – sometimes we crash and burn.

But you never whine, nor swear, nor do you take offence when your two partners are laughing like crazy at you – and someone will always ask, "Did you catch that rabbit?"

Table Scrap - Charlie (the rascal) pulls a fast one on me

Along with the supply plane comes letters. Jacob always gets a letter from his wife, and since he can neither read nor write, Charlie reads them for him – in Cree! They sit on Jacob's cot. Charlie reads a little and Jacob giggles, then Charlie reads some more and he and Jacob laugh and laugh. Hmmm – sounds like a good newsletter.

Another mail day arrives. We usually spend part of the evening plotting our day's readings at the map table, but on this particular evening Charlie says he wants to get the day's mapping done and maybe I could read Jacob's letter for him.

"I can't read Cree," says I.

"She writes in English," says Charlie, "I translate to Cree for Jacob."

Well, I soon find out why. Along with the local news and the kid's shenanigans, Jacob's wife also describes in lurid detail how much she misses Jacob. I won't go into specifics, but rest assured that the raunchiest stuff on the internet would have a hard time matching this. I turn red as a beet and I'm starting to sweat as Jacob giggles like crazy. I glance over at Charlie, and he is hunched over the map table shaking like he has the palsy.

From that time on I respectfully decline the offer to read Jacob's mail to him.

Table Scrap - Now Willie bites me – I am a slow learner

We keep our clothes and ourselves clean: ourselves with a dip in the lake and our clothes with a square tin washtub. For an agitator we use an upside down funnel kind of thing. You sit on a bench and plunge it up and down like an old cream churn.

One day it is windy with intermittent showers. We won't go to the bush today, so I decide to wash some socks.

Willie tells me that a good way to wash clothes when the lake is rough is to put the washtub with soap, water and dirty clothes in the bow of the canoe and drive around on rough water. The action of the bow rising and falling to hit the next wave supposedly acts as an agitator and washes the clothes. With Charlie's permission I borrow the canoe and set off to wash my socks. Well, the socks in the tub don't get any cleaner, but the stuff I'm wearing definitely gets washed by the spray kicked up every time the bow hits a wave. I return to the dock fifteen minutes later drenched to the skin – Willie got me!

Setting Lake is really rolling - we won't work to-day.

Table Scrap - Don't piss off the cook.

On another bad-weather day, sometime after breakfast I go to the cook tent to rustle up a cup of coffee and find Willie working at the washtub.

"Washing your shorts, Willie?" says Babe-in-the-Woods Bob.

Willie goes absolutely ballistic! It turns out he is washing dishtowels. "You think I'd wash my underwear with the kitchen towels, you SOB?! Get the hell out of my kitchen!"

I beat it back to our tent and we sit there listening to Willie rant and mutter. I tell Charlie what I said.

"Oh-oh" says Charlie.

For the next few days Willie is not his usual cheerful self, and it is cold shoulder for me. Finally Charlie takes me aside and says I should talk to Willie before we all starve to death, so I screw up my courage and go over to see Willie that evening. I tell him I made a stupid greenhorn mistake and I never meant to accuse him of washing his dirty underwear with the kitchen towels. Willie grudgingly accepts my apology and things get a bit better. Willie even starts baking pies and cakes once more, but it seems a little cool in our once warm, happy kitchen.

I am eating things I have never seen before: nothing exotic – just stuff not normally seen on my parents' table – sardines and smoked oysters for our noon snack or store-bought canned fruit and bottled pickles for instance. It's not any better than Mom's home cooking, just different. One thing I've never seen before is a Spanish onion thickly sliced and served in a water and vinegar mixture. What a taste treat it is to me!

33

Willie puts this out with our T-bones a few evenings later and I make a pig of myself. Willie even slices up another big onion. He enjoys feeding big eaters.

After supper I am relaxing on my cot when my stomach rebels. I moan and groan and then decide to take a walk. I pass through a fairly thick stand of mixed trees north of camp and I come out on a high rock ridge along the lakeshore. I find a nice spot in the evening sun about a mile north of camp and sit there looking out over the lake while my tummy decides it will digest the onion after all. In an hour or so I am feeling better and I hear the outboard chugging down the lake towards me. It is Charlie and Jacob, studying the shoreline as they motor along. They spot me and I go down to the shore.

"You better come back with us," says Charlie. "Willie is going nuts. He's crying, and he's sure you went off in the bush to die."

We go back to the dock where Willie is waiting. All is forgiven, but I have learned the first rule of camp life – NEVER PISS OFF THE COOK!

Table Scrap - How to keep your meat fresh

Our meat box is squirrel-proof but not bear-proof, as we find out one night when we hear a ruckus back there and the next morning find out meat cache has been raided.

Charlie has a single-shot 12-gauge that he keeps under his bed (every camp carries a bear gun) but he has forgotten to bring buckshot shells. I had found a lead weight from a commercial fishing net along the shore. Charlie pares it down to fit the shotgun barrel, removes the birdshot from a shell and replaces it with the lead slug. I am worried that it will hang up in the barrel and the gun will explode but Charlie says that he has done this before – no problem.

So Jacob and Charlie fabricate sort of a tree stand and hide out after dusk to wait for the bear while Willie and I drink coffee in the kitchen, keeping our ears tuned. Later on we hear some rustling behind the cook tent followed by a loud 'boom!' Back come our hunters, laughing like crazy.

It was too dark to see much so Charlie just aimed at the noise and pulled the trigger. Jacob says the recoil damned near knocked Charlie out of the tree. There is no blood the next morning, but it must have scared the bejeezuz out of the bear, because he never did return. I guess he figured berries and roots were a less stressful diet.

This particular campsite has been used repeatedly since the 50's when initial staking took place. A good expanse had been cleared on the point in front of the tents for a helicopter pad, and down near the floating dock there is an old steam bath, although we do not use it. It is a simple log structure about eight by ten with a good door made of plywood. There are two or three bags of bentonite (also known as drillers' mud) beside the shack. I experiment with it and find that if mixed with water it sets up sort of like plaster. It is used by drillers to seal off the hole if the rock is fractured. The bentonite keeps the sides of the hole from collapsing when the core barrel is pulled.

My problem is that I am not fully aware of all the properties of bentonite. I spend a couple of evenings mixing up my bentonite "plaster" and carefully chinking the gaps in the steam bath walls, covering the stovepipe hole with some screening, and making sure the door seals tightly. When our next meat order comes in we store it in the steam bath, hanging the summer sausage and kielbasa coils from the rafters.

34

The plot thickens; the weather has been nice early-August hot, but one night it rains. One of the properties of bentonite is its ability to soak up available moisture until it becomes a slurry. Unnoticed by us, the bentonite had turned to soup and run down the walls to the ground. A couple of hot days later Willie says, "Maybe you should look at the meat."

It is covered with flies who are chomping away and laying eggs like crazy! We salvage what little we can and throw the rest into the garbage pit behind the kitchen. It will be wieners and beans for a couple of days until our meat can be replaced and we are getting pretty expensive to feed. The next meat order goes back to the tree box, where it will remain safe until we break camp in September.

But our troubles are still not over. I don't know why people listen to me because it seems like every time I come up with a 'good idea' we all have to suffer.

A couple more hot days later we come back at the end of the day to find Willie jumping up and down on the dock waving his arms and babbling, "They're after me, they're after me! They're taking over the camp!" What the heck is this?!

When we had dumped the spoiled meat into the garbage pit, it never occurred to us to cover it with dirt. The fly eggs had hatched, and thousands of maggots – big maggots! - are squirming their way to the cook tent: not towards the bush, oh no - they are fanned out in the direction of the kitchen! The pit is about twenty feet from the kitchen, and the front echelon is over half way there, streaming and teeming down the path. Willie had discovered them when he took the garbage out in the morning, and was going more and more nuts as the evil horde advanced during the day. Willie can't swim or I'm sure he would have been halfway to Wabowden.

We try shovelling, but the maggots are so mixed up with branches and twigs on the forest floor that our efforts are ineffective. We spread six cans of lye (used for our outdoor biffy) in front of them, but they just ignore it and crawl on. We finally pour outboard gas on them and into the pit and stand by with pack pumps to contain the fire. It takes a couple of hours to fricassee them and back-fill over their corpses in the pit. We jump up and down on the dirt to pack it real good, let me tell you! The next day we dig another garbage pit well away from the first – another lesson well learned.

A Visit to the Dentist

It's midsummer, and my teeth are bothering me. I have two that are starting to keep me awake at night so I ask Charlie if there is a possibility of getting out to see a dentist. He talks to Moak about it on the next Sked and it happens that the leased Cessna is to go in for its annual check-up at the LambAir base in the Pas. The plane will be in the hangar for two or three days and I can get a ride back to camp on the return trip. Moak makes me an appointment and Keith Olson picks me up on his way by.

I get a hotel room and spend two days sitting in a torture chamber – eight fillings – and then he pulls an abscessed wisdom tooth with no freezing – Ouch!

The dentist is a recent immigrant from Korea and I think he is a little short on experience. Within four years every one of my fillings will fall out, but at least I get temporary relief. (I heard later that he left The Pas within six months, no doubt leaving behind a whole bunch of empty cavities to come in the future.)

35

I have some free time between my visits to the dentist. I hang out at the airbase and I meet Ron (Blokey) Martin, an Englishman flying the bush for LambAir and a great guy. He is outfitting his new Land Rover for a trip around the world! He is scheduled to

BRINGING A NORSEMAN UP ONTO THE HANGAR RAMP AT THE PAS. TOM LAMB'S LABRADOR/ASSISTANT BASE MANAGER SUPERVISES

head south in a few days to meet his partner in Winnipeg. They will circumnavigate the globe for the next two years, crossing every continent including Australia.

A National Film Board (NFB) cameraman is filming footage of the preparations and giving Blokey tips on camera stuff. They will supply Blokey and his partner with a camera and film to document their trip and will underwrite some expenses. Rootes of England is also on board, but if money

gets tight, they will interrupt the trip to earn some hard cash by doing some flying.

The NFB guy asks the doofus hangabout what he is up to, and the result is my first and last screen test. He films me climbing out of the 180 holding my jaw, hesitating before walking into the dentist's office, and emerging all smiles and thumbs up. I thought I played the part well, but I think I got left on the cutting room floor. Unlike Buck Owens, I failed to act naturally.

BLOKEY MARTIN AND HIS LAND ROVER AT THE LAMBAIR BASE

Was Blokey's documentary ever produced? I don't know. I did hear a year or two later that in Australia he traded in his partner for a good-looking, marriage-minded lady, and you know how women are. She no doubt decided that he should get a real job and settle down. Well, at least Blokey made it halfway.

All water in Manitoba flows north, and Setting empties at its northern tip into Brostrom Lake below Sasagiu Rapids. Brostrom is not much more than a large bay at the head of the Grass River, which is one of the many tributaries of the mighty Nelson. The Thompson Highway now crosses at Sasagiu, but in 1960 only the right-of-way was cleared. Brostrum becomes the Grass River within four or five miles and as the lake narrows the water picks up speed and Lynx Falls starts the river ever so dramatically! It is now a Provincial Park and Lynx is also the second highest waterfall in Manitoba but nowadays it is called Pisew (Lynx in Cree).

We have two weeks of work to do from Brostrum – to get there we have to deal with Sasagiu.

WWW.MANITOBAPHOTOS.COM
Sasagiu Rapids

The rapids are broad, shallow and quite fast, requiring a daily portage and our cumbersome freighter canoe requires two men to carry. With the five-horse outboard and 150 pounds of geophysical gear, it will be a time-consuming twice-a-day task.

After the first day Charlie realizes we have to go to 'Plan B' – we will build a railway! He and Jacob are the engineers. They walk the northwest bank in deep discussion. I have no clue and no input, so I sit in the sun and watch Mrs. Fish Duck and her brood. It's a pleasant interlude for me. Then, with the blueprint in their heads, we kick off the project.

First we knock down some good-sized trees and place them in two strings end-to-end and four or five feet apart between convenient docking spots above and below the rapids. These will serve to hold our cross-ties. We then cut a number of poplar saplings, peel them and fasten one end on a side rail with the other end on the ground, alternating left and right. This makes a v-shaped cradle for the canoe to slide on. The total change in elevation is less than 30 feet over a distance of 400 feet, and the slippery peeled poplars allow the canoe to slide easily. In the morning heading downstream with two men on the bow rope and the motor tilted up the canoe scoots down the railroad, and after a 400 foot trot we are back in the canoe and on our way in a matter of seconds. In the evening going upstream with one man pulling and two men pushing we still make good time. It takes a day to build, but Charlie's Railway saves us a lot of time and effort in the days to come.

Before we continue our bush work we cruise the lake down to Lynx Falls. It is a combination scenic tour/shoreline scope. Charlie wants to check out which picket lines we will follow. We pass a log cabin on our left. It looks like a cozy spot sitting in a clearing 60 or 70 feet upslope from the riverbank. Near the tree line are four doghouses, each with its own dog on a long chain.

Lynx Falls

They sound the alarm and a guy comes out and waves. Charlie waves back – we will stop and chat on the return trip.

When we near the falls the greenhorn gets a bit nervous. We can hear them roaring ahead of us and the river is speeding up, but I need not be worried with Charlie at the helm. Has he been here and done that, or is it just because he pays attention when the elders spin their tales? He knows exactly where the portage starts.

We tie up in a small cove and walk to the falls on a good, though seldom used trail. The falls are impressive, and we can hardly hear each other. The day is hot, but here it is a cool oasis. It is almost like a very light rain is falling. The trees grow tall and the whole area around Lynx is positively verdant - and wouldn't you know it, my camera is back at camp. We go back upriver, stop at the cabin and I meet John Pilog.

John is a trapper, and lives here all alone and all year long. He is of an indeterminate age, medium-sized and carries not an ounce of fat. He is clean-shaven, neatly dressed and he must cut his own hair because there are no scraggly tufts escaping his billed cap. He offers us tea which Charlie respectfully declines, but we do have time for a short visit.

We chat sitting in the afternoon sun on upended chunks of dry pine firewood. While the other three trade trapline stories I survey the homestead.

The log house is not large but the logs used to build it are pretty big. On the side facing the river is the door and a window and there is another window in the southwest end so John can pick up sunlight from early morning to late evening. The roof is shingled with wide, hand-split cedar shakes. It is obvious that all materials other than the windows had been bought locally at Mother Nature's big-box store. Was there more than a pound of nails used here?

The four chained dogs are quiet now. They lie in the warm sun and keep an eye on us. They are big ones (Charlie later tells me they are Malamutes) and they look pretty tough. One dog is running free. He is smaller – a half-grown Lab cross, full of adolescent energy and is certainly John's favourite.

I am disappointed that I never get to see the inside of the cabin. I guess maybe John had not had time to vacuum and make his bed that day.

We walk back down the path to our canoe, and the yard, of course un-mowed, is nevertheless neat and free of willow clumps. John has lived here for years.

On the way home Charlie patiently answers all my questions.

"Why are the Malamutes on chains?"

"That's because they are semi-wild, and if allowed to run free may be killed or adopted by a wolf pack."

"Yeah, right," I say to myself. I have yet to see or hear a wolf.

Charlie also says the dogs are let off their chains for an hour or so each day and that they are used to being chained up. In the wintertime when fur is prime, they will be on the sled trail every day – they like winters.

"What do John and the dogs eat?"

"They eat from Mother Nature's larder. In the wintertime John will cook beaver and muskrat and share the carcasses with the dogs. This fall he will shoot a moose or woodland caribou and hang them in his little log meathouse. He can also bag waterfowl in the summer months, ptarmigan in the winter, and there are always fish to be netted. Fish, in fact, is the main diet for the dogs."

"John has a small potato patch near the cabin. Potatoes are the only vegetable the rabbits and groundhogs won't eat. He will supplement his fresh veggies and fruit on a rare summer canoe trip to Wabowden. In the winter months Wabowden will be visited more frequently. Pelts will be shipped to a fur buyer, and on the backhaul will be flour, sugar, tea and tobacco."

"Where did John come from?"

"Nobody knows. John has never spoken about his roots, nor does it seem that he has a family on the outside. He neither sends nor receives any letters. He is not quite a hermit. He is always glad to have visitors, but he likes his dogs more than people."

(I wish I could have visited John myself, but the portage is too difficult for one man.)

We return to Setting via our dandy tramway and it is the end of a super day in the north. Tomorrow we will hit the muskeg again.

These are great days in late August/early September and we work a week or two off the bay at the head of Brostrum. Along the shore we see my first woodland caribou, and large flocks of waterfowl are mustering for the migration south. There is a family of fish ducks (cormorants) hanging around the rapids. The little guys can't fly too well yet, but they sure can swim! If we catch them at the foot of the rapids we pretend to chase them and they disappear under the water only to reappear above the rapids, going like crazy, kicking up a contrail just like a 200 horse bass boat!

One day we head down river halfway to John's cabin and walk inland with our gear. We have a couple of set-ups to do not far across the cleared right-of-way of the new highway. We pack through the bush to the right-of-way and as we cross it we scare up three good-sized timber wolves basking in the sun. It is my first encounter of the wolf kind! They lope off into the woods and we continue on for a few hundred feet to our set-up. As we rig up our tripod and hang our transmitter coil, the wolves start howling. What a thrilling sound! The hair on my arms and the back of my neck stand straight up! The weird thing is, we can't quite pin down where the howls are coming from. It seems one is to the left and another to the right, but where is the third? Their song circles around us and echoes back and forth, and it seems to say, "Go home, boys!"

The general drill is that I do my mag readings while Jacob runs the motor/transmitter at the set-up and Charlie does his EM. None of us wants to go out alone, and since three divided by two is one-and-a-half, there <u>has</u> to be one man alone. Charlie doubts that the wolves will tackle us, but they are obviously unhappy with our presence. In the end we decide to leave our gear and return tomorrow. The wolves moved on, I guess, because we never see or hear them again.

Towards the end of September and not too many days from our departure from Setting, we are working a couple of miles north of camp. We walk two miles to this job on an old muskeg tractor road, set up the transmitter coil and go to work.

It is fall and rutting season is in full swing for the moose population. By the time Charlie has taken three or four readings a bull moose starts to answer him. Now I would have thought nothing of it – I was too green to know that the moose was thinking that Charlie is another bull encroaching on his turf. Charlie comes back to the set-up and says, "Let's go home," which we do, very quietly. We finish the work the next day after the moose has gone on its way.

Charlie tells me that he was not really worried that the wolves would hurt us, but the moose is another story. Although I never heard of anyone suffering injury at the hands of a moose, I did hear of men being treed and even being run out of an area during rutting season.

So from then on, whenever Mother Nature said, "Go home," I went.

Table Scrap - Groundhogs, cabbages and rabbits

There is a groundhog burrow near the lakeshore 20 feet north of our dock. We have seen Mrs. Groundhog duck into her front door and by early July the three children are out and about. We keep bags of potatoes, onions et al in a tent with no floor. One day Willie says that something is eating our cabbage so I set a simple trap – a steel 45-gallon drum open at the top. I lay it on its side and use cabbage leaves for bait.

An hour later I hear something in the drum. I sneak out and upend it. Success! Inside are three terrified juvenile groundhogs. I reach in to pick one up but when I do so three sets of sharp teeth start gnashing – those woodchucks could chuck wood! I leave them in the hoosegow for a thirty-minute incarceration and then release them to serve the rest of their sentence in the community. They must have told Mommy because the grub tent was never raided again. I did my part in the rehab. Willie always throws away the outside leaves of the cabbage, so I donate them to the rodents.

Another evening just before sundown I hear a squeak and kerfluffle behind the tent, A mink is stalking a baby rabbit. I go out, chase away the mink and pick the little guy up. He's so terrified he can't move.

The rabbit must have been just weaned – he fits in one hand. I take him into the tent and he hops under my bed. I get a small piece of lettuce from the kitchen and put it beside him. He ignores it. I have thoughts of training him as a pet so I leave him alone at night and in the wee hours I hear him nibbling at something. At daylight I find he had helped himself to an apple (we keep some under the map table) and I figure he has graduated to hard food.

That cottontail is pretty agile. When we get back to camp that afternoon he hops off my bed – the little rascal had peed on my eiderdown! I am already having second thoughts about keeping him, so I return the little ingrate to nature. I left him an apple at the edge of the forest but it shrivelled up uneaten in a week's time.

Table Scrap - How to fool the Royal Mail and Moak Lake

(In 1960 our postal service was known as "Her Majesty's Royal Mail," before we got all politically correct and changed it to "Canada Post/ Postes Canada.")

One evening while having our coffee and pie in the kitchen, the conversation comes around to the unwritten "dry camp" rule and the ways that rule can be circumvented. Willie says it is easy – just buy a loaf of bread!!

Here's the deal: it is illegal to ship spirits by mail and if Moak finds out you are in double trouble, so you improvise.

First you need a friend or relative on the outside who is willing to aid and abet. You write a letter enclosing enough money to buy a 26 ouncer (US quart,) a loaf of bread

40

and postage – ten bucks should cover it. Why bread? Well, it's to cushion the bottle. (Styrofoam is neither widely used nor readily available in 1960.)

Your partner in crime buys a loaf of unsliced bread, cuts it longitudinally, hollows out a space for the booze and tapes it back together (he has already cracked the cap and filled the bottle to the brim – there must be no telltale gurgling coming from the package.) Now he packs it in an appropriate sized box and mails it to you. The booze will arrive safe, unbroken and incognito, and Her Majesty and Moak will never know the difference.

I sure enjoy Willie's story. It is a laughable convolution to get a snort of hooch but none of us are interested in having hard stuff in camp. We would like a cold beer now and then and we will have the opportunity to do so twice this summer.

Sidebar: You might think I am a little harsh vis-à-vis rough mail handlers – but listen to this. That summer my sister-in-law mailed me a care package of homemade oatmeal cookies. I wrote back "Thanks for the cookie crumbs."

Table Scrap - Swearing in Cree

I was told by Charlie that one cannot swear in Cree. There is no concept in the Cree language of blasphemy, either against their own gods (for wont of a better term) or their more recently acquired white man's gods. Of course they use descriptions of the various body parts to insult in Cree, but before the White Man's influence, these were merely words in everyday use. Hard to explain, but I hope you get my drift. For example, if someone pisses you off, you might say "Kechisk weet-say-gun!" (Your asshole stinks) or just "Weh-chisk." (You asshole) and then the other guy might say "Key-nah, oo-tay!" (No, it's yours,) Then they will both say "Tah-boy!" (That's right) and laugh like crazy.

Table Scrap - The Hudson's Bay Company/Northern Stores

We Canadians all learned about the Hudson's Bay Company in school: how they opened up Western Canada, their rivalry with the North-West Company, etc, etc, but my experience in the north gave me a different perspective – quite different in fact.

I started to hear a lot of stuff, like in the old days guns would be traded for beaver pelts – in fact they were called "trade guns." They were black powder muskets made of inferior metal, notorious for blowing up in one's face when fired and made intentionally with over-long barrels - the going price being a stack of beaver pelts as high as the gun was long.

I heard about a spot called Drunken Point on a lake that is part of the Berens River system, 75 miles or so upstream from where the Berens empties into Lake Winnipeg. Here the HBC would wait for the trappers to come down-river after spring break-up with their winter's catch of furs. HBC would have many kegs of rot-gut whiskey on hand and would get everyone drunk before ripping them off during the fur exchange, hence – Drunken Point.

As well, by meeting them up-river, the trappers would be relieved of their furs before reaching the big lake where they might meet Northwest Company traders and possibly get more value for their winter's work.

And, one day when Charlie read Jacob's mail I noticed Jacob looking worried.

41

"Trouble at home?" I asked.

"The Bay Store came to my house and took my Skidoo," he said.

"Why?" said I.

"They told my wife it was because I was behind on the payments."

"Why didn't you make the payments?"

"I don't use it in the summer," he said, "So why should I pay?"

Jacob had bought his family a fishing boat and outboard motor that spring. No doubt that will be repo'ed next winter after Jacob has bought another Skidoo. Can you spell rip-off?

I felt no sadness when the HBC Northern Division went belly-up a few years later.

The ultimate irony? The successor company is called "The Northwest Company," the same name as the one HBC had forced out of business 100 years before.

George Dram stepping out of the Norseman with our Fish Terminal reward.

Table Scrap - Fish Terminal

One day in early September we had just returned to camp when a Norseman landed and taxied up to our dock. Out stepped George Dram who owned Dram Airways at Wabowden. He talked with Charlie a bit and then we unloaded a number of fish boxes of fresh-caught whitefish. Seems that George, with an overload of fish from a commercial fisherman in Cross Lake, had spotted the Department of Transport Beaver at his dock as he returned to Wabowden. He snuck around behind a high ridge and flew to our camp to off-load some weight. Before dark he returned to pick up the fish and brought us a case of beer for our help. Cool.

Table Scrap – Archaeology

One beautiful July evening, with Charlie's permission, I take a short cruise with the canoe. I put a couple of good-sized rocks in the bow for ballast and head down the lake. There is a point about five miles south beyond our southernmost claim boundary and I want to see what is around the corner. As I round the point a large bay appears and on my right a beautiful beach curves off to the west in a long, lovely arc. The beach is about twenty feet wide and at least a half-mile long. It is clear, clean, golden-white sand.

As I chug along the shore imagining the tourist potential, I spot something sticking up in the brush just behind the tree line. I beach the canoe, tie it off and investigate. What I find is an old chimney, hand-made from local clay. It has crumbled to six or eight feet in height.

There are four upright poles sticking up on each corner, obviously part of the chimney and I assume, used to stabilize the clay until it hardened during construction. There is charred evidence of the building walls on the ground. The building was about 16' by 20' with a similar sized structure on the other side of the chimney. It had a hearth on each side and must have serviced a total area of 16' by 40'. It had obviously burned many years before. There is the remains of a birch tree that had grown and died where the building once stood. The rotting stub is a good ten inches in diameter and has been dead

The old chimney at Setting Lake. There is a duplicate hearth on the other side with both merging into a common flue.

for a long time. It is so interesting that I go back to camp and tell the others. They all want to see it, and I bring my camera this time. Charlie had never heard of the place, so on the 9pm party line he asks a guy who is from Wabowden if he knows what the deal was.

That guy writes a letter, and two weeks later we get a reply. His contact had asked his cousin, who asked an elder who remembered his granny speaking about a Hudson Bay fur trading post on that beach when she was a child. How about that, huh!

CHARLIE, WILLIE AND ME AT THE OLD CHIMNEY. THE TREE BEHIND CHARLIE IS THE DEAD BIRCH THAT HAD GROWN AND DIED AFTER THE BUILDING BURNED.

Table Scrap - Black Duck

One calm afternoon in early September we are motoring home on Setting Lake not far offshore when Jacob spies a black duck paddling along near the shoreline. When Charlie turns the canoe around, I'm thinking, "Now what?" – I should have known – Black Duck is going to provide us with some boredom relief.

So we spend a half-hour chasing a fat duck back and forth down the shoreline. When we get close to him he dives and doubles back. He can stay underwater for quite a while, but sooner or later he bobs up. We are never sure where, but when he does the chase is on again. After a while we tire of the game and go home, leaving Black Duck to his ducky ways.

I ask Charlie why the duck didn't simply fly off? He says that he is packing on the grub for the flight south and is so heavy that he can't take off without a good headwind. When he does fly south he will convert fat to muscle on his way.

It seemed like a mean game to play, but in retrospect I think we may have done him favour by giving him some exercise.

Charlie tells me these ducks supplement his people's diet in the fall. You can tucker them out and finish them off with a paddle, thus saving valuable shotgun shells.

We let him go because we have Mother Inco to feed us. Black Duck will return next year to raise another family.

a

On the Old Chimney beach. If you're going to sit on a big rock on a beautiful beach you might as well pretend you are quaffing a bubbly. Note the dead birch in the left background - the old chimney is to the right of the birch. Semi-naked dancing girls are offstage left.

On October 1st we say goodbye to Willie, pack our gear and move to Little Joe Lake, where we will finish out the summer season until freeze-up, `only weeks away. Willie is going home to Churchill and the Setting Lake camp will remain standing to await the winter season.

When we leave Setting we take all our personal gear, geophysical equipment and the canoe. From the kitchen we take all the fresh meat and vegetables and choose what we want from the canned goods. Anything partially used, for instance flour or dry cereal and some other stuff we don't want, goes with Willie to Churchill (with Moak's approval – better to use it rather than toss it.) All kitchen hardware stays to be used next winter.

From our tent we take only the mattresses, the gas lantern, the Airtight wood stove (with pipes,) the water pail and the basin. We don't button the tents up – the Moak Lake carpentry crew is coming tomorrow to winterize the camp.

It takes two Beaver trips to move us to Little Joe even though Willie has already been shuttled to Cook Lake. It wasn't that we have that much to move, but with the freighter canoe tied to the float struts Norm carries less weight in the cabin.

I am on the last load and as we lift off into a south wind and circle back over the camp, I get my last look and I think how great my first camp has been, and I also think I may never see this idyllic place again.

Little Joe Lake is not overly large but is more than big enough for a floatplane. It is situated about 25 miles south of Thompson and only four miles from the south end of the partially-constructed Thompson Highway.

The camp is in a state of disrepair. One tent frame has survived and Charlie and Jacob have already stretched a tent over the rafters. The kitchen frame is in shambles and the dock is not much better. No fancy floater here – it is made of two sheets of plywood nailed to a frame and sits on some posts driven into the offshore mud.

We button down the tent, set up our bunks and the tin heater and have just enough time to fry some chops on the Coleman gas stove before dark. It's a makeshift meal and I miss Willie. In fact I miss the four-star Setting Lake accommodations even more.

We are up before dawn, and when the sun comes up at 6:30 we are already at work. The kitchen frame has to be almost totally rebuilt and there is a lot of trash to pick up

44

and bushes to trim. Charlie tells me this was a four-tent/double kitchen camp in the old days. It has not been used for two years and it looks like it.

By late afternoon we have the kitchen spiffed up, our groceries put away, and we have a good supper. We have also set up a rusty, left-behind Airtight in the kitchen. It gets cold up here on October nights.

The kitchen at Little Joe. We must have had visitors. Jacob Brightnose centre rear, Henry Linklater on his left, Choochoo on his right, Landis Standlake in front.

We have not brought any kitchen stuff from Setting. Moak has supplied us with a "wanigan." (Cree/Ojibway for kitchen.)

A wanigan is a three-man fly camp's Swiss army knife. It is made of plywood (neatly varnished) and is perhaps 20 inches square and 30 inches long. On one of the long sides there is a strong, hinged lid. On the opposite side are pack straps. Inside the wanigan are nesting pots, a frying pan, a three-place setting of Melmac stuff, kitchen knives, utensils and cutlery. After everything has been unpacked, it becomes a table to eat on or a chair to sit on. They are very, very versatile.

The pack straps make me nervous, but in the years and wanigans to come I will never pack one further than from the plane to the tent or vice-versa.

The next morning we hit the bush to get our feet wet (literally) and return at 1pm to dry out. In the afternoon we cut up a big batch of firewood. We are keeping an eye on Jacob – yesterday when we were pulling the kitchen tent over the rafters Jacob's rope broke and he plunked his bum on a board with a rusty nail sticking through. Jacob just jumped up, rubbed his bum and we all laughed like crazy, but if he shows any sign at all of tetanus Charlie will call Moak for a quick flight out.

Progress is slow. It is wet every morning 'til 10 am and it's hard to get two non-rain days in a row. Our walks are not long, thank goodness, but the days drag on so slowly. We decide that we need a cabin fever reliever. Dr. Charles prescribes a case of beer.

Just across the bay is a muskeg tractor road which skirts the lake. Four miles north on that road is the south end of the graded 20 miles of #6 heading into Thompson. Thompson has a bar with Off Sales and Wray Dayson is in Thompson and Wray has a company truck – so two plus two plus two equals 2-4.

Contact is made – I don't recall how – maybe on the 9pm Sked or more likely by a note to Cook Lake. At any rate we are to meet Wray at 7pm to trade cash for bubblies.

Beer-run evening comes around and we eat a little earlier than usual. We have only four miles to walk on a good (?) trail and we figure on two hours. What we don't figure on is the muskrat.

J-5 Muskeg Tractor at a fuel dump near Little Joe. Jacob is in the centre, a driller on the right. The driver sat low in the front between the tracks. Note the bags of bentonite clay on the side racks.

At nightfall we estimate we have one mile to go. There is no moon and Charlie uses the flashlight sparingly – we did not bring extra batteries.

There is an animal on the road ahead of us! Sounds like a big one – hissing, spitting and gnashing its teeth. We stop, Charlie turns on the flashlight and there sitting in the middle of his road is a muskrat – five pounds of man-eating fury and he is not about to let us pass! If we go left, he goes right. The rat is a one-rodent Spartan army guarding the pass at Thermopylae. It takes us fifteen minutes to find a stick in the dark, and we use it to keep the foe at bay while we sneak past. We arrive at the rendezvous to find Wray waiting impatiently. We tell him about the rat and it's like trying to convince the teacher that the dog ate your homework. Wray doesn't believe a word.

We don't have time for much of a chat with Wray. He has been bucking deep mud for the last two miles and he hopes he can make it back to the gravel. We have to get going too – it's now a total blackout and we have four miles yet to walk.

Right off the bat we realize we should have used a smidgen of foresight before leaving camp. Two experienced, trap-line-hardened northerners and one now-not-so-greenhorn and we had neglected to bring a packsack! If you're dumb enough to carry a 24-bottle case of beer in the dark on a rough trail, you might want to try it some time. Muskeg tractor roads are like mini-roller coasters, and in the fall rainy season they are muddy and slippery. We don't dare carry the beer on our shoulders. A wipeout might break a precious bottle - so we tuck the beer under one arm and swap off often. We do have each other's back, though – if the mule stumbles the beer must be saved. Forget about muddy pants and shirts, they can be washed, but a broken bubbly is gone forever.

We pass the Muskrat Security Point but he has punched his time clock and has gone home to bed. We take a break and Jacob tells us a trap-line tale.

A few years ago he was harvesting muskrats just before break-up. He always carried a packsack, but to speed things up he would stuff rats in the side pockets of his cargo pants. When his pockets were full he would transfer the rats to the pack. One day a rat Jacob thought was dead revived and decided he would chew his way out of the pocket. Jacob says he didn't dare put his hand in to grab the rat and he did an impromptu ballet while dropping his pants. The rat got away, but Jacob's thigh still holds the scars. As he tell us this story, he is laughing like crazy – these guys will laugh in the face of certain death I am sure.

We make it back to the canoe and carefully motor back to camp. By now it is almost midnight and we crawl into bed exhausted, the beer unopened. It had been a long, hard yet interesting evening's work.

You might wonder, as I do now, why we would go to all that trouble for a 24 of Labatt's. We did not want to party. The beer – one bottle a day each before supper, will last for eight days. I think it is just that having that daily beer makes us feel that we are somehow connected to civilization.

Freeze-up is just around the corner and I learn that at this time of year the lake water gets heavier. I'm not kidding – usually our five-horse can kick up some spray from the bow, but as we cruise to the day's target the water just parts and flows past silently. It is usually dead calm in the mornings and even the underwater exhaust is muted. The moose are gathering to find a place to yard up for the winter and we pass a group of six, feeding along the shoreline. One day we catch sight of a big-antlered bull swimming across Little Joe. We circle him twice and let him go on his way. There are no ducks, geese or loons on the lake – they have gone south. The only birds left in the bush are chickadees and whiskey jacks.

Sidebar: In 2003 I was working at a tourist camp on Lake of the Woods, halfway between Kenora and the Northwest Angle. My main duties were to pick up and deliver guests to Kenora or the Angle. I drove a Lund Baron Magnum with a 200 horse Merc ETF on the transom. In between trips I carried a 2x4 and a hammer around camp carefully avoiding any area where there was actual work to be done. It was a tough job but someone had to do it.

Scooter was the camp in-house guide – in his mid-30s then, and had been on the lake since his high school days. Lake of the Woods covers one million acres and Scooter had tilled almost every acre – what Scooter didn't know about the lake was not worth knowing. He told me about the phenomenon called "turnover."

Here's the deal: northern lakes turn over twice each year – spring and fall. In the spring the fish turn over – in the fall the lake turns over.

In the spring the sun warms the upper strata of lake water and the walleyes and perch come up from the deeper parts to the reefs and shallower bays. In the fall the water turns over. The surface cools and the deeper water changes places, rising to the surface while the cold water sinks.

Good guides know when the turnover time arrives – the fish tell them.

At our latitude, Little Joe probably turned in September. In October the lake was preparing for freeze-up and the fish were already below, waiting for spring.

It's time to pull out. The last two or three days we have had to break some thin shoreline ice to get to open water. Moak gives us the word on the 7pm Sked – tear down tomorrow. The planes have to go back to base soon to switch from floats to skis. Keith and the 180 will go to the Pas and Norm will take the Beaver to the Austin Airways base at La Ronge Saskatchewan.

The next morning we are up before daylight to pack our personal stuff, and at daybreak we start to tear down the tents. There will be no breakfast today and no coffee either. It doesn't take us long – we are pumped. By the time Norm lands we have our baggage and the geophysical gear on the dock. Everything else is piled near shore, covered with the folded tents. The canoe is upended with the motor underneath. We have drained the water from the bottom end of the Evinrude. It and the canoe will be taken out by muskeg tractor before the snow flies.

We are on the first trip out. Norm will return for the leftover groceries and will fill the Beaver with as much as he can carry to Moak. What he can't take will stay with the canoe. The other camps have already folded their tents and Norm will leave for La Ronge tomorrow.

Norm drops us at Cook Lake and we load our stuff into Wray's 4x4 crew cab. I had thought we'd all be moving into a Midwest tent but Wray says there is only room for Charlie. Jacob and I will have to stay at one of the H-huts for three days until an empty tent is available.

Sidebar: The H-huts are aptly named. They are built in the shape of an "H" – two long sides, each having 50 double rooms – thus, 200 men to a hut. The crossbar holds sinks, toilets, showers and laundry facilities. There are four huts. Three of them accommodate up to 600 men who are working for the prime contractor building the mill, smelter and headframe. We stay in the fourth which is designated for salaried personnel only and which is seldom full. Jacob and I, who probably make less in one month than the rest pull down in a week, nevertheless qualify for VIP private room treatment.

In the middle of the four-hut quadrangle is the dining facilities, a huge one-story deal divided into four quarters. The kitchen is in the centre, and food is shuttled out cafeteria-style – choose your poison and fill your tray. It's not up to Tom's nor Willie's standards but there is a variety to choose from. H-hut "apartheid" is maintained. We salaried men have our own corner, so I'm filling my tray while straight across from me an ironworker fills his with identical food from identical stainless steel hot-table bins. Now, he knows the drill and he knows I am in the Ivory Tower section, and I am ashamed to say I feel smugly entitled.

There is a commissary here, somewhat like a US Army PX deal, and before we leave I buy a carton of Belvedere cigarettes. I am a sucker for a come-on. The only reason I bought that brand was because a spiffy Belvedere lighter came with. From the first drag I figure out why the lighter is free – after that carton is finished I never smoke Belvederes again.

Table Scrap

I didn't smoke before I got the Canadian Nickel job. I had tried it once in a while and it didn't appeal to me, but in the bush more than half my co-workers had the habit. Now, with snow on the ground, we would build a noon tea-fire and after lunch the tobacco would come out. Cigarettes would be rolled to be smoked while having our tea. I started practicing making roll-you-owns for Charlie and Jacob and then it was "Why not roll one for yourself?"

Before long my tobacco order joined the others being relayed to Wray on the Sked.

Sidebar: I still smoke (tailor-mades these days) and although I don't quite have to hide behind the barn, our lawmakers may yet force me to do so. It is even more difficult to roll my own. Players Fine Cut (if you can find it) is priced right up there with Maui Wowee. And if you ask for Vogue papers, the folks behind you at the checkout roll their eyes and nudge one another.

But those far-off memories lay soft and gentle on my mind – a soup can of sweet tea in one hand, a cigarette in the other, and the tea-fire turning to embers. Soon we will kick snow over the coals and strap on our snowshoes for another four hours.

For two days Jacob and I have been picked up at the hut and now there is room for us to move to the Midwest camp, which is halfway between Cook Lake and the H-huts.

This camp is a tent village with tents similar to Setting Lake. The main street leads off to the north of the Cook Lake gravelled road. Four or five tents line each side. The street climbs a gentle slope and beyond the tents is a conventionally framed kitchen/dining room. Beyond that there is a flat area with a large shop and lots of diamond drill stuff – this is a busy little town.

There are two geophysical crews working here this fall – Charlie and Henry Linklater (the other party chief) share a tent. Jacob and I share another with Henry's two crewmen. Both tents are at the bottom end of Main Street, and when we walk up to the kitchen we are expected to sit at the rear. Salaried personnel don't cut much ice at this camp – we know our Totem Pole status.

The first morning is a culture shock for this spoiled brat. There is no iron bar/triangle or wooden spoon/dishpan alarm. Grown men should know when breakfast is ready, so the first morning I go to the cookhouse and find that breakfast was ready at 6 am! There is no personal service here, and by 7 o'clock the precooked eggs are cold and rubbery, the toast is dry, and the pancakes are coffee table coasters. There is lots to eat, but I am disappointed with the set-up.

But the camp itself is not at all uncomfortable. Each tent has an oil stove and one light bulb – all we have to do is pull the string. There is a water pail for sure, but we don't have to fill it. The camp has a bull-cook who pulls a water wagon behind a muskeg tractor on his daily rounds. He even sweeps the floors!

Charlie and Henry have the map table in their tent and although I still do my corrections and my own mapping, progress is slow this time of year. However, this is semi-civilization, with a lot of activity round and about, so boredom is seldom a factor.

We are working off the new highway south of Thompson. Wray Dayson takes us out every morning and picks us up in the pm. Because the crew cab is too small for two crews, it's two trips out and two trips in. Wray has little else to do – we are the only crews working here. One other party is working out of Moak and the 180 is still on floats (Keith is just shuttling personnel in and out of Moak.) Cook Lake is still open – the coming winter has backed off for a few days and Cook is always the last to make ice – it must be spring fed.

Every day we pass through the mine site. On to the edge of the townsite we turn south on #6, and from day one the drive itself is an education.

Headframe at Thompson. Mill and smelter under construction nearby.

The huge, mine/mill/smelter complex is on high ground, of course, and is surrounded by a chain link fence. We pass through an open gate, pass the H-huts and construction hubbub and exit through a manned security gate on the east side. The west gate is unmanned and never closed. Why would anyone even want to escape to the Midwest camp?

Beyond the east gate a broad gravel road leads two miles to town, dipping into a black - spruce/muskeg area before climbing to high ground again. This muskeg curves around the south side of the mine complex towards Cook Lake. I will see it up close and personal in a couple of days.

The town is not much to write home about yet, but is expanding rapidly. Downtown is mainly the Thompson Inn, seating, I am told, 400 in the bar. The short main street has a smallish supermarket, a drug store, and a movie theatre. We went to a movie one night. In the lobby I could see where an eight-foot section of floor had sagged, leaving a six-inch gap at the base of an outside wall and insulation had been stuffed into the crack for a temporary fix. Some southern architect had a bit to learn about permafrost.

Northwest of Main Street, residential areas are under construction. As soon as houses are completed they are filled. Apartment complexes are also underway and there is a waiting list. Thompson is a zero-vacancy rentalsman's dream.

We dodge cement trucks, gravel trucks, shuttle buses and taxis, and to our left as we head south there are other camps belonging to contracting firms. One of these camps caters to small contractors and I hear it is definitely not up to H-hut standards.

Table Scrap

From time-to-time as we passed through the edge of the townsite I got the occasional glimpse of two tandem gravel haulers from Fort Frances. Both trucks were owned by a high school acquaintance, Dennis Busch. As a teenager Dennis' nickname, naturally, was "Sugar," The two trucks were designated as "Big Sugar" and "Little Sugar" – cool, thought I.

I didn't run into Dennis or his hired driver and the trucks disappeared before Christmas. About the same time we heard that there had been a riot at the sub-standard camp. It would take me 29 years to connect the dots!

In 1989 Pete Foster stopped in at our restaurant in Mine Centre every morning at 4am. He was pulling two loads of pulpwood every day out of Flanders Road South for Leon DeGagne. I had known of Pete when I was younger and now that I had met him I found that he also had many stories to tell.

Our combined reminisces came around to Thompson in 1960 – Pete was Dennis' co-tandem driver. I brought up the rumour of the camp riot, and Pete said, "That would be Dennis and me."

"What?" My big ears folded outward – I had to hear this.

What had happened was that the Thompson haul contract had gone south. Too many trucks were cutting the ton/mile rate and the sub-standard camp didn't help matters. Big and Little Sugar were loaded on a flatcar, the train would soon pull out and the two villains went back to camp. It was suppertime but they did <u>not</u> plan to eat – they were going to say goodbye. They waited until the food was on the table, stood up, dumped supper in opposite laps and coolly walked out the door to catch the train, leaving a riot behind.

A few weeks after the camp incident during my Christmas break I stopped in to see my brother-in-law at a Chev car/truck dealership in Winnipeg. While chatting I spied Big Sugar in the shop. Various guys were attending to him, laughing like crazy and having a great time. What's Up?

Turned out that Big S had a flaw. The 409 block had been poorly cast and one cylinder was more oblong than round. The big boy had been hard on oil since day one.

Dennis wanted a new block on warranty. GM would only kick in if the block was cracked and this is what the boys were doing – performing a big block crackectomy,

They alternately heated the block with acetylene torches, then poured cold water on it until the block cracked. Dennis got his non-oil-burning transplanted 409 back and I got a lesson on how some dealerships will go that extra mile!

I learn that while one can buy a case of beer one cannot take it back to camp. If you are caught trying to sneak beer or any spirits past the security guard you can be barred from the mine site, thus making it a virtual firing offence.

I learn that a school bus shuttles between the mine site and the townsite, with the final run at 10:30 pm. I decide to take a flyer with this method of joining the fringe element of Thompson society with mixed results, you might say.

I make a bus trip with Ian M. a member of Henry Linklater's crew. The bus is free, either owned or subsidized by Inco, and gives the mine complex workers a chance to hit the stores, or perhaps the bar at the Thompson Inn.

We must have missed a shuttle or two because when we get uptown there is a line-up from the Inn snaking back down the sidewalk almost to the movie house. We join the queue and at 7pm it starts to move. (Manitoba liquor laws mandate that a men's beer parlour must close between 5:30pm and 7pm. The Liquor Board calls it suppertime, others call it "ruining a good party.")

Ian and I hang in there and just as we make it into the lobby the line stalls. The bar is full, the doors are closed, and two burly bouncers stand on guard. If someone leaves the bar, another is let in – and it's not happening. We are still 50 feet from the bar doors and there is no point, so we do a 180 and join the line-up to the theatre. "NOW SHOWING – SPARTCUS." No kidding! After five months of 100% male companionship, are we going to watch Charleton Heston get all greasy while he gladiates? We catch the next shuttle back to the mine site vowing we will do better tomorrow.

We are determined to beat the system, so the next day we catch the bus right after supper. We hustle over to the Inn, and by golly, we are first in line. No wonder – it's just after 6 pm and we have an hour to wait, and is it ever boring.

The lobby fills up behind us and the line snakes outside. When I look back all I can see is thirsty eyeballs. The double doors are plate glass and the lights come on in the bar! Behind us there is an expectant murmur as hundreds of wristwatches are checked – five minutes to oasis time!

The two burly bouncers unlock the doors and swing them out. I am behind one – Ian behind the other, and we have a hope in hell of breaking out. The bar fills, the doors close and we are once more first in line.

We eventually did wet our whistles that night but it is obvious that plan B was no better than Plan A. The next morning, before leaving for work we hatch plan C.

The black-spruce/muskeg lowland circles the south side of the chain link enclosure. On the west side it becomes Cook Lake and on the east side it touches the road leading to the townsite. I have flown over this area at least three times since coming north and I know there are roads through the muskeg – winter roads to be sure, but roads none the less.

So let's recap. We can buy a case of beer, but we can't bring it through the security gate, and there are no security gates in the muskeg. Plan C is simple, needing only a packhorse to carry the 2-4 through the swamp. I volunteer to be that simpleton. I have survived the Little Joe deal and this will be a mile and a half at the most – what can possibly go wrong?

Today we are first out and first in, so at 3:30 Wray stops at the Inn and I buy a 2-4 of Labatt's Blue. Everybody has chipped in except me – mules drink free. I am dropped off well before the security gate and head off into the swamp, telling the boys I will see them soon. I disappear into the spruces and I mean disappear! Within 50 feet I learn that what looked so firm and walkable from the air is anything but. The snow has held off, and what little frost has arrived has not penetrated into this deep woods. The fall rains have done their job – the underlying peat moss is saturated, and I am already soaked to the knees. With 50 lbs of wobbly pop on my shoulder I sink no matter where I chose to step and there is no high ground.

I don't have my compass, but I don't need the mirror to tell me who is lost. I had thought the road would lead directly to the Midwest camp but it turns left, then right, and I have to choose a route at an intersection. It is a cloudy day, so I have no sun for a reference. I used to read Mark Trail's Helpful Hints for Young Woodsmen in the funnies – moss grows higher on the north side of trees. Not here it don't – every spruce has three feet of moss on every side. I thought I would be able to hear the noise of construction on my right but I am packed in deep cotton. The only sound I can hear is me as I huff and I puff and my feet make a sucking sound with every step. The only good thing is that there are no flies to add to my discomfort.

It will soon be dark and I am almost exhausted. I find a dry spot to set my cargo down. Surely I am lost but I am not worried – I just have to wait for nightfall. The bright lights at the 24-hour mine site will reflect on the underside of the low overcast and I will have my bearings. My only real worry is that the thirsty boys back at camp may hit the panic button.

Sidebar: One may be lost in the bush, but the real danger is in being lost in your mind. The woods will protect you – panic is your enemy. Four years later I was working for a different company and for a young man who told me this story.

He was eighteen and his uncle had hired him to help stake a claim group near Bathurst, New Brunswick. It was his first crack at that type of work and he was turned loose on his own line after lunch. He immediately got turned around in his head

Jim Cole has stopped in at the Dock on Cook Lake with Sioux Narrows Beaver ITW. The far shore behind him is where I emerged from my 2-4 trek and whistled at Wray.

and he panicked. He started to run, his uncle heard him hollering and cut him off. The rookie had run almost in a full circle. By the time his uncle tackled him he had no jacket, no shirt and they never did find one boot – PANIC KILLS.

So I decide to rest a while and just then I hear the 180 fire up on Cook Lake. I pick up my load and head toward that joyful sound. Within 100 feet I break out of the spruce and into the grassy swamp surrounding Cook just as Keith lifts off on his last trip to Moak for this fall. (He will leave for the LambAir base in the Pas tomorrow.) I am up to my hips in water now and as I part the last shore reeds I see Wray locking up the freight shed and heading for his truck.

I whistle, and my whistle can shatter plate glass at 100 yards. Wray, startled, looks across the lake, spots me, and I am saved. There is a canoe upended by the shed. Wray puts it in the lake and paddles over to pick me up. Of course he is laughing like crazy, but I am too relieved to take offence. He delivers me to the Midwest camp and I deliver the beer to the boys. I guess they enjoyed it, but as for me – I enjoyed dry clothes and a warm sleeping bag. (And I missed supper that night.)

The weather turns, the temperature plummets, the lakes are frozen, it snows daily, the snow gets deeper, and I am introduced to snowshoes.

Now snowshoeing is an art, and any budding artist should be mentored by a master – Charlie is such a master. I have now worked with Charlie for five months, and while I have learned so much from the man, I have also learned a thing or two about the

53

man himself. He is not big on lectures unless it is important to lay down the foundation, and this snowshoeing is surely an important issue because he gives me a tutorial.

Moak has delivered a selection of shoes and first we choose my winter underpinnings. There are two basic styles in storage, OJIBWAY and HURON.

Ojibways are pointed at the front. The birch frame has a canoe-like prow to prevent the shoe from digging into the snow with each step. Charlie says they are easier to handle in the bush but they are for men smaller than me.

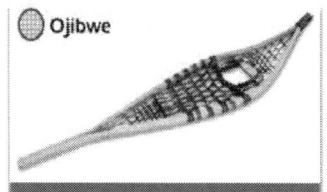 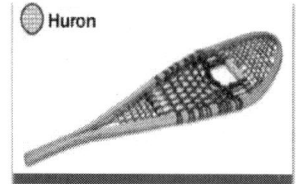 Hurons have a round tip, curved slightly upward, and being teardrop-shaped, are wider at the front. These are the ones for me.

Length is important. Forty-inch shoes are nimble but I need square inches, so we choose a pair 48" long. This is the type that Charlie uses. At 180 - 190 he could do with a shorter version but he has spent his winter-walking life on shoes and can handle the length. At 230 lbs I could use more than 48 inches, but it would take 40 acres to turn that rig around.

Next comes the harness – attaching same to the shoes and fitting my feet into the deal. CANICO supplies leather harness. They are easy to attach to the snowshoes and the user's feet if the user actually has normal feet. I don't – I have huge feet and have already had to deal with this issue.

A couple of months ago while still at Setting Lake I had pretty much destroyed the work footwear I had packed in June. I needed an upgrade and I needed help. Eaton's had work boots in many styles but only in sizes up to 11. Wray was asked to find me a larger pair at the lone Thompson haberdashery and thanks to Wray I was already a widespread source of amusement on the Sked. He had suggested a couple of small canoes (haha.)

But Wray did find me a pair of size 13 felt packs. These are Logan types – rubber bottoms, leather uppers, and a heavy felt liner boot. By removing the liner and cutting proper sized insoles from the felt I could fit my size 15 clodhoppers inside if I kept my toenails clipped. They worked well in the summertime – on snowshoes not so much.

The leather harness is "one size fits all." The toe strap and the heel strap are adjustable with little steel buckles and holes in the straps. I use the last hole but now my feet don't fit the shoes. If I have my foot too far ahead my toe hits the front crossbar and if I move my feet back, my heel hits the rear crossbar. One's feet should rest between these bars cradled by the rawhide webbing. When walking one does not really lift the shoe with each step – it sort of follows naturally. I do a field test – I have to lift the shoe, and this is an unnatural way to walk. It will have to do for now and the next day we go into the bush on snowshoes.

It is a brutal day for me. I have to lift my toes with each step and by noon the muscles on the front of my shins are screaming. I am slow and clumsy, and the worst thing is that I watch Charlie and Jacob handling their own shoes as if they were born with catgut beneath them. I am holding up production, but the boys are patient with the clodhopper millstone far behind them.

At noon we build a tea fire, and I welcome the break. After toasting our sandwiches we have a cup of super-sugared tea and Charlie tells a story.

It has taken me months, but now I have this crafty guy figured out. Once Charlie has given me a basic run-down on things that I must know, he changes tack. He never says "You should do this," or "You should try that," – he tells a story. Sometimes they are just stories, but I have learned to pay attention. Today's story is about a relative who, when working away from home, got caught by winter without his moosehide moccasins. He made do with a pair of rubber overshoes.

After supper that same day I catch the bus uptown and buy overshoes.

Now – with insoles and multi socks to keep my tootsies warm, I have flexible footwear. My feet are cradled properly and I don't want to lose these shoes so I write MINE on them with a black magic marker. (You may not believe this, but those snowshoes were waiting for me in camp on Jan 2, '61.)

The only remaining issue is the harness. Charlie and Jacob use lampwick, which has been the trappers' choice since the fur trade days. Before that (I am told) the people used moosehide strips. Lampwick has many advantages over manufactured harness. It is light, adaptable to any snowshoe, and is easily replaced. A spare leather harness can be carried in a lunch pack, but there's enough stuff in there already. Spare lampwick rolled up in your parka pocket is there when you need it.

Charlie shows me how to make a lampwick harness. It is so easy to do that I cannot possibly describe it in words. Like tying one's own shoelaces, tying a lampwick harness becomes second nature.

The next day in the bush, the change is dramatic. I'm still no speedster but at least you don't have to set up pickets to see if I'm moving. My shin splints don't hurt as much today but I'm still working on my thigh muscles. As I take each step, the inner part of the shoe tends to whack my stationary ankle so I do an unnatural swing-out step. I am watching Charlie and Jacob ahead of me. They tilt the shoe sideways with each step, so I try it. Good – no more whack, and within hours it becomes an automatic thing to do.

I am envious of my mates' moosehide moccasins. They look so darned comfortable – light, but warm. Today there is another teatime story.

Charlie and Jacob talk about moccasins – who makes them, different styles, etc. I'm up to speed.

I ask Charlie if he can get me a pair. He says he will ask a lady at Cross Lake at Christmas. He says we will draw an outline of my feet on a piece of paper so the lady knows what size I need. When we part a week later he takes the paper and $20 to Cross Lake with him. When I return on January 2 the moccasins are waiting with a note, and for one last time Charlie bites me again. The footwear and the note went to Wray Dayson – it would not do to send it privately to me – Oh, no!

The note, which Wray gleefully reads on the World Wide Radio Network on the first Sked of the New Year, says, "Tell Bob the lady had to shoot two moose to make his moccasins."

The Bigfoot legend builds.

I finish out the last two weeks at Moak chasing anomalies with Irwin Wilson. Tom has gone to dry out – his replacement would rank high in any other bush kitchen, but Tom is a hard act to follow. Charlie has gone home to build his pool hall – I will never see him again. Jacob will spend next winter in a different camp. I will pass

Hoooty the Owl

greetings to him on the 9pm party line, but we will also never reconnect. Willie has retired to Churchill, and the old gang is gone forever.

It's hard to feel bad when you're feeling good. We fly out of Moak every day in the 180 or the Beaver, and the common room now boasts an evening card game. At a penny a point one has to be very unlucky to lose more than a buck and a half.

Table Scrap – Hooty the Owl.

One evening I am heading to the kitchen for our 9pm pie and coffee. It is dark as hades and from just above my head comes a loud "Hoo!" It startles me and I duck into the kitchen saying I was attacked by an owl! Everybody laughs like crazy!

"That's Hooty," they say, "He lives here."

Hooty was not taken from a nest and tamed – he just showed up one day and never left. Voles and other small rodents are the bane of any camp, and Hooty took care of that business. When the rodent population cycle hit a low point, Hooty was put on a subsidy program. He was not a scavenger, but he appreciated the odd donation.

One lovely May afternoon the cook was having his siesta, and with no flies yet out and about, he left his door open. Hooty was jonesing for rodent replacement goodies, so he flew into the cabin and perched on the cook's footboard. The cook was sound asleep and Hooty was impatient – he woke the cook up. "Get out!" said the cranky cook.

"Hoo!" answered Hooty,

"You, that's Hoo!" and he threw his slipper at Hooty.

So I got my flash camera and no doubt temporarily destroyed Hooty's night vision.

And on December 3rd I climb into the 180 with my suitcase. I am going home! My packsack, eiderdown and axe stay behind. I don't know yet where I will be in January, but wherever it is, my stuff will be waiting.

I overnight at the Midwest camp. Wray will take me to the station bright and early tomorrow morning. There are two other guys who are also leaving. We decide to visit the Thompson Inn one last time this year.

It is busy, but not packed – construction slows down during the cold winter months. One of my tablemates is a three-year Canico veteran. He tells me that when the Inn first opened the capacity rules were even more loosey-goosey. If a chair was not available, you bought one – a 2-4 case of beer on its end became a chair.

On one end of the bar sits a board – three feet long with four spotlights attached. I have seen this before and had wondered what the deal was. The vet tells me these are trouble lights – what trouble?

Now, you have to understand that Thompson, in its infancy, was lacking in electricity. Light plants – many of them – were the main source of power. A dam had been built on the Burntwood River and the grid is being switched to water-generated power, but outages are not uncommon.

Tonight the Inn has such an outage, and the bar goes basic black. The battery powered spotlight board comes to life, but does not light up the entire 400-seat bar. The vet tells us to stick our heads under the table. I am still wondering why, when, from a dark section I hear a bottle smash against the wall. A bartender shifts the spotlights to that direction and now our section is dark. Another bottle hits the wall not far away, and believe me – we are definitely under the table now.

The lights come on, the spotlights go out, and the bar chatter resumes – not a guilty party in the crowd. It's just a pleasant, normal evening at the Thompson Inn.

We only have a couple and catch the shuttle at 9pm. There are not many on the bus with us and they are mostly middle-aged gents. We are feeling Christmassy and sing a carol or two, and others join in. We are not doing a bad job of it, either. I look back as we pass under a streetlight. Two of the older gents behind us are singing up a storm with tears rolling down their cheeks. I am surprised and touched by this unexpected display of loneliness.

Sidebar: I was told that when the Inn was built it straddled the original Thompson geophysical baseline. If you were doing base station checks on a mag survey you could walk in with your Sharps A-2, take a reading, have a cool one and continue on out the back door.

I'll believe anything, but I found that one a bit iffy. Common room/coffee klatches can get pretty inventive.

Table Scrap

In 1960 Canada was emerging from a mini-recession. Jobs were becoming more plentiful and not everyone wanted to brave the isolation and winter cold in Thompson, so immigrant labourers and tradesmen filled the gap. A lot of these guys were from Portugal and their families were still back in the old country. No wonder they teared up when the glee club sang in the bus.

I had seen these gents uptown or walking to and from the H-huts. When the weather got cold they still wore a fall type of coat with an old parka hood sewed onto the collar. I thought this was carrying economy a bit too far until I was told they were saving their money to bring the family to Canada. Another plan was to amass a nest egg at home. A small apartment unit could lead to easy living in a simple, sun-drenched Portuguese town.

I <u>did</u> hear about one guy who was a common room legend, economically speaking. He was a Canadian – a recent geology graduate on his first full-time posting. No one knew why he was such a cheapskate. Perhaps he owed money for his schooling, or perhaps he was a natural-born miser – but he refused to spend his cash. They say he came into Moak with two shirts and two pairs of pants and left with the same clothing. We all used canvas sample bags to patch worn jeans, but he even carried that to extremes – his pants were all sample bags, and if someone tossed a pair of boots he salvaged the leather uppers to re-sole his own. They say he bought toothpaste only – perhaps twenty bucks in two years. They say he refused to take the seasonal break until they had to fire him to get him out of the bush. They say he left with six months of accumulated time, and because of him, Canico had to write the seasonal time off rule. They say no one knew where he went and no one cared.

I go to bed a Midwest but I have a hard time falling asleep. It is like when I was a kid and Santa Claus is coming. I wake up early, too wound up to eat breakfast, and I'm ready to go long before Wray picks us up. I find a day coach seat on the crowded train and before long the permafrost shakes the nervousness out – the Bay Line will do that for you.

It is a long, old, boring haul by rail to Winnipeg. There is plenty of time to anticipate nearly a full month of family love and Christmas visits. There is also plenty of time to think of the last six months. So I anticipate and I think and I reminisce and I become introspective.

Table Scrap

The tea-fire tales can get wild and woolly. To settle things down someone will say, "You B.S. your friends, and I'll B.S. mine, but let's not B.S. each other."

So I'll tell you no lies. I was 21 years old and not a philosopher by any stretch of the imagination. I have a few things on my mind as the train pulls out of Thompson; the main thought being, "Let's move it on down the road." I had money in the bank – not a whole lot by today's standards, but enough to buy Christmas presents and perhaps a party or two. I looked forward to spending a few days in Winnipeg with my sister's family. I will rent a new Chev Biscayne for three days, so I won't have to impose on my brother-in-law to chauffeur me around. I will bus it home to spend three weeks with family and friends. I look forward to Dad's approval. I won't have to bum money from him and I can tell him that I am actually going to keep a job for a change. (He doesn't say much – he never does say much.) I also look forward to my little brother being impressed. And, of course, that's also hard to tell.

But now, more than 55 years later, I can pretend that a younger me actually did get introspective on that train ride, so let us introspect and reminisce together.

When I look back at that season at Setting Lake so long ago, I marvel at how steep the learning curve was. I thought I was a man of the world, or at least of North America. In 1946 my family had moved from New Brunswick to Manitoba, half-way across Canada, with all nine of us and our worldly possessions (including the piano) stuffed in the back of a 45 Ford two-ton and a '37 Plymouth 4-door sedan, and just one year before boarding the train for Thompson, I had hitch-hiked from Toronto to Winnipeg, then driven to Texas in a '47 Chev! I then followed my thumb to the Peace River country in Northern Alberta and back to Toronto - over 10,000 miles on less than $200 – that's two cents a mile!

I had swum in the warm Gulf of Mexico, baked in the hot sun of Western Texas, and spread my bedroll beside a grease pit in Roswell, New Mexico. I had hit the harvest in Nebraska. I had written my name in hot asphalt in Wyoming with the snow-capped mountain peaks off to the west so inviting to my parched throat, and marvelled at the depth of the Peace River Valley at the Dunvegan crossing north of Grande Prairie, Alberta.

I had been there and done that, I thought, but when I stepped off the 180 float at Setting Lake, I was a babe in the woods.

I was so lucky in so many ways that summer. The camp was almost ideal for a starting point in my bush career. Subsequent camps would run the gamut from B to Z, but Setting Lake was a straight A.

I was lucky to have been introduced to flying by two most excellent bush pilots. I didn't fly often that first summer but I learned a thing or two, one being how to load an aircraft. (Heavy stuff to the front – lighter stuff to the rear.) Another thing I learned was to always trust a good pilot. In the future I will learn it is pretty darned easy to separate the good from the not-so-good. In fact, many chapters onward (in Book Two) you will read about how I almost got out of a 180 before the idiot started the engine.

And – was it blind luck, or was it predestined that I would start my bush career with Charlie? I was plunked in Setting Lake in June. Charlie, Jacob and Willie had been there at least two and a half weeks before I arrived. Summer seasons always started following spring break-up, and while the ice would have hung around at that latitude later than back home, May 15 to May 20 would have been the kick-off date. Canico always planned ahead and crews would have been set up in April. Had a summer student failed to show? Did someone bolt after the first deer-fly bite? Whatever the reason, I drew Charlie out of the deck and thus I was on my way.

There were some tough days that summer, but as the weeks and months went by it became routine. The flies were the hardest to get used to. I had expected to deal with mosquitoes, but they were only an irritant. What I didn't count on were the clouds of black flies. Who invented these evil insects? From day one they did their best to drive me nuts. I ingested them on the picket line, I shared my lunch with them – it was impossible to eat a sandwich without eating a black fly or two. They liked to draw blood – not to fill up like a mosquito – they just took a gulp. I learned to keep my shirt cuffs buttoned and to filter my oxygen through my teeth, and I learned to live with them.

The deer flies were something else – I had never seen a deer fly before. They were fast and they were mean, about the size of a large housefly with dark stripes on their wings. They operated as a hit-and-run surgical strike team – get in and get out, and they really drove me nuts. They would bite through a sweat-soaked shirt and would even crawl into the short hair on one's head. When it got really hot in July I fooled them. I shaved my head and doffed my shirt, preferring to put up with a few scratches rather than deal with them. With repellent slathered on from my beltline northward those deer flies just skidded off the runway. Charlie and Jacob thought I was goofy but I had to do something.

What really pissed me off was that those two had little trouble with bugs, and I noticed one very interesting thing – natives don't sweat – at least not those two. Perhaps it was because I had to work twice as hard just to keep up with them, but I seldom saw them sweat.

(A few years later I developed a theory fly-wise. By my third summer I was not using nearly as much repellent, and although the flies were still out there I seemed to have built up immunity. Charlie and Jacob had dealt with flies every summer of their life – perhaps they had been inoculated years ago.)

Other days were tough for different reasons. We might get caught in a cold shower. Warm showers on a hot day were a relief – a cold shower on a cold day was miserable. Then we would beat it back to camp, build a fire in the tin heater and soon we were comfy in dry clothes.

Some days the dreaded sunspot diurnals would mess with me and my magnetometer. After evening notebook corrections were made, Charlie would tell me my day's work would have to be redone. I learned to take more frequent base station checks. A little extra work today saves a re-read tomorrow.

But the bad times were never a deal breaker, and that was because Charlie, and to some extent Jacob, always tempered adversity with humour.

Charlie made a man out of a malleable hunk of clay that first season. Had I worked for a lesser mentor I might never have made it. In the coming winter my party-leader would be a non-Charlie – a sometimes grump who thought he could push me around. Thanks to Charlie I was so well trained and secure in my own skin that I could deal with the man. Once he learned that I knew my place in the Northern Sun he backed off, and the rest of the winter ran smoothly.

So Charlie set the bar, and I am forever grateful. I would try to emulate Charlie in the years to come with mixed success. I am not Charlie, but I came close from time to time.

Table Scraps: Other notable memories.

The Noon-hour Message Period.
The kitchen always smells good: steaks or chops may on the stove, fresh pie cooling on the plywood counter, and tea or coffee in the pot. Wonderful! On rainy days we have lunch in the kitchen and listen to Willie's radio.

"This is CKDM Dauphin Manitoba, 800 on your radio dial – here's a song by Georgie Jones, the Beaumont, Texas Lad:"

"Open up your heart and let me come in,
I've got no hair on my chinny-chin-chin.
I love you and I want you and I need you too,
I'm not the Big Bad Wolf,
I'm just your Little Boy Blue."

(Thirty-five years later I would attend George and Tammy's "Together Again" tour at the Winnipeg Arena. I'm not a concertgoer, but I just had to see the Beaumont Texas Lad.)

Now it's time for the Northern Message Period, brought to you by Ostrowski Furniture with outlets to serve you in Swan River, the Pas, Flin Flon and Churchill. Remember, Ostrowski pays half the freight.

First of all – to Nelson Longstreet at the Two Bridges sawmill: Harry twisted his ankle and can't come to work next week,

To Erasmus on the trap-line: Mary had a fine baby boy this morning. Meet them at Thicket Portage next Friday. Signed, Auntie Helen.

This one goes out to Belle at Cormorant Lake: Dad is in jail in the Pas for ten days. He will released next Tuesday. I will bring him to The Narrows. Don't forget to lift the nets. Signed, William Jr.

The next one is to James on the cat swing, somewhere between Ilford and God's Lake Narrows: Betsy has a bad cold. I will have to take her to the doctor. Please send money. Love, Verna.

Our last message goes out to my dear Daisy: I will be on the train tomorrow. Keep the bannock and beans warm, if you catch my drift. (This one is unsigned.)

The message period was always great to listen to, and what amazed me was that no matter who the message was from, or to whom it was going, Charlie, Jacob or Willie knew the people, were related to the people, or knew someone who knew the aunt of the sender or recipient.

Every day, seven days a week, the messages went out. Any community with an HBC store or a church had access to a Manitoba Telephone System radiotelephone, but the calls were not cheap. CKND accepted collect calls for the message period.

Now, I can imagine that little Dauphin station with its wide listener base must have generated a fair bit of advertising revenue. The last time I listened in 15 years ago, they had a Saturday Night Polka Party program. Do they still do messages? I doubt it, but perhaps they do.

I go back to the kitchen in the afternoon to have a cup of coffee and play cribbage with Willie. We don't expect a big meal on weather days. We snack, sleep, do a little map work, and tell lies. Willie has switched his radio to short-wave and is monitoring Moak Lake's channel. A junior field geologist is reporting a hole from a drill camp and we listen in. It's probably a hot prospect and the Moak geologist wants an up-to-date reading on the core. The airways have a thousand ears so the report is in code.

Hole S-12 – sixty-five degrees - collared at 50 feet; 25 feet of mixed pickles, three feet of peaches, four feet of cherry pie, six inches of pears, followed by potatoes and gravy. Core is fun to listen to, even though it means nothing to us.

Sometimes we pick up a skip. Skipping, as explained to me, is a phenomenon of sorts. Most radio systems have their range limited by power output, but atmospheric conditions can result in skipping, whereby a signal can travel unusually far. When I was

61

at Lambair in the Pas, they talked to one of the Lamb boys at Baker Lake, at least six hundred miles to the north.

So we are listening on our static-filled company frequency when a voice booms out clear as a bell.

"Buffalo Narrows, La Ronge, Buffalo Narrows, La Ronge, do you copy?" back to static, then, "Buffalo Narrows, roger on that, what is your weather?" more static followed by, "Roger on that, Buffalo, we are CAVU (ceiling and visibility unlimited,) winds southwest at 29 knots." Then we lose the skip.

The boys tell me that Saskatchewan Government Airways services the northern part of the province from their base at Lac La Ronge, 300 miles to the west of us. They use the same frequency as Moak, but because of the wide area monitored, they have oodles of power. Moak, by comparison, needs to cover an area of 100 miles, so our power output, as mandated by the CRTC (Canadian Radio and Telecommunications Commission) is pretty feeble.

I think this is wonderful, and as I snuggle into my eiderdown that night I can once again envision thousands of concentric airwaves connecting remote communities, trap lines, and commercial fishing camps throughout the north.

Another weather day sticks in my memory. The wind picked up at midnight and awoke us. Charlie says we are in for a good one, so we go down to the lake in the darkness. We make sure our floating dock is lashed tightly to the shore, pull our canoe up on to the dock, stow the life jackets and gear in the old steam bath, and then return to our bunks.

All night long the wind increases, coming straight up Setting Lake, and by morning there are five-foot waves crashing on our rocky point. Before breakfast we check the ropes on the tent flys and make sure everything is snug. The wind howls all day long, and today we don't do much except read and nap. An odd shower hammers the tent fronts, and when we hit the kitchen for a snack, we don't dawdle. In fact, we have to lean south during our 25-foot dash; otherwise we may end up in the bush behind camp. There is no radio today. The storm has reduced it to static and garble, and tonight there is no Sked. No one has any work to report, anyway. Even the camps on smaller lakes take the day off – it's not worth it to chance falling trees, and besides, the wind makes voice orientation impossible.

That night the wind dies down and by morning there are just some residue swells left on the surface of the lake. No damage done, although we lost a couple of trees behind camp. Back to work, doggone it.

Coleman Lamps

The Coleman lamps were constant, faithful friends in any camp. They used white gas (naphtha) which when pressurized by a simple pump in the lamp base, fed through a carburetor, where the fuel vaporized and burned at a mantle, giving off a warm, bright, cheery light. As a bonus, in addition to giving off adequate light for map work, reading, letter writing, or just plain story-telling, it emitted a comfortable hiss that told you it was working correctly. A full lamp would always last an evening and bedtime would roll around before the lamp was empty. Nails in the ridgepole were our lamp hangers, one over the map table, and another in the centre of the tent where everyone could share the light after the homework was done.

Camp protocol, unwritten but tacitly understood, was nonetheless fairly rigid. The low man on the totem pole had two main responsibilities: keep the water barrel in the kitchen full, and keep the lamps full. So, in Setting Lake in 1960 I was lamp and water boy. Keep things topped up was our motto. Every day after changing into our formal hang-around camp clothes, I would fuel up the lamps while Jacob gassed up the outboard motor and checked the spark plug. If Willie had been hard on water, Jacob would help me carry pails up from the lake.

The white gas was treated with great care. Water contamination would plug up the lamp's carburetor, necessitating a change-out, not a big job, but why chance a shortage of spare carburetors? Our naphtha gas in those days always came in ten-gallon drums, which we always stored in a dry place. We built a simple cradle to hold the barrel on its side, put a tap in the small bung hole and filled the lamps through a small funnel with a felt insert filter, which caught any contaminants, including water.

The long summer days were not big lamp-time-wise unless the evenings were dull and overcast, so the fuelling-up process was not as critical as when days got shorter in the fall. By the time the winter of '61 rolled around, I only had to maintain our lamp. The larger kitchen accommodation for the cook and bombardier driver made two lamps necessary over there, and the driver kept them operational.

In later years the Coleman would always be used in fly camps and snowbank set-ups, but in larger camps light plants started to arrive. The hammer of a two-cylinder diesel was and is completely devoid of romance.

I miss that soft hiss of the lamp, accompanied by the quiet turning of pocketbook pages, punctuated by the splash of a jumping fish or the call of a loon on the bay out front answered by a far-off echo.

Sidebar: In 1969 I was running my own show and decided to go upscale with a propane light system. I learned a lesson: you can't make an axe-carrying bushwhacker into a propane technician. After buying double fixtures, 200 feet of copper line and routing everything meticulously, I would arrive in camp to find one disaster after another. Someone would forget to check the tank which would run dry after dark, the tank changer would neglect to change the regulator resulting in blown mantles all around the camp, or just to be inventive reroute my system and leave a fitting loose then wonder why the propane didn't last.

Eventually the propane system was donated to a normal person and the Coleman lanterns returned.

Necessity is the Mother of Invention

This story I got from the horse's mouth in the winter of '61. He was Bob Tilden – a party chief at another camp who had been around for a few years – this was not his first rodeo.

First we must set the stage: Christmas break 1960. The Bay Line meets the east/west CNR mainline at Hudson's Bay Saskatchewan, just over the Manitoba border. When I go to Winnipeg, our train hooks up with the eastbound here, and there is a good connection. Bob and his group want to go to Prince Albert Sk. and there is an eight-hour wait for the westbound, so Bob had left his car in the Pas – he and four others will drive to P.A.

It's a dicey deal. A new road has been built between the Pas and Carrot River, Sk. where it joins an existing highway. Bob isn't too worried. He knows the road is new but it's wintertime, and it will be plowed and relatively smooth.

They had left Thompson a day or two ahead of me, arriving in the Pas well after dark. The car, a '52 Meteor, had been sitting all summer at the train station. Now it was a brutally cold evening and the six-volt battery was stone dead, so they called for help and a push started the car. It ran well with the generator supplying the spark plugs. It was too late to buy a new battery so they shoved off into the wilderness.

It is over a hundred miles to Carrot River with nothing but white highway and high snowbanks ahead – zilch for habitation of any kind – so of course, fifty miles later the Meteor died. The battery had thawed, lost its water due to a cracked case, and had shorted out. They can run on the generator but they need a battery in the circuit – what to do?

Keeping warm is not a problem. They have axes, and there is bush here – nothing but.

The problem is time. They want to get home. It is after midnight and the road sees little traffic even in the daytime. It is clear and cold – snowplows will not be running.

Two of the boys have radios powered by two 1.5 volt "D" cell flashlight batteries. The radios are multi-band and only partially transistorized. They are energy hogs, so the boys have made up battery packs – four "D" cells wired in parallel pairs give the radios longer life.

The solution is simple and the operation is performed by the light of a campfire in – 45 degrees "F" (for freezing.)

They rewire the "D" cells in series in two tape-wrapped bundles of four, and hook the two bundles in parallel to the disconnected battery cables. It isn't enough juice to roll the starter – they have to push-start the car – but the batteries supply the initial spark to the coil until the generator takes the load.

They make it to P.A. with "D" cell power to spare.

Bob said those two boys were certainly inventive and their mothers were gratefully acknowledged.

Procrastination can also be the Mother of Invention

This is told by my retired OPP friend – therefore it will hold up under cross-examination.

He was based in Moose Factory on James Bay and one day he was flying up to Fort Severn, farther up the west coast of Hudson's Bay. The aircraft was a conventional government Beaver – the chauffeur a conventional bush pilot. (The term "conventional" draws a broad brush stroke.)

My friend noticed a jury-rigged tube leading from the firewall to a funnel near the pilot's right leg. On the floor was a jug of engine oil. Halfway to Fort Severn the pilot uncapped the oil and poured some into the funnel – what's up?

64

The pilot said he had an oil leak. He didn't want to land every 150 miles to add oil so had developed an in-air oil maintenance program. When the oil pressure dropped, he gave the Beaver a transfusion.

"I'll have to get that fixed one of these days."

The next morning, 24 hours after leaving Thompson I detrain at the Union Station, blaze a picket line to the nearest Avis office and I am back in the land of indoor plumbing until January One, 1961.

Chapter II Thompson - Winter Season '61

I arrive in Thompson at 9 pm January 2nd, having left Winnipeg on New Year's Day. Canico men have joined the train en route but I see only two or three familiar faces. There may be a few others I know but we are spread out in the six passenger cars. There are no empty seats – the first shift of the year is on its way back to the Nickel Belt.

The old-timers know what to expect when we pull into the station, but I am downright impressed. Lined up beside the platform are five Canico Bombardiers, numbered 2 to 6. On the platform, Wray Dayson stands holding a clipboard – he is directing traffic. When Wray spots me he simply says, "Bob – number 6 – Setting Lake," and I am checked off – present and accounted for. I feel pretty good that I will be hitting a familiar campsite.

A convoy of three loaded Bombardiers pulls out of Thompson and heads south on Hector's Highway #1. Two units take men to Moak on #1 North.

Table Scrap: Hector's Highways

The highway system had been in use for years, evolving as the Nickel Belt claim group expanded. The northern terminus was Moak Lake – Land's End was at the camp on Setting Lake. Along the way feeder routes (secondary class B) led to present and past camps. Like any rural highway, other side roads accessed smaller, temporary whistle stops such as drill camps.

Hector's #1 South was pretty much all-weather and stuck to higher ground, thus did a fair amount of twisting and turning. Naturally travellers had to know where they were at, and where they were going to, so signs were nailed to trees at intersections.

Between Moak and Thompson the Mystery River crossing isolated Moak in the summers. From Thompson to Setting no major rivers cut the highway, although there were small creeks along the way. These were no impediment to winter traffic, and only muskeg tractors used the highway in frost-free months – they could handle the occasional damp spots.

There was a fuel dump approximately halfway to Setting Lake: no Esso sign – no pump attendant – it was gas-and-dash.

Hector was the transportation guru at Moak. He could be a bit autocratic, some folks say, and to honour his exalted (in his own mind) position, his men had named the highway after him.

Hector also had his own personal Bombardier. Naturally, it was Canico #1. It was never used for freight. It was as clean as a whistle inside and the blue exterior was polished every fall. He usually had a chauffeur, and if he drove it himself he had valet parking. Hector travelled business class.

(I had seen seven Bombardiers lined up at the Moak Lake shop last summer. Where was #7? I am told that unlucky 7 had met some thin ice before Christmas. No lives were lost – only a soaking wet, shivering driver's pride took a hit. #7 was retrieved, but the motor was running when it went into the lake. When the running motor took a drink of water the result was a bent crankshaft. Now # 7 was used for spare parts.)

I had been told last summer about the highway system, and had seen glimpses of it from the air, and had walked sections at Setting and Little Joe. Now I am travelling on #1 and soon will be using some feeder routes.

Actually we do not leave on Hector's #1. The first twenty miles are on the finished stretch of Manitoba #6, then we pick up #1. For the next 50 miles we will criss-cross the cleared #6 right-of-way before finishing the last six bush miles to our camp.

We are third in the convoy. Ahead of us Neil McAskill's machine peels off at the Grass River North exit (yup, road sign.) Eight miles onward we lose Bob Tilden, on his way to Grass River South and we motor on alone. Bombardiers are no speed wagons. They might hit 30mph going downhill with a tailwind, so we arrive in camp at 2 am. A light is on in the kitchen, hot coffee is on the stove and two fresh apple pies are on the table.

We hadn't talked much on the trip south. Conversation is difficult in a dark, bouncy, and slightly noisy snow-runner, but now we can be properly introduced.

My party chief is Harry from Preeceville, Saskatchewan. This is not his first season, but it is his first crack at running a detail/review camp. He has four men this winter. Besides me there is Art the axeman, and a new kid – Harold H. Art is from Oxford House, further up the Bay Line, and Harold is from Prince Albert The Bombardier driver is Keith Devine from Carrot River, Sask. and the cook is Karl Muzyka from McCreary, Mb. I am the only Ontarioan on the Nickel Belt this winter.

We don't chat long. We are all tired and we go to our tent and hit the eiderdowns. The tent is toasty warm – Keith had fired up the oil stove before going to Thompson to pick us up. Tomorrow will be a day of organization, and we will be on our snowshoes the day after.

Keith Devine and I share some absolutely true stories on my cot at Setting Lake. Note walk-in closet behind us.

The Setting Lake camp has been tuned up and waiting since before freeze-up. I will now describe the modifications to our camp and I assume the others are similar.

The only real change to our tent is the installation of an oil-burning space heater to replace the wood-burning stove.

The oil tank is a 45 gal drum outside the tent. It has been placed on a cradle lying on its side and a half-inch copper line runs under the tent and up through a drilled hole in the plywood floor to the stove's carburetor. The line is wrapped in burlap to prevent freezing and if a line does gel on a bitterly cold night we can switch to a small tank attached to the stove. This is for emergency use only. Because the stove burns virtually

24/7 the small tank will only last four to five hours. Keith tops up the outside tank every three days. (These tents – though never cold, are not exactly up to R20 code.) The only other mod required for our quarters was sealing the screened window over the map table. We won't need cross-ventilation this winter.

The kitchen has undergone an extensive renovation – it is now 16'x32'. This was accomplished by grafting a second tent onto the rear of the summer kitchen. With some alterations to the canvas it is now one large room. Keith and Karl live at the back and extra shelving now holds all the groceries. There is still a storage tent outside that is now only used for our work gear and Bombardier stuff.

Karl Muzyka and me at the kitchen door.

There is an oil stove in the back part of the kitchen. It helps heat the front part, aided and abetted by the ever-busy propane range.

Keith has had a lot to do since his arrival just after Christmas. First he has had to break trail to camp. (Bombardiers are not great at breaking trail in heavy snow - even worse if they are heavily loaded). He has hauled barrels of stove oil and gas from the fuel dump. He has cleared snow from the tent flys. (Snow does not slide off the canvas and we don't want the fly touching the tent roof.) He has fired up the oil stoves for a field test. He has transported propane cylinders, snowshoes, axes, and the Lancaster radio including batteries and the one-cylinder motor/battery charger. He has brought our gear from Moak – the EM and mag, and also eiderdowns and packsacks left at Moak way back in December. (We own our own sleeping bags, but Canico pays the shot to have them dry cleaned once a year. They send them to good old Woods Tent and Awning and they return daisy-fresh – Woods even replenishes the goose feathers.)

Keith has cut a water hole in the ice in front of camp and will keep it open every day. He has packed a runway on the lake for our weekly supply plane and marked it with small spruce trees. While packing he has eliminated any chance of slush. All lakes can be susceptible to slush conditions and slush is an anathema to Bombardiers and aircraft on skis. A large lake like ours is generally slush-free but when the snow gets too heavy, water can come up through cracks in the ice. The snow cover is an insulator and the slush will hide below.

And two days ago Keith had brought in Karl with the initial grocery order - so when we arrive, the camp is ready to go – It is a turn-key deal for us guys.

No one expects us to be up very early and we start to straggle into the kitchen at 9 am. Karl has heard us shuffling around next door and he has the coffee on. Breakfast is not a big deal today – we have to clear the cobwebs and get up to speed. We will have a late lunch and a light supper. Tomorrow we will hit the picket lines and our strong appetites will return.

We spend the morning in the kitchen, common-room style, establishing credentials and getting better acquainted. We will be spending the winter together, so we have to scope each other out.

Keith Devine – I like him right from the get-go. He is perhaps 25, single, has a cash crop farm at Carrot River, and has been coming to Moak for a few winters to earn off-farm income. He tells us he puts on 30 lbs every winter and works it off every summer. I am impressed, but I have reservations. Does he lose the pounds because he works hard, or is he just a rotten cook? (He will prove to be a bun-warrior.) Keith is a pleasure to work with, eternally pleasant, and does his job well.

Art is a cipher – a silent presence in camp. He is an excellent man with an axe and can handle the EM receiver if necessary. He speaks perfect English but seldom speaks, and is also a whiz on his Ojibway snowshoes. Art is our trail-breaker, but the little bugger is only 150 lbs and I still sink halfway to China when I follow him. Occasionally I will take the lead just as a matter of pride, then I power out after a hundred yards and Art flies past. He can break trail all day and never break into a sweat.

Harold H. is the new kid on the block, a tall, raw-boned country type. He is puppy-dog friendly and puppy-dog clumsy – always puppy-dog happy and willing to work hard. He won't be with us long, though. Within a week or three a man will drop out at one of the Grass River camps, to be replaced by Harold, but by that time Moak has realized that with Art and me on the job Harry does not need a third man. Harold's bed will remain empty until spring, occupied only by the occasional visitor.

Side bar: Harold was a magnet for good-humoured teasing. One night on the evening Sked his new party chief told Moak that Harold would like to have fresh milk in camp. We didn't need to hear the result on the radio – the whole Nickel Belt erupted in thigh-slapping laughter. The Cree word for milk is Choo-choo-sa-boy. Harold became Choo-choo, and by spring no one could remember his real name.

Karl is a good cook and a pastry specialist. He owns a locker plant/butcher shop back home in McCreary, which the family runs every winter while Karl is on the Belt. He likes to cook for hungry men and we will all put on some suet this winter. (That is – all of us except Art – he is an ectomorph.)

Karl is also (uncharacteristically cook-wise) always happy – in Karl's life the sun always shines 24/7.

That leaves us with Harry, and I can tell that this man will need an attitude adjustment.

He is pretty much full of himself and wants to show us he is the top dog. It seems that Moak has told him that I am only a mag man – I tell him I am no slouch on the EM. Once he finds out that is true he will start to off-load some of his own work onto my

69

shoulders. I think he is a bit lazy. He will work five days and day-off for a day, saying he is not feeling well, or saying he has map work to do etc. etc.

The thing is, on those days he only has EM set-ups for the daily progress report on the Sked. He will report mag readings but to do so he has to borrow from the bank. I don't like my savings withdrawn just so he can have a day off and I finally have a chat with the lad. We reach a truce of sorts. I am always willing to help him out but I do my work first. We become co-bosses in some strange way and I never usurp his authority or try to make him look bad. He always stays in camp when the boss is coming in – that is understandable. Does he try to undercut me on those days? I neither know nor do I care.

After lunch Harry and Art go out with Keith to check some areas we will be working. There are no flies on Moak – the maps were waiting when we arrived.

I spend the afternoon with Harold checking out the gear. First we do a "pretend" set-up in front of the tents. This serves two purposes. We want to know if the motor and associated paraphernalia works before we pack it into the bush. It also gives Harold an idea of what will be on his agenda. This doesn't take long to do – one hour max.

Our next job is to calibrate the mag. It a bit technical, but yet rather simple. A Hemholtz coil (sort of like a hamster wheel) is placed over the mag head. By using hocus-pocus magic (I am no technocrat) I can induce, and increase by steps, bogus magnetic fields around the mag sensor. This is done with an induction box hooked to a 12-volt battery borrowed from the Lancaster. I step it up 100 gammas at a time and plot the results on a graph – the range being 800 to 1800 gammas. This is within most of our magnetic activity range. We often get higher action, but a 2000+ high is not important for calibration purposes. What I am looking for is an out-of-sync reading – the graph should rise in a straight line. A blip may indicate a damaged quartz knife-edge, and the quartz bar will have to be flipped to a new edge or even replaced.

The graph line is straight – good – this means the previous operators of this unit had been careful. We pack everything up and the gear is ready for tomorrow.

(Of course this is all Greek to Harold and I tell him not to worry. Seven months ago the only Greek I knew was "pi" and I am still far from fluency.)

We have an hour left to kill – time to get Harold onto snowshoes. There is a pair with leather harness already attached. Harold has normal feet and he tries out the shoes – not too bad – he will be OK.

The Bombardier boys come home and after supper we do a final check of our go-to-work clothes. Harry gives me a hint of where we will be working. I am a dentist – information is not shared by this man – I have to extract it.

We will start at good old Soab Lake for the first week. This time there will be no long walk – a tributary of #1 leads to Soab. We had not used it last summer, as the winter road runs through boggy Labrador tea. It is virtually unwalkable when not frozen (N.B. Early Nov. '60 Cook Lake 2-4 expedition.) Keith will not venture onto Soab itself. That little pothole is sure to be hiding slush beneath the smooth, innocent snow-cover.

70

We do not have a large area to cover here. Our job is to extend the picket lines across the lake and on a ways over the marshy west rim. There is some preparatory axe work to do so we cut pickets and run five or six lines across Soab. The lines extend a half mile into the bush and Art cuts them with no help from us. It's not thick bush and Art is a line-cutting machine. I am gaining more respect for this man's ability day by day.

The reason we are on Soab this winter is because last summer Charlie, Jacob and I had started to pick up some very interesting readings near the lake. This winter the lake survey shows that something is going on beneath our snowshoes, and a diamond drill program will confirm that an ore body exists. Within ten years Soab will be drained and the ore extracted from an open pit. (Of course the lake had undergone a re-brand by that time. No ivory tower executive in the world would ever dare to brag on his Son-of-a-Bitch mine.)

Sidebar: Years later Cook Lake would also be drained and open-pitted. With an all-weather road into Moak Lake, Cook was already retired as a transportation hub. (Times change – Moak Lake was later put back into production also.)

And this is downright eerie – it seems that Canico has disappeared! I am no whiz on the ethernet but Google as I might I can find no reference to my old exploration company. Did it ever exist? It must have – I cannot have invented this stuff. Maybe I was in a parallel society in '60–'62. (A shiver runs up and down my spine – the wolf hairs on the back of my neck are bristling – Brrrr!)

One more thing, and this refers to both Soab and Cook Lakes. In the years ahead I will often lead a crew through veritable parkland to hit an airborne anomaly. We trot along thinking that this will be easy-breezy only to find that the response is dead centre in a fly-infested, tangled cedar swamp.

It is not a hard and fast rule but the Canadian Shield can sometimes guard her wealth very well.

It's my first winter in the Great White North and it's a whole lot different from my first summer. The loons no longer call from the little bay west of the tents, and woodpeckers no longer smash their faces against a dry pine. Big, black ravens (Thompson Turkeys) arrive to check out our garbage pit, and though their normal raucous caw is anything but pleasant to the ear, they can surprise you. Listen closely when they are having a private chat, and you will hear little warbles and trills as they trade private thoughts. If they spot you, they revert to stridency – probably embarrassed that you caught them being soft and cuddly.

Whiskey Jacks (aka Canada Jays) abound. They have their camp robbers' reputation to protect and they work hard at it. They have little fear of us and join us at the noon fire where they accept bread crust donations, and if we are very still, will take it from our fingers. Each tidbit is taken away to be cached, never twice in the same direction. We figure those birdbrains can't possibly keep track of that many locker numbers – no doubt the semi-hibernating red squirrels will also benefit.

The hum of the outboard is replaced by the grumble of Keith's Bombardier. Aircraft fly over almost every day, and an old hand like me can identify the plane before I see it. The 180 hums by at 2150 rpm, the Beaver is more business-like at 1800, and once in a while an Otter will fly over, hammering like a John Deere, with the big three-blade prop cranking over at 1600 revolutions per minute – just a fast idle. If the Otter is up there at 3000 feet, we have to plant some pickets to see if it is moving.

Last summer Charlie knew them all: "That's George Dram's Norseman," or maybe his Stinson, or the government Beaver from Snow Lake. The Thompson Airport is almost completed, and community strips are in the works, but in the early sixties, floats and skis rule the northern skies.

In the mid-winter bush all sounds are muted somewhat. Snow build-up on the evergreens muffles our sound orientation until a breeze clears the boughs. On the lake the wind is always there, stronger on some days, but even a light wind at minus 40 means you must keep your parka hood wrapped around your face. We always swing our arms like power walkers between lake stations, and when we meet we always scope the other guy's nose and cheeks for the telltale signs of frostbite.

The winter lake work is done at a trot with no lolly gagging. Art has the worst job to do. Harry and I are always moving, but Art has to stand at the transmitter coil. He wears an extra layer of clothing and is pretty warm, but the cold wind still gets him. As for me, I now keep up to five or six readings in my head, stopping every 500 feet or so to write them in my notebook. I don't want to take my right hand out of my mitten too often.

It's no use to wait for warmer weather – so what if it's cold today, tomorrow may be colder. If we are close to camp it's nice to see Keith come out a little early to pick us up for lunch, and on really, really cold days he will be on the lake by 4 pm – good old Keith. When the lake work is done we will hit the bush again, and it will seem almost balmy by comparison. The wind swirls around the tree-tops, but it can't find us. We are always moving and pass the open swamps quickly. Mr. Wind shakes the trees and snarls at us. Tough cookies, pal! Go to Grass River North – maybe you can find some lake victims there.

The noon tea-fire is a treat. Sights, sounds and smells – I can remember them as if it were yesterday, not 50-some years ago. We pick a sheltered spot in the lee of an evergreen grove and cut dry pine logs about three or four feet long. A little fire starter (a.k.a. mixed gas) and soon there is a warm semi-forest fire going. A fallen tree or a pile of balsam boughs serves as a seat and the tea-pail is hung from a stick poked in the snow angled over the fire. Our cans of soup are lined up beside the fire next to our frozen oranges and a sharp stick holds our frozen peanut butter and honey sandwiches over the open flame until they are charred on the outside and semi-thawed on the inside – they sure taste good.

When the top of the soup can starts to bulge it is grabbed with a mittened hand and opened with an axe. A 2½ lb Walters is the bushwhacker's swiss army knife – keep 'er sharp son. I can shave with mine (no kidding.)

Now the oranges are thawed on one side and frozen on the other – squeeze a few times to mix the hot and cold, suck out the juice, eat the pulp and toss the peel to the whiskey jack.

The tea is poured, or more accurately, dipped. Sugar is added from the glass pickle jar in our lunch packsack, and I roll a cigarette.

72

Table Scrap: Thompson Tea.

Utensils: One tea-pail: Canico always supplies us with a tea-pail – a three-quart aluminium pot with a bail handle. The pail is carried <u>outside</u> the lunch packsack, dangling from one of the flap straps. (The pail is black and sooty and our lunch pack is kept relatively clean – we are a fussy bunch, as you will soon find out.)

Sometimes we will lose the tea-pail on the picket line. A branch may snag it and the lunch carrier guy will be pretty unpopular at tea-time. The pail may be half a mile behind us and may never be recovered so we have to steal a pot from Karl and make a haywire handle. Diamond drillers always have tons of haywire and Keith has his own purloined stash.

Cups: We carry Mel-Mac cups. If we forget them an empty soup can with the stray noodles rinsed out will suffice.

Spoons: We carry one. It gets pretty grungy but can be polished with snow on a pant cuff. If a spoon disappears, a twig becomes a swizzle-stick replacement.

Now you melt snow in the tea-pail until you have ¾ pail of water. Snow is always clean in the bush. Before you bring the water to a boil, scoop out the rabbit turds.

Bring the water to a boil and add four tea bags. Roll the tea three times – this is very important and is easy to do. Each time the tea boils you swing the support stick away from the flames and after three rolls the tea is ready to drink. Stray pine needles add flavour, and if we have missed a turd we don't know and we don't care.

(Picket-line tea is absolutely delicious and is impossible to duplicate in your kitchen.)

Now, for a little after-dinner ambience some green balsam boughs are added to the dying fire. Balsam is an aromatic tree to begin with, and a whiff of burning balsam is like pot-pourri. As the congealed sap expands needles start firing from the branches, and like miniature missiles they arc in all directions trailing little contrails of hissing smoke until their fuel is exhausted. They land on our pants, parkas and hair, and bounce off our noses. The little buggers are hot, and we dodge, yelp and laugh like crazy. I prefer what is called a Hydro parka – heavy cotton twill impervious to the incoming attack. Some guys wear nylon or some other synthetic material, and the balsam missiles leave little melted dots on the fabric. If you see a guy wearing a parka that looks like it has survived a bird-shot salvo, you know he has spent some time around a balsam fir fire – that guy is a player!

On cold, clear, windless days the bush is alive. My snowshoes crunch as I walk the packed grid lines, and off to the left I hear Harry holler for orientation followed by the hum of Art's motor generator. If this cold weather holds out we will hear gunshots behind our camp tonight as the expanding sap splits a jackpine or spruce. If it warms up tomorrow a sleepy squirrel may give me a half-hearted lecture. He has come out to hit his pine-cone/mushroom stash, or perhaps to raid the whiskey jack's hoard.

Sometimes I will come across the signs of a ptarmigan or spruce hen – a hole in the snow and a couple of tracks and wing-marks where he has taken off from his overnight bivouac. They don't always make it and evidence of a kerfluffle is left behind – a few feathers with fox or lynx prints leading off into the surrounding bush – the cycle of life goes on.

The ptarmigan are invisible, white as the snow and very wary. If I am lucky I spot one - and of course he has seen me first. We study each other and I leave him alone – he doesn't need any hassles from me.

The spruce hen is no rocket surgeon for sure and I don't know how they survive – good luck rather than good management, I think. One day we have our tea fire right beside our picket line, and as we enjoy our last cuppa, a spruce hen ambles by on the snowshoe trail not five feet from our fire. He is in no hurry and nods to us. "What's up, boys? Are you working hard, or hardly working?" and disappears down the picket line. Art says they have sort of a wild taste but if you are hungry they will do in a pinch. To catch them you attach a loop of rabbit wire to a slender spruce pole, find a covey of spruce hens in a tree, and pick them like apples. They seldom fly away, and if they do they only go 50 feet to the next tree. They must raise large families every summer, or perhaps their gamy taste is a last resort for a predator.

We are doing review/detail as per the previous summer and the two Grass River camps are doing the same. This is the last big push to fill in the gaps in the huge Inco holdings. Our camps are situated far enough apart to avoid interfering with each other, and of course there is a limit to how far a Bombardier can reasonably travel. We go out each morning with Keith who then picks us up in the afternoon, leaving each day at 8 am to return by five. It is seldom more than a half hour ride each way, and with a one-hour lunch break around the tea fire, we have a good seven hours of fieldwork.

Mapping and field note corrections take up an hour or two each evening followed by a late cup of coffee and piece of pie and then it's off to bed – no time to get bored.

Keith's job, besides driving and servicing his Bombardier, is camp maintenance. He keeps the water barrels full – a little more work than in the summertime, as the ice hole in the lake has to be cleared daily. Keith also keeps the propane tanks fresh for the cook stove and from time to time might even make an extra trip to the "Esso station".

We continue with the review work where we had left off the previous fall but with a new dimension. In addition to the bush work we are also extending picket lines onto the lakes and open swamps, as winter freeze-up enables us to get into areas too wet to tackle in the summer. Keith runs the lakes with the Bombardier, but is always very careful to avoid getting stuck in the slush. It will build-up on the skis, and there is also the danger of getting "high-centred" with the tracks in the air. You then have to shovel out the slush and cut poles to put under the tracks – hours of hard, hard work. Keith never breaks trail on the lake alone, and when a trail is established we mark it with small spruce trees stuck in the snow in a line, the same way he outlined the airstrip.

I don't have very many miles in my snowshoe logbook yet, and the first two weeks just about do me in. At night my thigh muscles and calf muscles are so sore I get very little sleep. Am I going to go through this every winter? After two weeks the pain goes away, and never-ever comes back in years to come.

After Soab is finished we extend the grid lines out onto Setting from the west shore across the north end of the lake. We get a good sightline from the west shore, chain across the lake and plant pickets every hundred feet.

The grid covers the lake from the north end by the rapids to about two miles south of our camp, and we put in 66 lines. The lines at the north end are fairly short but become up to three miles long as the lake widens. Sixty-six lines with an average length of two miles, and with a picket every 100 feet, means we have 7,000 pickets to plant. In the bush, pickets are made using any nearby small tree. You chose one two inches in diameter, trim it to four feet with a point on one end, make a flat blaze to write on, and you are done. Seven thousand pickets will denude a large area, so Moak buys cheap 2x4 lumber, has the maintenance crew saw them into laths four feet long, tie them in bundles and ship them to Setting Lake. We sit in camp at night and pre-number smaller bundles to carry out on to the ice, planting them in order as we go along.

We are able to walk to the closer work, then use the Bombardier as we get further from camp. During the day Keith also dumps bundles of pickets at appropriate points along the west shoreline, and picks us up to go back to camp for lunch.

As soon as the grid is done we proceed with our instrument work. I have over 132 miles of mag work to do and Harry has to cover the grid with the EM survey. We can easily double our bush output, and so by mid-February we are able to get on with the bush sector of our winter's work.

JOHN PILOG AND FRIENDS. IT IS LATE WINTER – THE LAB IS HEAVIER NOW AND HAS BEEN PROMOTED TO SECOND SWING. NOTE BOMBARDIER TRAIL INTO BUSH IN BACKGROUND. THIS IS KEITH'S ACCESS/EGRESS TO THE LAKE ICE. THE ROAD LEADS TO THE TENTS OFFSTAGE LEFT.

Every two days or so John Pilog passes by with his dog team and sled. One day he checks the traps north of his cabin, and the next day when he does his south run he sometimes stops for a chat. His five-dog team consists of the four big, tough looking malamutes and the Labrador cross, now almost full grown. He runs in third place – two malamutes ahead and two behind. It is funny to watch them take off when they get the "mush" from John. The Lab almost comes off the ground when the traces snap taut, but once they are in travel mode he regains some traction. He obviously enjoys being part of the team but he has not yet shed all his puppy-dog ways – he is easily distracted.

One day I am on shore when John passes by about 500 feet out on the lake, and for some reason I whistle. The pup looks over at me, turns right, and what a train wreck! Dogs pile up, the sled is on its side, and John is mad as hell! I hide behind the trees until John leaves and I never do that again.

75

Table Scrap

One lovely March afternoon the boys are sitting on the steps of the Hudson's Bay Store. A trapper pulls up with his dog team and goes in for supplies. The dogs lie down for a while but they decide he is taking too long, so they go home. One old guy says, "Heh-heh, must have left 'er in gear."

One day we wake up to silence. It is not much below freezing outside and a heavy, heavy snow is falling straight down – not a breath of wind. The kitchen is eerily quiet, just the faint sound of Karl's radio and muffled clinks of enamelware. We walk to the kitchen like soundless wraiths and no one talks very loud. It could be Sunday in Church and there is just the odd slurp or cough as we eat our breakfast and pack our lunches. Even the Bombardier is respectful, the exhaust burbling quietly as Keith warms it up. As we head down the bush road, even the tracks rumble a bit more reverently.

Valet Parking at Setting Lake
George Dram has stopped in with his Bombardier Taxi, a brand-new top-liner - 313 Chrysler power w/automatic transmission.
Note headlights higher on the cowl.
Also Note the A-2 mag with its protective sample bag cover.

On the job the silence is deafening - I can barely hear Harry and Art is off the radar. I glide soundlessly over the snow, trying to keep the instrument eye-piece clear, and protect my notebook as I jot down readings.

Harry asks Keith to come out early, as the soft, wet snow may eventually cause problems with the gear, and at one o'clock we hear the horn toot – we didn't even hear him coming. We return to camp for afternoon coffee, hang our gear to dry on the ridgepole clothesline and lay on our bunks. We didn't even build a tea fire today – a gallon of fire starter would have been wasted.

The next day it's a whole new world outside. It's like Mother Nature has erased the slate. Now we have an extra foot or so of snow to deal with, but we don't care – we are over the hump. Winter's back is broken, the days are getting longer and it's a downhill grade 'til spring and another period of renewal in the north woods.

The weather has been very cooperative and the work thirteen/one-day-off deal has become routine. By mid-February the Grass River boys are looking for some R&R.

Arrangements are made on the 9pm party line. Grass North will motor to Grass South where the two groups will all pile into Neil's machine and continue on south to Wabowden. Keith, with Harry, Art and Karl will go in our Bombardier. Karl does not drink, but he is a sociable guy. Besides that – a day off for us is not a day off for him and he wants to eat in a restaurant for a change. I decide to stay in camp to keep the home fires burning. It's not that I'm too cheap to party, although I am paying off some old bills and sending a buck or two home: I just think I'll enjoy an evening of solitude.

Plan A had Neil and Keith meeting in the evening at Wabowden. Keith was to head down the lake after supper – a drive of less than one hour on the firm, windswept lake (we have already determined that Setting is now slush-free as long as Keith stays away from the deeper snow near shore.) The boys can whoop it up all night long if they wish, and return to their respective camps the next day (which is our earned day off.).

So we are on the lake in the morning doing our thing when a Bombardier appears from the south! It is Neil! Unbeknownst to us Plan B has been in effect since the night before. They had pulled out after the 7pm Sked and had slipped by while we were fast asleep. Both Grass River camps had cashed in their progress savings accounts to have an extra day off – they are on a two-day party!

We also have progress in the bank, so we pack up and go back to camp with Neil. Karl and Keith are surprised to see us back so soon. They hadn't seen Neil come up the lake.

While Harry and Art and Karl change into their town clothes I have coffee with Neil and he has quite a story to tell.

They had intended to hit our camp the night before but had missed the turnoff and continued on down the #6 right-of-way to Sasagiu Rapids. They were already half in the bag thanks to a sleight-of-hand 2-4 (these boys are experienced party-goers.)

Faced with the open-water rapids they decided to ford the river. They knew that Sasagiu is not that deep in the summer – probably even shallower in the winter months. They also knew that the upper part of the rapids is rough, but fairly level.

Neil was not driving. He was on the passenger (upstream) side.

Sidebar: The Bombardier is a people/ freight carrier. It has two steering skis in front and the tracks start about a quarter of the way back, extending to the rear. There are passenger benches on each side with another across the rear in front of the engine. There is a little louvered door in the back wall which can be opened on really cold days to allow engine heat into the cabin. There is no forced air heating system. There are two little fans mounted on the dash to keep the windshield defrosted, and the one windshield wiper is hand-operated. It is only needed if wet snow builds up on the windshield. Bombardiers can run on dry roads with the skis replaced by wheels, but why would anyone want to run one in the summer?

Bombardiers have no seatbelts, air bags or power windows. In fact, you can't roll the windows down, although the front doors can be easily removed and tossed in the back. When you order a new one you don't have to delete the radio – what radio?

They do have an overhead escape hatch which is large enough for a little guy like Art (if he sheds his parka.) If a bombardier loaded with people were to go through the ice, it would probably float long enough to allow egress through the front doors or the large side door on the right-hand side.

And, to complete the scene in this epic, the floor between the two front bucket seats is six inches lower than that in the passenger compartment.

Sidebar: I put a lot of Bombardier miles in my log book that winter. Like putting my trust in competent pilots, I had complete confidence in Keith's expertise. Keith never took risks but those boys at Sasagiu, full of bubbly bravery, took a chance.

So Neil's driver heads out into the twilight zone of Sasagiu, and he feels the current pushing him downstream. It's not moving him sideways yet, but the pressure is there. He asks Neil if the water is deep on his side. Neil can't see out of his frost-covered side window so he opens his door – and Setting Lake rushes in! The driver opens <u>his</u> door and Setting Lake rushes out! They are in midstream, and they can't back up, so it's pedal to the metal – and the good old girl climbs out on the south side with a clean floor. They continue on to Wabowden laughing like crazy – just another minor incident on the Nickel Belt behind them.

The two Bombardiers pull out for Wabowden at noon. I make myself a light lunch and retire to my tent. I figure I will have at least ten hours to myself before the crew returns and I have been instructed to skip the 7pm Sked, which I do. In fact, I even forget to monitor the airways that night. Had I done so I might have been able to avoid some embarrassment the next morning.

I have also been told to charge the Lancaster batteries. It is not a difficult job – we have a little Iron Horse twelve volt battery charger. I slide the batteries out from under Harry's bed, take them outside and charge each for one hour, then I hook them up and slide them back under the bed, forgetting to put the piece of protective cardboard on top of the batteries. It is not a fatal error, but certainly a fateful one.

SPUDS AND SUDS AT OUR WINTER KITCHEN AT SETTING. THE SUDS ARE LEFTOVERS FROM THE WABOWDEN TRIP/PARTY. KEITH DEVINE IS POSING – HE IS FAR FROM BEING A CHUG-A-LUG ARTIST. SOMEONE GAVE KARL MUZYKA AN EMPTY BOTTLE TO HOLD. KARL SMILES IN HIS SLEEP I THINK – HE DOESN'T NEED ALCOHOL ENHANCEMENT. NOTE SPIFFY ROYAL DOULTON (SERVANT'S TABLE) DINNERWARE. ALSO NOTE ABOVE KUSTOM KITCHEN KABINETRY – THIS IS WHERE THE TWO 14X16 TENTS WERE JOINED.

I do some map work, write a couple of letters, turn off the lamp and hit the sack. At one am the boys come back – two snowmobiles full of happiness. Gas lamps are lit, lunch is put on the stove and beer is on the table. I get up, get dressed and join in, have a couple myself and listen to the evening's tales. After a snack Keith and Karl, sober of course, kick everyone out of the cook tent, so what is left of the party gravitates to our bunk tent. A few more beers are opened and things start to slow down, bodies are stuffed into #5 Bombardier and it is sent home. I go back to bed only to be awakened at 5am by a mild ruckus. Harry has lit the lamp and is pouring water on his mattress!

When he went to bed his weight pushed the springs down, they contacted the unprotected battery posts, shorted out and lit his mattress on fire. Harry starts to get really warm and luckily the heat wakes him. He realizes what the problem is right away, douses the fire (or so he thinks,) replaces

78

the cardboard cover over the batteries and goes back to bed. These mattresses are pretty basic – just a cotton/wool sort of fluffy stuffing in a heavy denim cover, and once on fire they are hard to extinguish. After we go back to sleep the mattress starts smouldering again, so Harry throws it out the door and crawls back into his eiderdown on top of the springs.

Harry and the Lancaster radio. Note the spiffy interior wall treatment. The two 12v batteries are tucked under the bed.

This is our day off so we are sleeping in. Suddenly Karl comes banging on the door.

"Get up! Getup! The boss is here! Clean up the mess quick! I'll try to keep them in the kitchen a while!"

We jump up, dress and quickly hide the empties and sweep up the labels and bottle caps on the floor. There is a knock on the door, and in walks Glen T. – Inco's Chief Exploration Geologist (Canada.) He has come to Moak Lake and is doing a camp tour. We leave Harry to show him our maps and hustle over to the kitchen.

When I step out of the tent, what to my wondering eyes should appear but the mattress – smouldering away! Our doorstep is a 3' x 3' piece of 3/8 plywood nailed to a couple of blocks of wood. When Harry threw the mattress outside it didn't clear the step and has slowly burned its way back towards the tent taking half our step with it! I quickly throw the culprit into the deep snow beside the tent, kick snow on the charred step and go to the kitchen to await a dressing-down. Norm is having a cup of coffee and I say, "Jeez, Norm, we never even heard you come in."

Norm feels terrible. "Bob," he says, "I always shoot up a camp before I land, but today I didn't. With the south wind, I came straight in."

Glen finishes his map inspection, has another coffee, chats a bit and goes on his way to the next camp, and not a word is ever said about the unusual aspect of his visit. He probably had a good laugh about it after. The only up-shot of the whole deal is that I have to give my mattress to Harry while I wait for a replacement

And we have to charge the batteries again.

It is now March and the lake work is completed. We have finished off frozen swamp areas and the lake work. Art and I are cutting picket lines on a gap in the grid four miles north of camp while Harry is mostly resting.

Cutting picket lines is uncomplicated but can be strenuous. We are extending existing lines, so Art and I place two or three pickets in a straight line and then work off

79

the end, keeping a watch behind us to stay on course. We are cutting 12 lines 4000 to 5000 feet long, which is a piece of cake for Art, but I am new at this. Once the lines are started we work alone. To cut a picket line you merely open up a path in the bush wide enough to see your back pickets and thus keep the line straight. When you hit an area of thicker bush or swamp alders there is more cutting involved, as the operator must be able to get through on snowshoes. We are on high ground with very little underbrush, mostly small jackpine and spruce with the occasional larger tree to knock down. We also cut a blaze on the larger trees to make it easier for future crews to visually pick up the line.

So we start cutting. I am on one line, Art is on the next one 400 feet away and I can hear his axe snick-snick-snicking as I work – he is going much faster than me. When the sound stops Art is probably having a smoke. If he hears me catching up the snicking starts again until he moves ahead far enough to take another break. This goes on all day until Art hits the boundary, then he comes over and helps me finish my line. I watch him as we work together and start to pick up a few hints and tricks of the trade. Had Art been working by the mile for a line cutting contractor, he could easily have cut three-plus miles a day. But he is paid by the month as I am, and besides he doesn't want to show up the Moniass (Cree for greenhorn.)

By day three I am moving along a lot better and we are even able to chain the completed lines before walking out the mile and a half to the lake for our ride home with Keith. On day four I am full of piss and vinegar. There will be no smoke breaks for Art today. My axe is flying! I am like the wind! The snick-snicking stops once, and when Art realizes I am right beside him he starts up again. He doesn't say much at the noon tea fire – he never is very talkative anyway.

It is a great March day, sunny and barely below freezing. I am heating up, so I hang my parka on a tree branch and continue on my way. By the time we finish the day's work I am drenched in sweat and soaked through and through. We are a little early for Keith and the temperature is dropping with a brisk wind blowing down the lake, so I am thoroughly chilled by the time Keith picks us up. After supper I go straight to bed.

The next morning I don't feel so hot, but decide I will tough it out anyway. A little hard work never killed anyone, right? By 10am I power out, build a fire and sit there shivering. Art, hearing only silence from my line, comes over and finds miserable me by the fire and immediately walks the four miles to camp. He returns with Keith, helps me out of the bush and by the time we drive back to camp the 180 has come in to take me to the new hospital in Thompson. I have pneumonia. I feel much better in a couple of days, but they keep me in for a week. It is a nice break, but when it is time to be released I am ready to go back. I don't care much for the needle in the bum twice a day.

Sidebar: Would you believe that in the Thompson Hospital there was a cigarette machine in the hall just outside our ward door! It was a standard cigarette dispenser. You put in two quarters (it <u>was</u> 1960) and choose your brand from a row of buttons. Some choice – every button is Buckingham – Buckinghams only. It is a four bed ward and all of us have to smoke Buckinghams – the strongest cigarette known to mankind.

Table Scrap

Perhaps I should blame my pneumonia on Billy Pronteau. He was Moak's claims specialist and had come to Setting a week or so before my air-evac. We were living in semi-isolation that winter and were never sick, but Billy brought in a flu bug. He only stayed two days but he left the bug behind and we all got the sniffles. We worked through it, but I am sure Billy bought my plane ticket to the hospital.

The reason Billy came in was to check out the claims at Soab Lake. Due to our most excellent survey earlier in January Moak realized that the Soab claims would eventually be patented – Billy's job was to make sure the claims were properly staked.

He was a welcome visitor – a happy, energetic guy who entertained us in the kitchen common room each evening. Because he was essentially a rover with no fixed camp address, he knew everyone and had many tales to bring us up to date anecdote-wise. Billy knew everyone on the Belt, and everyone knew and liked Billy.

He was a tea drinker in the evenings – gallons of tea. Then, all night long, every two hours, Billy would crawl out of the sack and visit the yellow snowbank behind camp. (Anyone of my advanced years will know where Billy was coming from in 1961.)

So – after Billy left camp Karl had his own "Billy story" for us.

It was a hot summer night – one of those sultry August dog-day evenings and the tent was stifling, so Billy took his bedroll outside to sleep under the stars. Nature called and Billy – half asleep – crawled out of the sack, stumbled into the tent and pissed on the floor. We thought it was surely a funny story but we found it a little hard to believe.

On my release from the hospital I am flown to Moak Lake where I will join another crew for a week or so. Harry had borrowed a man to replace me and Setting was about to shut down anyway. They didn't know how long I would be laid up so they had packed my gear and it is waiting for me at Moak.

We will be chasing anomalies out of Moak, flying out every morning and back every evening. It's pretty hard to take, what with central heating, 24-hr power and more than the same four guys to talk to every night

Anomaly chasing is similar to what I have been doing but without the benefit of a cut grid. An airborne survey has identified areas of possible conductive and/or magnetic activity. These are anomalous to the general background readings; hence the term "anomaly."

The deal is this: We have an eight by eleven copy of a ½-mile-per-inch map showing the day's target. Our three-man crew flies out to the lake nearest the target (Keith or Norm.) We unload our gear, establish a compass azimuth and snowshoe in on a direct line to the anomaly. If the response is actually on a lake, it's easy-breezy. The area is dotted with lakes large and small, so few anomalies will be out of reach snowshoe-wise.

Once we reach our location we set up the EM gear and read "search squares," 800 feet to a side, making a square around the set-up. If the readings show nothing then we move to each corner of the first square and repeat the process, thus covering an area of 1600 feet per side. Lack of any type of response writes off that particular anomaly and the next day we hit another target. If we get an indication of a conductor on any square then we establish the strike, pull out our ever-present compasses and put in a small grid. It's nothing fancy – we blaze the trees and clear the underbrush as necessary to

81

maintain a straight line. We do not chain. We count paces and write the station number on a blaze.

Once we have established the base line for a distance of 800 feet on either side of our initial conductor indication, we blaze a cross-line every 400 feet to a distance of 500 feet either side of our base line. Then the boss and his helper set up the EM gear and trace the conductor while I do the magnetometer survey. When the first four lines are completed we look at the notes and decide which way to extend the grid. The EM gear is then moved 800 feet left or right, the base line is extended and two more cross lines are put in. With another cross line at the original set-up, this gives us a compass grid 2400 feet long by 1000 feet wide. This is our basic work plan. The map is then turned over to the brass and we move on to the next anomaly.

The work described takes three days, and if we find a parallel conductor (very interesting) we may extend the lines, but if the grid gets too big control suffers due to the fact you are using a compass and pacing the distances. The more interesting areas may require a cut grid and detail work in the future, although quite often our grid is the one used for follow-up diamond drilling.

An airborne crew is working out of Moak and has been for the last two months – we are chasing their anomalies. They are flying a Anson from an ice strip on Moak and are almost finished. I meet the crew and establish a good rapport with Jim Lee, the crew leader. This friendship will last for years, and within two months will lead me to new horizons in Ontario.

There are five men on the aircrew – Jim, a navigator, two draftsmen, and George Charity, the pilot. Jim and his three men are based at Inco's exploration headquarters in Copper Cliff, Ontario. George, like Norm Kearns, works for Austin Airways. Inco owns the Anson – Austin supplies the pilots.

Table Scrap

George was another legend in his own time. (I'm telling you – back then I could hardly turn around without tripping over a legend.) He did not fly airborne surveys exclusively – it was a seasonal deal. Austin had many bases in the north including one at South Porcupine near Timmins, Ontario. They were used by many mining companies working in the Shield and thus I would cross paths with George more than once in the next decade. I would never fly with him but the mining exploration fraternity on the Canadian Shield operates in overlapping circles like Olympic rings. I wonder now – how many times did our paths intersect? It was a matter of altitude – I may have been trudging on snowshoes while he was flying overhead.

I was at Nighthawk Lake in the winter of '65 working for another company. We had flown out of the Austin base in South Porcupine, and although George didn't fly us, our pilot knew him well. He said George had been flying out of South Porcupine, but had recently pulled the plug after a set-to with the base manager. In 1964 Texas Gulf Sulphur had pulled a glory hole. A huge staking rush had kept the Austin pilots hopping – both George and the base manager were stressed out.

Now ordinarily when one quits, or heaven forbid gets fired, one kicks one's lunch bucket down the road, but not our George. He had a project underway building his own plane in Austin's hanger, and he had almost finished the build. The fuselage had been painted and was sitting on its wheels waiting for the motor and wings. When George quit, he picked up the tail and pulled the plane behind him as he walked home up the main street of South Porcupine.

In 1964 the legend grew. George was co-pilot of an Austin Airways DC-3 in the Nemiscau area of Northern Quebec east of Moosonee. They were heavily loaded with fuel oil in 45 gallon drums when they lost both engines. Unable to reach the nearest frozen river they had to ditch it in the bush. George, as the senior pilot, took the controls, brought it in barely above stall speed and just as they hit the treetops he kicked the DC-3 sideways. On impact the barrels went through the side of the plane instead of piling up on George and his co-pilot. George had intentionally hit the bush with his side of the aircraft and thus received two broken legs and a broken arm. They had to wait 30 hours for a rescue, but the grim reaper had to wait for another day.

And – the DC-3 sat in the treetops for two years before a forest fire brought it to the ground, where it remains to this day.

It was November of 1969 before I ran into George again. It was another staking rush – this time the Mattabi rush at Ignace Ontario. Staking rushes are like a swarm of bees - the queen bee is the discovery and the worker bees gather around the queen.

I walked into the bar at the HI-WAY Motel and there sat Jim Lee and George Charity – big as life and happy to see me. It had been eight years since I had last seen George and perhaps six for Jim, so we had a lot of catching up to do.

They were not in Ignace because of the Mattabi find. They just happened to be there – flying an airborne survey from the Ignace strip.

So we caught up – where we had been, and what we were up to now. George was near retirement and since the Nemiscau mishap had been sticking to airborne. I asked him how he was doing health-wise.

"Bob, he said, "I've been flying those Ansons for so long that every time I get over 400 feet in the air I get a nosebleed."

Just in passing – I had met his son in Chibougamau years earlier. He was a chip off the George block and was Fecteau Airways chief pilot.

In the Canadian Shield those Olympic rings may shift like weather patterns, but they always and forever interlock.

So in the evening common room at Moak I chat with Jim Lee and the boys. I learn about airborne surveys and I like what I'm hearing. These guys don't break trail on snowshoes in the wintertime, nor do they fight flies and Labrador tea in the summer. They never live in tents. They always work out of places like Moak or out of a motel, so they never have to cook their own meals. They don't have to patch their tea-worn pants with sample bags and their footwear can actually be worn downtown. They make better money than I do (I am now considered trained and as a qualified mag-man my monthly stipend has been raised fifteen whole dollars to $275!)

The only drawback, which I think is inconsequential, is that their work hours are what you might call flexible. They never fly if the winds exceed five miles per hour –

83

therefore, to take advantage of the usual calm periods, they will fly two hours in the early morning and/or two hours before sunset. So in the long days of summer they may lift off before 5am and end their day at 9:30 or 10pm. If the day is totally calm they never fly more than two three-hour stints. The Anson must be refuelled and more than six hours at 400 feet is too much for the pilot and navigator to remain sharp – and sharp they must be.

Another thing – if the winds blow continuously they sometimes sit for days on end. I don't consider this a negative but will change my mind later on.

The draughtsmen work a normal day transferring the results to maps. I would not like a desk job – I've been there and it's no fun. I'd like to be a navigator, and I broach the subject to Jim Lee and he says he will talk it over with the power-tower. I don't recall writing a formal request to Copper Cliff although I may have done so.

The upshot was that before I left Thompson at break-up I was told that I would be transferred to Copper Cliff the following summer season. I would not go directly to airborne but would stay in the bush until there was an opening. Cool!

Sidebar: A condensed overview of Inco's Airborne Division

They flew Avro Ansons for two reasons: the Anson was a two-engine aircraft with good performance characteristics on one motor – an important safety consideration when flying at 400 feet. Also, the Anson was built of molded plywood, and with little or no metal to interfere with the signal, the transmitter coil could be installed inside the fuselage.

The receiver (known as "the bomb" for obvious reasons) was lowered on a winch cable and trailed the aircraft – the Anson at 400 feet and the bomb at 200 feet! George had to be very careful in the driver's seat.

The navigator also had to be careful. He had a rolled strip of an air photo montage with the flight lines drawn and had to keep George on track. This could be difficult at 400 feet. If there were lots of rivers and lakes below it made his job easier, but in thickly forested sections he had to navigate in 50 shades of grey – darker evergreens and lighter broadleaf trees.

Jim Lee rode in the back, keeping an eye on the instruments which included a camera that recorded the flight line as flown. Sounds complicated but it worked well. (Actually, Jim's toughest job was to keep the crew half-assed sober during the inevitable downtime days.)

2400 Ansons had been built in Canada before the design was retired. Few remained and now Inco flew the largest Anson fleet in the world. They had three in the air and one on the ground for back-up. The hardware was no problem – the engines were widely used – it was the darned plywood.

They bought one from a guy in Riverton, perhaps hoping to modify it as another daily flyer. I don't think they flew it back to the basin – at least I hope not – because when it arrived they pulled off some after-market aluminium to check the plywood and found sawdust!

With the end of the Anson era in sight Inco decided to switch to the Twin Otter. This involved some re-thinking as the all-metal aircraft would need to have an external transmitter coil. It would form a triangle – wingtip to wingtip to rudder. There was some structural re-engineering involved and early designs had airworthiness issues.

84

Wheelbarrows of hard cash were dumped into the project before a workable set-up was found.

I don't know how many air miles were flown with the Twin Otter. I do know that the technology was undergoing quantum changes thanks to the magic of computers.

Nowadays, what with GPS and Laser Altitude Control, one man plus the pilot in a chopper does it all in the air and computers do the mapping. The feet on the ground are now meters, and the romance is gone – replaced by one night stands and speed dating.

Yet I can't help thinking that there are still common room legends out there, and I hope they haven't been miniaturized.

Table Scrap (true story)

The prez is coming to visit Sudbury from his corner office high above Bay Street in downtown Toronto. He wants to see and experience an airborne survey first-hand.

The boys at the Frood strip have prepared the queen of the Anson fleet for the demo. It is squeaky clean and shiny. The crew is squeaky clean and shiny. Ditto for all the survey stuff, which has been checked and double-checked – even the cable that pulls the bomb has been replaced.

They have picked a previously-flown line where they know a response exists. As they line up on their flight path the crew chief (name withheld for obvious reasons) releases the bomb.

The winch turns. The cable unwinds and disappears! They had forgotten to secure it to the drum!

The prez raises his eyebrows and says, "Does this happen often?"

Harry has closed Setting, has moved to the Midwest camp in Thompson, and I am on my way back to Harry. Spring break-up is not far away and once again we are cleaning up odds and ends. Keith is still with us and we go out each day by Bombardier. Two other crews are working off the highway by truck. Our job, accessed by a Hector side road, is to stake 20 claims and cut a grid. I am to do the mag survey once the grid cutting is under way. Harry and I are joined by Joe Wenusk and Willie Wood. Art has gone home and I will never see him again – I liked Art.

We stake 20 claims along the east boundary of the Nickel Belt. They are necessary to make sure an interesting zone is protected, so they are staked side by each. This will extend the east boundary ¼ of a mile. Picket lines had been cut on the claims we are tying onto, but they were cut years ago. They are ill-defined and the chainage is almost indecipherable on most of the pickets. Good control is a priority, so we will cut a new baseline and because a claim is ¼ mile square it will be five miles long!

We find some readable pickets and establish our starting point. The work is done with Harry on the transit – Willie, Joe and I are on the axes. It's open bush and we make good progress. We turn cross-line angles at 400 foot intervals as we go along and by day two we have a half-mile completed – but we have a problem.

It is a 35 mile drive for Keith. We use a short-cut across Cook to pick up a Hector Hwy to #6, but our last seven miles is on an old road which Keith has reopened. Even if we leave Midwest 15 minutes early we don't arrive before 9am and we have to leave the bush shortly after 3:30.

85

So after our second day of work Harry, Keith and I sit down in the evening and do the math. Our baseline will be five miles long. We arrive at the north end and walk south and the walk will get longer every day. By the time we are halfway down the baseline it will be 10am before we start cutting and we will have to head back by 2:30, leaving us with a short lunch break and four hours of actual work. We need a plan B – and there is always a plan B.

Keith knows of another old road that branches off before the north end of our new baseline. He says it passes near the south end and he figures he may possibly be able to open up a 1/4 of a mile of new road to meet the end of the baseline. It'll be tough slugging for the Bombardier – there is now over three feet of snow. Keith may also have to clear some trees but he thinks it will work. The only problem is that he does not know where the baseline will end. Someone will have to be there to meet him – I will be someone.

I don't have a whole lot of compass time in my log book, only a couple of weeks chasing anomalies with Irwin Wilson. I have the pacing down pat, but outside of the two weeks on chase duty I had always been on cut picket lines. Harry is sure I can handle it – I hope I can.

Keith drops us at the baseline the next morning and leaves to open the road. He estimates it will take him at least five hours to do so – we will rendezvous, s'il vous plait, at two-ish. I work with the boys until noon and strike out south with no lunch. I have three miles of heavy straight-line snowshoeing ahead of me and two hours to get there.

I am very careful trying to mix accuracy with speed. I take no breaks and have no mercy. As I have said, it is open bush, but if a tree is in my way I am ruthless – "Sorry about this, little jackpine, but you must fall to the axe."

With 200 feet left to go I hear the rumble of the Bombardier, and while I make a fancy four-sided blaze on a big jackpine, Keith makes a turn-around. We go back on the newly broken road to pick up the boys. Keith and I are both pretty proud that we had made the connection. I will have some trouble sleeping tonight – I must have pulled a muscle while patting myself on the back.

When we finish the baseline we find that my four-sided blaze is forty feet further south and only ten feet off-line. I pull a back muscle again.

With the baseline finished Harry goes home. The poor guy was married a week before Christmas, and has spent the last three months in the bush with a bunch of yahoos. No wonder he tended towards crankiness.

So Willie, Joe and I are now on our own and also share a Mid-west tent, but neither of them speak a word of English. I'm not kidding – not one word! No problem on the job, as each one knows exactly what to do and does it very well. However, the tea fire and evening gabfest are sort of two-sided. I had picked up some Cree from Charlie and Jacob and I can count to ten. The only number I remember now is "six" (gah-toss-ik.) I could say, "It's raining" (no-tem-a-si-wan,) not much use in wintertime, also "It's cold out" (kay-nis-i-tayo) and I might even have these vice-versa. Criminy, it's been more than fifty years ago, after all!

So I say "Kay-nis-i-tayo," they reply "Eh-hey" (yes) or "Tah-boy" (that's right,) always accompanied by a smile and a nod. I never hear them say "no" or shake their heads. I love these guys.

I have never heard a native say "No." Perhaps, like blasphemy, it is not part of any native tongue.

Willie and Joe have a small battery powered record player and one – I repeat, one – record, "Country Classics;" Good songs by good artists, but Jeez! Every night?

Two other party leaders, Bob Tilden and Henry Linklater, have their own tent and I spend a lot of evenings there, mapping, shooting the breeze and catching the sked, during which I also report my progress. One night Moak asks if Joe and Willie are in my tent, which I confirm.

"Well tell them to send some money home. The Hudson's Bay Store cut off their credit and the families are hungry."

Henry goes back to my tent with me and passes the message on in Cree. Willie and Joe dig into their packsacks and pull out every check they had received that winter! Neither one smokes, drinks nor even goes to town – they just never sent the checks home. I get an envelope, they put their "X's" on the back (which I witness,) and I mail them to Split Lake the next day.

"Megwitch," (Thank you) they say, nodding and smiling. What an enjoyable pair!

Winter is coming to an end and the last couple of days our return trip to camp on the mile of ice road across Cook Lake is

6 BOMBARDIER – SKIS OFF – WHEELS ON – READY FOR A THOMPSON BEER RUN. KEITH SAID HE WOULD TAKE US UPTOWN IF WE HELPED SWAP TO WHEELS.

through a foot of water on top of the ice. Keith tells me that as long as the water stays on top of the ice we are Ok, but when the water disappears it means the ice has candled and is now floating on the water.

We only have a couple of days left anyway. The Bombardier will return to Moak and I will go home. The second-last evening we talk Keith into taking us uptown in the Bombardier. We pick up a couple of 2-4's and we have a mild going-away party. One more short workday and I will be heading south on the Thompson Special to Winnipeg. Woo Hoo!

Table Scrap: Charlie McLeod Sr. and the 55th Baseline

While working out of the Midwest camp I meet Charlie McLeod Sr. who is axe-man on Rusty Cordell's land survey crew. By 1961 the claims in the vicinity of the new mine were patented and surrounding claims were maturing to patent status. Rusty's four-man crew is doing the legal survey on these claims.

87

Charlie's dad is an old hand at this. Charlie had told me that in the 30's his dad was part of a crew who cut the 55th baseline (55th parallel) from the east boundary of Manitoba to the Saskatchewan border just north of Flin Flon. It was this baseline that established the benchmark for legal surveys to come.

It was all winter work, and all done with axes and crosscut saws – no chain saws in those days. The camp was mounted on sleighs and followed the progress of the baseline. The crew went out after freeze-up and returned home before break-up, with a week off for Christmas.

Manitoba is 400 miles wide at the 55th parallel, and I'm guessing that the crew, besides surveyors, kitchen crew and support personnel would have had up to fifteen men cutting, and two miles a day would have been a pretty good average. There would be a certain amount of manpower changeover, as they passed near Oxford House and other settlements along the way, but Charlie said his dad stuck with it from east to west and was away from home for three winters.

(I have flown over the 55th baseline in summer and in winter, and it is easy to spot from the air, even from an airliner at altitude. It appears to be at least 50 feet wide and stretches out like a belt across Manitoba's tummy, fading off in the distance as far as the eye can see.)

In the '60s the corner grocery store landed at our camps every week, and fresh meat, fruit and other goodies disgorged from the cargo door of the Beaver. Not so simple on the 55th - flour, salt, sugar and other non-perishable staples were fairly easy to keep on hand, but a two or three-man hunting crew kept moose and woodland caribou on the menu. Dried fruit would hit the spot for apple and raisin pies after dinner, and breakfast would be oatmeal and homemade bread. They must have had an occasional rendezvous with a cat swing and when they crossed the Bay Line they no doubt had fresh eggs, apples, and oranges for a treat, but some stretches of uninterrupted bush must have made the dinner menu same old, same old. When crossing lakes, nets were set up and fresh pickerel, northern pike and baked whitefish would be a cabin fever reliever.

So every morning us young bucks semi-stagger up Main Street to the Midwest kitchen for our rubber eggs and greasy bacon and pass by Charlie Sr. as he smokes his pipe, sitting on a small bench by his tent door. He always rises early for a quiet breakfast and smokes a pipe or two before the day's work begins. I get up early a few days also. The eggs are a little hotter and the bacon not so greasy, and I sit with Charlie Sr. from time to time. We never talk much – maybe discussing the weather a little bit. I sit back and suck a couple of Rothmans King Size, smelling the pleasant aroma of Charlie's pipe and listening to the camp wake up. It is easy to figure out where Charlie Jr. got his calm, unruffled, patient demeanour.

Wonderful, wonderful mornings: but tonight someone will suggest a bar trip, and tomorrow it will be rubber eggs again.

Table Scrap

The camp was OK, but I really disliked the Midwest kitchen and the men therein. We had to sit at the back and were totally ignored – it was Apartheid on the Belt. So one early morning I pick a spot on a bench in business class. I have just started to dig in

88

when a guy walks in and stands beside me. I know who he is (I have filed him under S for "snotnose") and I know what he wants. In fact, I had picked my spot on purpose.

"You're in my seat." He says.

I feign surprise and concern, look left, right, and under my plate. "I don't see your name anywhere," I reply.

Will he take a poke at me? I hope he does – I'm leaving tomorrow anyway. I slow down, savouring every greasy mouthful, and he eventually chooses a different spot in the almost empty kitchen.

So I had the satisfaction of poking Mid-west with a pointy stick before I left, and I think the snotnose wasn't too popular anyway. As I got up to go the cook gave me a little nod and smile.

Manitoba and Ontario use similar systems to bring claims to patent. When a claim is staked it remains in good standing for a year. You can then re-stake it, but if it is a hot prospect, you stand a chance of losing it to a competing staker. Thus, you perform certain assessment work on the claim, file the work with the Department of Mines, and the claim remains in good standing for another year.

In Ontario, 200 man-days per claim are/were required to bring it to patent. The first year you needed 20 man-days, year two through four required 40 man-days, and in year five, 60 man-days completed the assessment work. Work performed could be blasting, trenching, sampling, assaying, and in some remote locations, just the fact that you made it in to the claim group in one piece deserved a certain amount of man-day credits. In the old days a small 8 x 8 foot exploration shaft might be sunk, all by hand-driven steel and shovel and sometimes to depths I shudder to think about, with cross drifts to reach a vein to take a bulk sample.

In the early days, building a cabin was accepted as part of work performed, but this was eventually dropped from the list as cottagers realized they could use this short-cut to pick up a 40-acre lot on a not-so-remote lake.

Almost any reasonable amount of work or expense was/is acceptable to satisfy the claim block assessment. Flying to remote lakes has always been expensive, and a decent trail has to be cut to the showing. With the introduction of geophysics to the mix, line cutting and EM or mag work became a credit – so many man-days of credit per line-mile. Diamond drilling, always relatively costly, was credited at the rate of one man-day per foot drilled – pretty straightforward.

Larger companies and syndicates had office staff to handle the paperwork but the individual prospector had to supply hand-written reports, hand-drawn maps, and naturally, receipts and assay bills. Assay results were not required by the DOM – some things have to remain confidential.

It was, and still is, common practice to stake enough land to protect the perimeter of a prospect, and the Department of Mines (DOM) recognizes this as a necessity, therefore the Mining Act does not stipulate that each unit of assessment work be performed on each individual claim.

To clarify this, suppose you have a nice showing on an east-west strike with a contact/fault line on mostly outcrop, surrounded by heavier overburden or swamp. You may stake a nine claim cube with your main showing on the centre claim, and this is where you will be performing the bulk of your assessment work, which can be credited to contiguous claims in the group. Common sense in a government agency – how refreshing is that?

So now you are in your fifth year. A legal survey is needed to patent your claim, and legal surveys are also expensive. Once again, common sense kicks in. The legal survey is credited at 60 man-days, and now the ground belongs to you and your partner in perpetuity.

But wait a minute! Partner needs to be pluralized at this juncture. You and your buddy never did have deep pockets, and you have been pounding in steel, blowing off fingers and shovelling rock for four years and your overalls have patches on the patches. Every winter you return to your long-suffering family, find an honest job to put food on the table and presents under the Christmas tree, and you hit your grubstakers up for enough cash to do the next assessment round. Five percent here, five percent there, and by year five you are hanging on to the last 20 percent and hoping – always hoping – to see that golden road off in the distance.

Some of the old-timers had to be pretty astute and sometimes a bit shifty. I've read about Sir Harry Oakes and Ernie Martin, partners in the Lakeshore Gold mine. They did a pile of work themselves and hired men to bring the mine to production. In the early days of Lakeshore, the money was always tight, yet they ended up with a huge share of the loot. When the dust settled, Sir Harry was mysteriously rubbed out in the Bahamas, and Ernie eventually passed on with less than $20,000 left. Sad, but true.

So this was why Rusty Cordell, Charlie Sr. and the crew were surveying claims at Thompson. Another outfit, Canadian Engineering Services, (CES) was also on contract surveying claims for Inco, and we would cross some of these claims during a trek to work. The lines were always well-cut and skylined (clearing all branches to open the line skyward) and were, of course, straight between claim corners. By this time I knew that running a compass line meant that the line was not exactly dead on all the time, and I asked Charlie how they established their target before cutting the transit line.

"Easy as pie," said Charlie.

They set the transit up at the #1 post and sent a guy on to the #2 corner with a flare gun. The transit guy took a shot at the vertically fired flare and the azimuth was established. Then they cut the line to the #2 post, repeated the process to #3, north to #4 and back to #1. The claim was not exactly square, but it was surveyed as staked – a simple solution for what I thought was a big problem.

By the 70's the DOM realised that after probably a century or more of claims being brought to patent that a problem existed. Claims could be handed down to surviving family members and larger companies were buying up old patents and sitting on them, essentially removing them from possible development, to be held at no cost and no further work required. Once again common sense prevailed. A land tax system was put in place – in other words, poop or get off the pot. A lot of the patents are still held, but there is now some incentive to get things moving.

In 1968, for instance, some deep pockets bought up the stock of the old Jackson-Manion gold mine east of Red Lake, Ontario. They now own the whole deal, shaft, antique mining equipment, and lock, stock and barrel. A big private lodge was built, and now I'm sure that surface improvements trump any possibility of the Jackson-Manion ever being re-opened. Maybe it's not a bad thing – I just don't know.

90

Chapter III   Summer of '61

At home on my spring break I get a letter from Herb, head supervisor at the Copper Cliff geological/geophysical field office. Instead of reporting to Copper Cliff I am to join John F. and his crew near Shebandowan. This is good news – I am paying my own way and Shebandowan is less than 200 miles away, saving me a one thousand mile ticket.

The letter contains other news – not so good. At Moak I had started at $260/mo and on Jan 1 had advanced to $275. At Copper Cliff I would have started at $300 and should now be pulling in $330. But Herb, (who I will find out loves Nickel and dimes) tells me that $300 is actually a raise.

This is not a deal-breaker. I like the job too much to bark, but for the next four years I will be sucking on the hind you-know-what. Herb and I will sometimes choose our corners in the future, culminating in an unnecessarily bitter parting. (And don't forget – Henry Levac had given the same Herb a swimming lesson once.)

I catch the train to Shebandowan where I am met by John F. the area supervisor. He is set up at a tourist camp on Greenwater Lake, and I am to join Alec Godfrey who runs a packsack drill crew. The next day I fly in by Cessna 180 to join them at their camp.

This will be my first experience with the packsack drill and I will never be a great fan of that vibrating little bugger. Alec's crew (I make three) has two holes to drill before we all move to Marathon.

For the second hole we have to carry the drill a mile down-shore. The bush is very thick along the shoreline and we know it will take at least two days to cut an adequate trail to the new drill site. Lacking a canoe, we decide to build a simple raft, load the drill gear and push the raft down the shore to the spot where we will have to carry the drill only 150 feet to our target (my idea.)

It works fine. The water is knee deep and we push the raft to our spot keeping our boots on since the lake bottom is mostly softball-sized rocks. We unload the drill and take our boots off to wring out our socks. Egads! Our legs are covered up to our knees with leeches! We discuss so-called tried and true methods of leech removal. Alec says liberal doses of salt will dislodge them. We have salt in the lunch pack, but the nearest salt mine is a thousand miles away in Southern Ontario – plan A is out.

The other guy says that if you pee on them they will drop off. Now we stand there, laughing like crazy, peeing in our own legs. The hydrants soon run dry and the leeches are undisturbed. We jump back into the cold water to rinse off.

Plan C works and we sit there for an hour pulling those slimy rascals off. They are latched on pretty good, and it takes quite a pull to unlatch them. No harm done, just lots of red spots on our legs – but we cut a bush trail back to camp after all.

Sidebar: In 1988 my wife and I opened a restaurant 40 miles east of Fort Frances in logging country. We had a fine dog – a Blue Merle Collie.

Trapper was a diligent parking lot attendant – he marked every tire that came in.

Day two after grand opening festivities, two wood haulers pulled in. Tandem tractors and five-axle trailers – Trapper went to work.

He never made it down one side of the first unit. Like Peeaire, the dehydrated Frenchman, Trapper ran out of magic marker.

He was a smart dog, so he went to plan B – one tire per truck.

With the hole completed we await the 180 to pull us out. The plane arrives, is tied up nose in by our little dock, and we load the drill and rods along with the camp gear. So we load, and we load, and we sink the Cessna tail first – no kidding! We quickly off-load until the plane re-floats. Luckily, the plugs in the top of the float compartments held, and an inspection shows that no water has seeped in. Once again, no harm done and we all laugh like crazy.

By the time the 180 returns for Alec and myself, the wind has died down, it is dead calm and the lake surface is a sheet of glass. Our little lake at a mile and a half long is certainly adequate for a loaded Cessna 180 to take off, but on glassy water it is difficult to break the surface tension. The pilot can get the plane up on the step, but it refuses to lift off. With the hot afternoon, the air is too thin to give either prop or wings the bite they need to do the job, so the pilot does a high speed taxi in a large circle and then makes his take-off run. At the crucial lift-off point, he runs into his own wake reflected from the shore, the plane bounces off the waves, gains the few mph of airspeed needed and we are airborne. Once again, no problem, and I am learning something new every day. (Glassy water can pose a problem also when landing. Lack of any surface ripples hampers the pilot's depth perception.)

We are packed and ready to head to Marathon the next morning. John has bought us a 2-4, but with six of us thirsty guys it disappears pretty fast.

"Let's hit the Shebandowan Hotel!" (Not my idea.)

One guy is a non-drinker, so John lets him take us in the company truck, a Dodge Power Wagon panel. Alec has his own Triumph TR3 – his baby. It has recently come out of the body shop where repairs were done to rectify some damage caused by an upside-down incident. Alec, with Mike Easton riding shotgun precedes us. We head out the ¼-mile driveway to the side road, and there in the far ditch is the TR3, upside down again, wheels still turning. Alec and Mike crawl out miraculously unhurt and the party is over before it really gets started. We all laugh like crazy and go back to bed.

Table Scrap

In 1968 a tale of extreme fortitude and presence of mind would unfold on Greenwater Lake. Inco was doing a diamond drill program in the area and had a Beaver on lease. One fall day the Beaver was on an empty back-haul from a drill camp. It was goose season, and the pilot, an avid hunter, was carrying a shotgun in the aircraft. He spotted a flock of geese having an evening snack on Greenwater, landed downwind and drifted toward the flock. When the geese finally took off he bagged one, which fell into the lake. He fired up the Beaver, taxied over to the goose and stepped out on the float, leaving the engine ticking over. Without thinking, he moved to the front of the float to retrieve the goose and the spinning propeller took his arm off above the elbow and knocked him, stunned, into the water. The cold water immediately revived him and also served to slow blood loss. He grabbed the back of the float, unaware that he'd lost an arm until he had trouble climbing up onto the float. He made it back to the cockpit, removed his belt, cinched it around the stub of his arm and, holding the end of the belt

92

in his teeth, was able to keep pressure on the makeshift tourniquet while he flew one-handed back to base. The aircraft mechanic was luckily on hand, and although not a

pilot, took over and flew the plane directly to Thunder Bay, 50 miles to the east, under the pilot's direction. An ambulance was waiting at the floatplane base to whisk the pilot to the hospital. One life saved – one bush flying career nipped in the bud.

Point of interest – the Beaver was CF-FHX – the same Beaver that flew out of Moak and Cook Lake in '60 – '61. (Norm Kearns was not the pilot in 1968.)

We move to Marathon May 1st without the drill – Mr. Packsack has gone on to another area. I will be working as mag-man for Mike Easton, a third year geology student and in his third year of summer employment with Inco. The third man in the crew is Larry Marshall, a geology student and a nephew of Freeman Marshall (who will be appearing later on.) Alec heads a second party, also made up of students. They are a great bunch of guys.

Marathon is a pretty little paper mill town on the north shore of Lake Superior mid-way between Fort William/Port Arthur (Now Thunder Bay) and Sault Ste. Marie. We check into the Everest hotel, store our equipment in a shack at the airport and prepare to spend five weeks in the dirtiest bush I have ever worked in. You can find it rough going in many parts of the Canadian Shield, but the North Shore never gives you a break – hills, hills, and more hills on top of hills. The whole area had been logged off years before and moose-maple, alder and rabbit bush is prevalent. (Rabbit bush is thickets of small balsam firs growing so close together only a rabbit can get through.) But life is good. We sleep on a real bed, eat decent restaurant food, meet lots of pretty girls, and I get my first helicopter ride.

Room assignments are arranged, and I am bunking with Mike. I notice a couple of sidelong glances and grins but think nothing of it, but they know Mike and I don't. Turns out Mike is prone to vivid nightmares. It scares the poop out of me the first night but I find out that if you holler at him he will come out of it and quickly go back to sleep. I soon get used to it and it never really bothers me. The thing is, Mike is never violent during his regular nightly episodes – he's just trying to be helpful. For instance, one night I am awakened by his hollering. I turn on the light and there is Mike, standing on his bed with both hands pushing on the wall. "Get out quick," he's yelling, "I'll hold the car up before it falls on you!"

I say "Okay, I'm out," and Mike goes back to sleep.

Our chopper is a leased Bell G-2. Our pilot is Van Den Bos, ex Dutch Air Force and a real character. The G-2 is underpowered, and combined with laminated wood rotor

93

blades, the result is poor performance. On takeoff the chopper will lift off only as far as "ground effect" allows. (Up to a few feet above the surface of the ground the air displaced by the blades is compressed, giving extra lift.) On a cleared runway Van lifts off ten feet, tilts the machine forward to gain airspeed, and then has enough lift to climb. As airspeed increases, power can be reduced. Stationary hovering for any extended period of time is inadvisable as it will result in overheating and possible engine failure. Nowadays of course, with turbo-jet power and advances in rotor design the modern chopper can hover indefinitely, even out of ground cushion. (Point of interest – a Beaver will out-climb a Bell G-2.)

Two crews of three men are flying out every day: two men to one location, then two men to another spot, and then finally two men to be dropped off one at a time at each work site. Our gear, weighing just under 200 pounds total, is tied to cargo racks over each pontoon. (Our choppers were always equipped with rubber pontoons rather than skids.)

Van taking off from Marathon Airport

Van decides how much gear goes with each load. I weigh 240 pounds, and one kid on the other crew is over 200, so we never fly together. Even the fuel load is critical. Van makes all the decisions about flying. He has the experience, and we don't argue.

Once in the air we fly to the location of our intended work target, (we always have a small air photo with us) and on arrival over the designated area we look for a decent landing spot as nearby as possible. There are very few lakes - mostly little ponds. Occasionally we find an abandoned lumber camp near a river with a reasonably sized clear area for landing.

94

Sidebar: Neys Provincial Park is situated on the north shore of Lake Superior west of Marathon on Highway 17. One of Ontario's most pleasant parks, it nestles amongst the pines on a sandy plain almost devoid of underbrush. My father, veteran of most of the gruesome land battles of WWI, spent WWII as a Captain in the home guard. Posted to many camps, he had spent time as Officer-in-Command at POW Camp Neys, the site of the present-day park. When I told him that we used old logging camps as landing sites, he was of the opinion that some of them no doubt housed POWs. The good guys were given the opportunity (not forced – he stressed this) to break the monotony by going to these logging camps to work, and even receiving a small stipend for doing so.

We used one of these old camps near a feeder creek to the Pic River as a landing spot for a couple of days. A few tumbled-down log buildings were all that remained, and while waiting for Van one afternoon we poked around. Only one building had a roof left, and in it we found a few receipts and stuff. This must have been the timekeeper's shack, and had been built with a little more care than the others, accounting for its better state of repair. Not much was left behind, but some of the receipts were legible. One was for an orange – one orange – price, thirty cents. The timekeeper always ran the commissary on his own nickel, and apparently the men had to buy their own fruit. I doubt if the POWs' pay was more than fifty cents a day, and Canadian cutters probably didn't make big bucks either. Did they merge the work force in these camps, or were the POWs working in separate camps?

No matter how you cut it thirty cents is a hefty cut out of a day's wages – pretty harsh!

On the first landing at a new target Van sets her down very gently, with the tail rotor perilously close to small saplings or willow clumps. When he leaves, our first job is to dress up the landing spot. We clear away all small brush in the near vicinity to make sure nothing will interfere with the tail rotor and we also cut down nearby trees which might impede the evening's loaded take-off run.

Van always does his best to save us a long walk, but sometimes if we do have a long hike we cut a chopper pad in the bush. On these occasions only two men will be dropped off the first morning. I am usually one of the two, and carrying our faithful axes we compass in and spend the day cutting a pad. (Our third man joins the other crew for the day.) When we find a good spot we start knocking trees down until we create a circle of about 100 feet. Some fallen trees are cut into eight or ten-foot sections and carried to a level spot in the centre of the circle to build a rudimentary platform for the chopper to land on. This also raises the height of the machine enough to keep the tail rotor out of trouble. We then cut down any tall trees for 200 feet in the direction of the prevailing wind (and hope it doesn't change.)

Usually Van appears just as we finish our hard day's work. Now the pad will be used for the next three or four days until we move on to another target. You may wonder why we go to all this trouble, but we never know if a target will prove to be interesting enough for extra work, and a three-mile walk in and out every day leaves little time on the job.

One such pad is surrounded by very high trees – no problem with a decent headwind but one day it is dead calm. Van merely flies to one end of the pad, swings the chopper around, flies back to the other end, swings around again and flies out. It's tough on the engine and it gets pretty hot.

Van makes the mistake of telling us about "auto-rotation." Works like this: at a safe height the pilot simulates an engine failure by cutting the power and the chopper immediately drops two hundred feet, and the increased airflow over the rotors speeds them up. (In normal flight the rotors spin at 300 rpm, the motor at 3000 rpm – a ten to one ratio.) So the rotors speed up to 340 – 350 rpm and, in the case of an actual engine failure, the pilot looks for a landing spot within gliding range. He gets one chance only. Pick your spot, flare out at ground cushion height and set her down quickly before the rapidly decelerating rotor blades lose their ability to generate lift. These guys earn their money!

So now, every trip it's "C'mon, Van, let's do an auto-rotation!" We drive him nuts.

One day flying back to the airport with a light load, Van cuts the power with no warning. We drop like a stone, leaving our stomachs 200 feet above us until the chopper settles into its auto-rotation attitude. Van powers up and we fly home.

Everyone in the group gets to enjoy an auto-rotation experience and we never rat Van out. (Practise auto-rotations are only to be done on check flights with a DOT observer on the ground – never, ever with a paying passenger on board!)

For the rest of my life I have used Van as a benchmark for my own grading of chopper pilot skills. Many would measure up to Van's flying ability, but few had his unique attitude: always cheerful, optimistic, and curious about everything, with a never-ending exuberance and enjoyment of life's ups and downs (pun intended.)

Some of our targets require longer flights, and when possible, we rendezvous at a designated spot. The first two men leave the airport with Van and the other four go by road to a site near the day's work. One particular day the meeting-place is at a former highway construction campsite 30 mile east on Highway 17. We park the truck, Van picks us up and we chopper to work.

Highway 17 had been completed to Sault Ste. Marie in 1960, so the tourism department has been flogging the "Circle Route" around Lake Superior. Many folks are doing the circle tour, but provincial park infrastructure is still not complete and travellers must do their overnight tenting at any spot available.

So when we return to our truck in the late afternoon, our landing spot is plugged with campers. Van chooses a place to land, comes in downwind over the horde below, banks and flares into the wind in one easy motion. For a moment the chopper is on its side in midair, right above a family preparing their evening cookout. It's summertime, and we have taken the doors off the chopper, so when the lady looks up, she sees a helicopter on its side above her with us three yahoos looking straight down. She screams, grabs a kid under each arm and heads for high ground. Van feels sheepish for being a show-off, but what the heck – no harm done and she has a good story to tell the grand-kids.

Sidebar: The following year I flew in Northern Quebec with a pilot who never banked. On landing he would turn the chopper in a flat attitude using the tail rotor. I told the mechanic about Van's method and asked if it was dangerous. He said that Van's use of the main rotor as a sort of pendulum was easier on the chopper. Using the tail rotor to turn puts added stress on the relatively lighter sections of the rear airframe.

Table Scrap

Van told us a tale of his days flying seismic crews out into the desert in Libya. The camp was on an escarpment and when returning to camp they faced a sheer rock wall rising 200 feet above the desert floor. Every day a strong prevailing wind would come off the flat desert and flow up over the escarpment. The chopper pilots (bunch of mean rascals), when flying newcomers, would head right for the wall. The new guys would start looking over at the pilot and you could almost hear them thinking, "Does this guy have a death wish? Pull up! Pull up!"

Then the wind swooped them over the cliff, and they proceeded on to camp for a change of underwear.

Rarely is anyone seriously injured on the job. We are young, agile, full of piss and vinegar, and every day is a cardio workout (probably why my misspent life has lasted longer than I would have expected.)

However, one day Carl R, a real good summer student (whose father is an Inco VP in New York) is driving a compass line through some rabbit bush. Those little balsams always have small, dead branches sticking straight out. You have to tunnel your way through, and when Carl turns to check his back line one of those branches goes into his ear and punctures his eardrum. Wow – painful! There is nothing we can do but wait for the chopper, and Van takes Carl straight to the Marathon hospital. From there he is transferred to Fort William and then on to expert care in New York. I never did hear how it turned out, but no doubt it affected his hearing forever after.

We seldom lose days to rain. It rains mostly at night and the bush is drying by the time the chopper drops us off. Fog, however, is another thing. Lake Superior never gets really warm – it is the deepest of the Great Lakes. We often wake up to fog, and sometimes it extends to the airport, five miles inland and 200 feet higher than lake level. The surrounding hills hold the fog on calm days – no visibility – no chopper ride. We always have spare equipment, so six of us pile into the Power Wagon and chase road anomalies, easily accessible and saved for such days.

One foggy day we are off to Hemlo, 35 or 40 miles east on Highway 17. Hemlo isn't even a town as such, just a spot on the CPR line near the junction of the Manitouwadge road and Hwy 17. There are quite a few airborne responses within a radius of three miles or so, but most can be written off as being caused by a nearby high-voltage power line and steel railway tracks.

There are targets not far off the highway and with the combined crew of six young bucks, we knock off three of them. The first two conductors are immediately pinpointed on outcrops. A few strokes of a prospectors hammer shows Mike that the conductors are graphite in iron formation and are of no further interest. After lunch we hit a third response. This time there is no outcrop, so we do a quick four-line grid. The geophysical results are similar – a strong, well-defined conductor with a coincidental mag high writes this one off also.

97

That evening Mike and Alex discuss the day's results with John. The upshot is that the gaggle of airborne responses at Hemlo will be abandoned. We have other areas to cover this summer.

Sidebar: We were walking on GOLD at Hemlo (maybe that's why my feet tingled,) and we walked away from three gold mines!

I don't mean to be critical of anyone involved. Had they run sample assays the gold would have run grams per ounce, and gold was still pegged at $35/oz.

A few years later two prospectors from Timmins would cash in. They believed in the area and thanks to gold being allowed to float, and thanks to Murray Pezim, (a Vancouver promoter,) they would become gazillionaires with an unheard of five percent net smelter return (NSR.)

Table Scrap

The two prospectors were John Larche and Don McKinnon. I haven't googled them – this is common room chat. It's my story and I'm sticking to it:

Larche was into heavy equipment contracting. Some say he quit the contracting business but bought new stuff anyway. He just wanted to admire it and keep it shiny yellow.

McKinnon was younger, and continued to prospect, shifting his attention to the Yukon. Was he successful? Google will tell you.

But I like this part: McKinnon ran for public office in Timmins – perhaps provincially, and I don't know which party he was associated with – it doesn't matter.

He didn't need to raise campaign funds – he had lots of his own. His opponent, on the other hand, was struggling.

So he donated to his opponent's campaign!! He said it would be no fun if the teams were not competitive. He must have been a remarkable man.

We have two targets near the source of the Pic River. They are 60 miles from the airport, and as they can be accessed by an old Marathon Paper logging road, it is decided that our crew will drive in and camp for a week because the chopper will be leaving soon. This work must be completed and it is too far for daily flights in and out. Alec is working between us and Marathon. Van will deliver his crew, then pick us up at camp – it is a time and money saver.

We take our sleeping bags, a ten by twelve sail-silk tent, a wanigan and some grub and power-wagon in from Coleman, a small railway town 15 miles west of Marathon. The Pic River is a fair size, but is only about 60 miles long. It drains a good portion of the hills surrounding Lake Superior and thus is fed by many creeks along its way. A few miles in we cross the river on an old bridge at a narrow spot: a pretty basic structure – six or eight huge trees decked with 3"x12" rough planks span the river. It was never meant for logging trucks, as back in the day, all logs were floated down river to the big lake and taken in booms to the mill.

Another 30 miles up the road, we pass a large old logging camp, with buildings still intact, but by now, in various states of disrepair. Ten miles on we pitch our tent and go to work, and on the second day we decide to do a little exploring after supper. We drive up the road to find it ends, more or less, at a little bridge over the Pic at its headwaters, where it is fed by Kilalla Lake, a fair-sized body of water. Here the road becomes a causeway, curving out across a huge cattail-covered shallow swamp to the open waters of Kilalla. The causeway, set on pilings with a plank deck, has to be a half mile long and is obviously unsafe to drive on by now. We walk out on it a ways, marvelling at the vast volume of lumber used to construct the thing and trying to figure out its purpose.

I don't want to get up! Maybe it is because I've grown another foot (note three boots). This was a bare minimum set-up, as you can tell by the rudimentary canvas cot. we were not here for a long time, just a good time.

Just under the bridge, where the swamp becomes the Pic, there is a rapids. The pure, clear water, not too deep and not too wide, rolls over smooth rocks, dropping 20 feet in a distance of maybe 30 feet to start its way down to the big lake. We peel off our clothes and spend the next few evening hours sluicing bare-assed down the rapids: good, clean fun, and no body parts are lost, just bruised a bit. We laugh like crazy.

Table Scrap

Three years later I was working for Mining Corp of Canada at Manitouwadge. Glen McNay had been added to the crew for a three-month term of employment and I soon found out that Glen trapped at Killala in the winter months.

He had built a secluded cabin on the lake and his line covered Killala and down to the old logging camp. He tended his traps on two loops, one day touring the Killala Lake portion and two days on the south loop. He had a Samoyed cross as a pack dog and faithful companion. He skinned the larger animals on the spot and these pelts would join the smaller animals in the pack, to be cleaned and stretched on frames in camp that evening. Flour, sugar, tea, salt and tobacco would be packed in from Coleman at the start of the season, with one mid-winter trip to replenish supplies. Glen and the dog each carried their share, with the dog also pulling a small sled (Glen would help on the upgrades.)

And the dog was smart. On the Kilalla swing, take-off trails were used to work a couple of beaver dams. Coming back to the main trail, Glen had to get ahead of the mutt to make sure he turned left, and as they neared the junction the dog would be checking over its shoulder to see if Glen was gaining on him. If Glen sped up, so did the dog. If the dog reached the junction ahead of Glen, the dog would turn right and head the two and a half miles back to camp. Glen swore the dog was laughing at him. No harm done, it only added five miles to the daily walk.

It was a good trap line, with otter, mink, fisher, marten and of course beaver, which would also be their fresh meat supply (beaver, that is: I have eaten beaver and muskrat – very tasty.) Their diet would also include grouse and spruce hens, and fish netted in Kilalla. The south loop of the line was longer. Glen fixed up a small shack at the old logging camp and he and the dog would overnight there, eating from a small food cache and dressing out the day's pelts, which were stored there until break-up. The next day they would complete the loop, and in this way he serviced the traps every three days. It was a peaceful yet hard-working existence, with only the dog for companionship.

Three weeks before break-up Glen would pull the traps and concentrate on the marshes of Kilalla for his muskrat harvest. One year he took out 3000 rats – a good year at $3 per rat pelt. Normally the marshes would yield 1000 – 1500 rats, but that year was extraordinary and the rats represented his bonus check. In the spring four of five pack trips to Coleman would bring out the fur, to be bundled and sent to the buyers in Port Arthur. Then Glen would find summer employment while the dog stayed with a family in Coleman.

By the winter of '61 Glen had a Skidoo and sleigh to pull supplies and pelts. The dog, refusing to ride, still followed along and still packed on some of the more difficult trails on the line. On the last trip to Coleman in the spring, the dog would lag behind, Used to the winter supply trips, he no doubt expected to meet Glen on the way back to the cabin. The dog would overnight on the trail somewhere and show up the next day, wagging his tail, and no doubt saying, "What the hell happened to you yesterday?"

But in '62 the dog never showed up in Coleman. He was getting older, and maybe the wolves got him, or, Glen thought, he might have been stolen or shot. Glen went out in the winter of '63 dogless and terribly lonely. Then in the summer of '63 the Department of Lands and Forests burned the Kilalla causeway, burned the old lumber camp and ripped out the log bridge over the Pic on their way out. The loss of his dog had already taken the starch out of Glen. He would never again trap Kilalla.

From Kilalla we move to Jackfish Lake, farther west and ten miles inland from Lake Superior. Van has gone on to another contract and the work around Jackfish is easier to get at. We fly in by 180 (Superior Airways) and set up camp – two tents for two crews and a communal open-air kitchen beneath a tarp. We have one canoe. I wonder how this will shake out – and it does the next morning. At sunup the 180 comes back with a message from John. A three-man crew has pulled the plug in Chibougamau – Mike, Larry and I are on our way to La Belle Province.

We take our personal gear only – a camp awaits us in the bush in Northern Quebec. We will go to Copper Cliff to pick up our travel orders. To do so we will catch a train at Jackfish.

Now, this is pretty cool. Jackfish is a stop on the CPR Line right on the north shore of Superior. It is not a town, it is just a hotel, and a nice old hotel it is. When the railway was built in the late nineteenth century the North Shore of Superior was virtually unknown to non-native Canadians. During construction it was easy to figure out that this was moose country, so the CPR built a hunting lodge. It was in no-way on the scale of Banff et al – just a comfy ten-roomer with a nice view of the lake beyond the main line. There had once been a road to Jackfish Lake but with the arrival of float planes in the late 20s the road was abandoned. When Hwy 17 was built, five miles of road was resurrected to access the hotel.

We stay overnight and catch the eastbound Transcontinental the next morning. It is a flag stop – no kidding! The hotel manager stands on the little platform with a flag.

We load our stuff with our chests puffed out. The folks in the dome car probably think we are V.I.P.s.

Sidebar: We may have been the last guests at Jackfish. With the new highway finished, hunters and fishermen did not need the railroad. I heard that the hotel was closed and demolished in 1962. We are careless with our heritage in Canada.

On August first we arrive in Sudbury and go to the field office in Copper Cliff. We are given our train tickets and a small travel allowance to eat with, plus enough to buy our bus tickets for the last leg of our journey. I check with Herb S. regarding my transfer to airborne. "Not yet," he says.

The next afternoon we continue on by train to Montreal where we will switch to CNR to Roberval, on to Lac St Jean, change again to a day coach to St Felicièn, then on to the final three hour bus ride to Chibougamau. When we get off the train at St. Felicièn, there is a bus sitting on the street near the station. The sign over the windshield says St Felicièn – but we knew where we are at, we want to know where he is going. Our French is abysmal. Mike, who is taking geology at McGill University in Montreal can barely say "oui" or "non." He was raised in Barbados - Spanish is his forte.

Larry and I scratch up our high school French: "Chibougamau?" (Pretty lame, huh?) "Oui."

"A quelle heure vous parti?" We think that is not too bad. We want to know if we have time for breakfast. We are hungry!

"Nevair," says the driver. Our faces fall.

"Never?"

"Oui, nevair."

How in the hell are we going to get to Chibougamau?

Just then the driver cranks the windshield banner over. It now says "Chibougamau."

He takes our money and we get on the bus, where we sit for forty-five minutes with growly tummies. By ten to nine there are a few more passengers loaded, and at 9 am on the nose our bus pulls out. Finally the light bulbs come on. The driver had said, "Neuf heures," – nine o'clock. Let's face it – we are not the sharpest pickets on the grid.

So at "Nevair" we are off to Chibougamau, unfed.

Sidebar: Our lack of Francais had led to another glitch on our way to Chibougamau. It was not a serious foul-up, but was an indication that we were anything but savvy travellers. I could blame it on Mike – he lived in Montreal – but maybe he had a brain fart.

101

We are travelling fully loaded with packsacks, eiderdowns, suitcases and our trusty axes. So we hustle out of the CPR station to a cab parked at the curb. The three of us and our gear fill the car. We have ten minutes to switch trains.

"Where to?" (En Francais.)

"CNR station! Vite Vite!" (Hurry up) – Blank stare from the cabby.

"Vite! Vite! Dix minutes!"

The cabby shrugs, pulls a U-turn and parks at the opposite curb. The CNR station is right there on the other side of the street!

Fred Ellgring meets us at Chibougamau and checks us into the Hotel Monaco, where we eat, nap, and discuss the upcoming work with Fred.

The Hotel Monaco is owned by the family of the wife of my previous Canadian Nickel supervisor at Moak Lake. Small world, eh? It is a nice modern hotel with a nice bar, so we are told, but the doors are locked. They had run afoul of the liquor inspector (insufficient bribe, no doubt) and are sans licence for a while. We are tuckered out anyway and we spend the evening watching TV "en Francais" and have a good night's sleep.

Chibougamau is a modern, bustling mining town with two copper mines nearby, and is also supplying numerous exploration camps in the untouched virgin forest north, east and west of town. I don't know if the mines are still going. As the old saying goes, "The day a mine opens is just one day closer to the day it closes."

Saturday, October 20, 2012

Sudbury nickel mine to stop operations at year's end

THE CANADIAN PRESS

SUDBURY — Operations at a nickel mine in Sudbury are being suspended at the end of the year.

Vale, which operates the Frood site, says it is closing the mine due to market volatility, declining metal prices and falling demand for the product.

The price of nickel has dropped 17 per cent this year and 30 per cent since 2011.

Vale says the closure will not lead to any job losses.

The company says the 85 workers currently employed there will be reassigned to other jobs within the Sudbury operation.

The Frood site has been mined for more than 100 years, but the ore now has low value and the company had been mining the site at a loss.

This mine operated for 100 years!

But Chibougamau still exists as the terminus to the north, and recently gold has joined the base metal play. If you Google Chibougamau on the Internet, you will be amazed at the network of roads north of the town.

Our area supervisor, Fred, is highly intelligent but different, if you get my drift. He is never an "I am the boss" kind of guy, but you have to learn his quirks. I will work for him again the following winter, and we will clash from time to time, but when the chips were on the table, as you will learn if you stick with me, Fred was the only one willing to deal me a fair hand.

When Fred's crew quit, the camp had been left intact including non-perishable grub. Fred had given a fresh meat order to the local grocery and it had been delivered to the airbase. (Every northern town has at least one grocery store that will fill food orders to be picked up or delivered to the air service. HOT TIP: Use liquor boxes – they are very strong, and the dividers can be left in for glass jars.)

After breakfast Fred drives us down to the base five miles south. We will do all our Chibougamau fixed-wing flying with Fecteau Air, without a doubt the most well organized air charter service I will encounter. We jump into a Beaver and head to Frotet Lake, 70 miles north.

Frotet is a sprawling body of good clear water, over 30 miles east to west. With plentiful shoreline outcrops and many creeks and inlets, it is a prospectors' Xanadu. However, for us anomaly chasers it is Labrador tea; tea on the flats, tea on the ridges, tea everywhere. The only place it is sparse is in the low spots and there you walk on water-soaked moss. Labrador tea bushes are low, never growing more than knee high, but the little interlocking branches make for tough slogging, and it is hard on boots and pants. Some guys prefer leather boots but the tea wears out the toe leather in a short time. I wear "Logans," with leather tops and rubber bottoms. The heavy rubber withstands the beating better than leather – pants, however, are another problem. The tea soon wears out the knees, but we have sample bags and Jiffy Sew.

Sidebar: Every camp has a supply of strong cotton canvas sample bags in three sizes. A small one makes a nice cover for the mag head.

Sample bags are our Swiss Army Knives. They make a good ditty bag to hold your tooth paste, tooth brush, and razor, to be tossed into your packsack when you are in transit. Small bags, opened at the seams and liberally coated with stuff called Jiffy Sew, make excellent knee patches that stand up to the Labrador tea better than denim. If you are ever at a northern float base and see a group deplaning wearing sample bag patches, you know they are players.

The large bags will hold a slab of bacon, and when tied to a high branch out of bear reach, it keeps the flies and squirrels out yet lets the meat breathe.

If you are unlucky enough to hit a submerged reef and slash the canvas on your cedar strip Peterborough canoe – no problem. Mr. Sample Bag and a tube of Ambriod to the rescue.

I think you get my drift by now – endless uses. Nowadays the bags are made of plastic or synthetic fibre and are useful only as intended.

Oh, Yeah. Once in a while we might actually use a bag to send a sample of a mineralized outcrop in to the boss. I can't really remember that happening too often – we are anomaly chasers, baby!

The camp is standard issue. Every Inco fly camp is the same - blindfold me, put me down in any camp and I can stand on the dock and point out where everything is. This one is nestled in among some good sized jackpines in a slight shoreline indentation just deep enough to be called a bay, but sheltered somewhat from the west winds by a small point. We step out on a standard issue dock, two pieces of 1/4 inch plywood, 2' x 8' nailed to sapling stringers and supported on rock-filled cribs. It is just long enough to reach the cargo door on a float plane nosed in to shore.

The tent is standard issue 12x14 sail silk with four foot walls. The ridge pole is a four or five inch diameter pine held up at the ends by crossed poles that also support the side poles at the four-foot eaves. The tent peak is tied to the ridge pole using small ropes sewn into the peak by the manufacturer. The eaves also have sewn-in ropes to tie them to the horizontal support poles. Over this is stretched a canvas fly, slightly larger than the tent, and tied off at the sides. The fly handles the heavy rain, and any spray that comes through does not penetrate the inner sail silk roof. The fly also serves to keep the tent cool in direct sunlight on hot summer days. The front of the tent has flaps which can be draped over the sloping side support poles in warm weather, or closed and tied down on cool windy nights. Mosquito netting keeps the bugs out and can be rolled up out of the way in the fall. An asbestos stovepipe ring at the rear completes the deal. In the centre of the tent is an airtight tin heater for cool nights. In three corners are canvas stretcher beds and personal gear. In the other corner is a homemade makeshift counter for the Coleman stove, water pail and various condiments. We eat sitting on our beds. In the summertime we cook outside under a tarp, but in the fall we move the kitchen inside.

An old 8x10 canvas tent serves as a storeroom for canned goods, gasoline, life jackets, work equipment and a five horse Evinrude for our 17' v-stern canoe, which is overturned against a tree.

So now we go to work, and we lose a few days in August. Overnight showers may keep us out of the bush until 10 am, but by then the tree branches have shed their water droplets. The Labrador tea stays wet until noon or later, and since it is only knee-high, our instruments are not affected, but we are soaked to the knees most of the time.

When we circled the lake upon arrival, we had seen a tent on the northwest shore, five miles from our camp. One evening right after supper we motor on over for a visit. Sundown is 7 pm, but we are not going far from camp.

(It is common practise to check for submerged reefs on any arrival even if the pilot has landed there in the past. Changing water levels can cover a reef that had been above water only a few weeks before or vice-versa.)

They are glad to see us. We are their only visitors this summer and tea is poured.

They are two old prospectors (in their early '50s, I'd say – I'd call them young now.) Every year they form a grubstake syndicate with one or more investors and head for the bush. They must have a good rep, as a piece of the syndicate money is always supplied by a major mining firm. In the fall they return home to make plans and raise the grubstake for the following summer, saving enough syndicate money to pay themselves a stipend over the winter months.

Their camp is compact, neat, and oh-so-clean; an 8x10 sailsilk tent and a tarp-roofed patio with plastic sides that can be rolled up or down depending on the weather. A hand-made counter for cooking, a hand-made table for mapping and hand-made chairs complete the deal. The only lumber is new plywood for the tabletops. The camp is located on a broad flat outcropping exposed to the breezes to keep the bugs away. I file this for future reference. It is super clean – our camp is a slum by comparison.

They have no radio – Fecteau overflies Frotet almost daily and a signal fire or flag will indicate an emergency. Once a week a Fecteau fly-by to someone else's camp drops off a grocery order and mail. These prospectors would be charged pro rata by the weight of their order. This is called a split charter, one way of keeping the overhead down. The pilot always checks the tent for outgoing mail and next week's grub order is waiting on the bed. What a great system!

104

"Getting dark," we say, "Better be on our way."

One of the old guys looks at the other one, who nods and a bottle of rum appears. They have been saving it since May!

So we have moose milk: two ounces of rum, equal parts of condensed milk and water and a spoonful of brown sugar. What a smooth drink! We sip moose milk and trade stories for a few hours in the light of a hissing Coleman lantern. (Actually <u>we</u> don't have much to trade, we just listen.)

All good times must come to an end. At 11 pm we climb into our canoe and head home – not impaired mind you, twenty-six ounces of rum between five guys over four hours is not enough to get anyone in trouble.

It is the darkest night I have ever NOT seen. A low overcast and a dead calm leaves us totally blind. We know the direction of our camp so we take a compass shot before we depart. (We all carry a Silva compass hanging around our neck and stuffed into a shirt pocket – sort of like "prospector bling.") We chug along at a slow idle as Frotet is notorious for reefs. From time to time we light a match for another compass check.

Thump! We run into a small island that we know is a half-mile from camp. We pick up the shoreline and paddle silently home across black velvet water guided by the faint outline of the tree line to our right.

A Six-mile Hike

We have a two-way radio supplied by the air service - no more ex-Lancaster boat anchors. We are now in the transistor age and the radio is compact, reliable and easy on batteries. We check in every morning for messages from Fred (Fearless Fred to us, of course.)

One morning in late August Fred is at the airbase for our 8 am check-in. He tell us he is sending us maps and an air photo regarding an anomaly six miles off the southeast corner of Frotet. The maps will be dropped off at camp that afternoon and we are to prepare for the arrival of a chopper two days later, weather permitting. A Montreal charter service has a Bell G-2A coming by on its way from James Bay to Montreal. It will make a slight detour, deliver us to the target and continue on its way. "Hmm," we think, "One way only."

We are to take our 12x14 sail silk tent (very light) roll up our sleeping bags, pack a little grub and our Coleman stove, and fly in with our geophysical gear for a five-day stay. When finished, we will leave the tent and stove (to be retrieved the following winter) and carry our gear and sleeping bags back to camp. As our camp is a mile from the southeast corner of the lake, it will mean a seven mile trip! While we are chasing the anomaly Fred will have Fecteau drop off a new tent and Coleman stove at camp to await our return.

(You may wonder why we would go to all this trouble rather than walk in every day. Well, it is hard to make two miles per hour in the bush even on a cut trail. Three hours in and three hours out doesn't leave much time for work.)

The day of departure dawns clear and sunny, obviously a good flying day, so we prepare for our trip. We decide that Larry and Mike will go on the first flight. They weigh a lot less than me and they will take some stuff with them. I will load the remainder of the gear into the canoe and motor down to the end of the lake to be picked up there on the second trip. This means that the pilot has to wait a half hour for me to get to the meeting point, unload the canoe, overturn it on shore with the motor and life jackets underneath, and load the chopper. The pilot is cool with that – there will be lots of daylight left for the remainder of his ferry flight. As for us, we have cut our return walk to six miles.

It works out very well. We throw up a rough camp and immediately go to work, locate a conductor a few feet from the tent, and by the end of the fourth day we have worked up a good-sized grid. The weather is great, lucky for us, as we had left our tent fly at camp to save weight.

We can understand why the anomaly is a high priority. The area is a hotbed of conductivity with lots of good mag association. We had worked dawn to dusk, and although the area needs more work, we are running out of grub. We have a lot of info to be plotted on maps and forwarded to Fred. We are well satisfied.

On the morning of the fifth day we are done by 10 am and getting ready to return. We roll up the tent, tie it and the Coleman stove to a strong tree branch and head back to the canoe. We are each carrying at least a hundred pounds, and when we reach the canoe we are tuckered out. We still have to pitch our new tent and although we have less than three hours of daylight left, there should be no problem. Right? Wrong!

We arrive back in camp to find our new tent sitting on shore at the end of the dock, and what a tent! It is14 x 17 with three-foot walls, made of heavy canvas and no fly provided! It weighs at least 200 pounds! Our tent poles were designed to support a light 12x14 sail silk. Now we have to reconfigure everything and daylight is waning. We had left our 12 x 14 fly in camp (luckily for us, you will soon see why) to cover our stretcher cots, kitchen stuff, stove and stovepipe. We pull our fly to the side and wrestle that heavy beast of a canvas tent into place and tie it up well enough to give us some shelter, then crawl into our sleeping bags too exhausted to cook supper.

We spend the next day and a half rebuilding the tent. The ridgepole is long enough, but too small to support the new tent. No problem – we put a support pole under the centre by the stove. We move the crossed support poles at each end, cut longer side poles and move our stretcher cots out to the new walls. More room – not too bad at all!

Black September

By noon of the second day we have our new camp in shape and the Beaver comes in to take Mike and Larry out. They are returning to school and I have a new party chief and helper, Jack and Hal. I have heard about Jack, he is a good man, well experienced and easy-going. Hal is a buddy of his and though green, he is a willing worker and fits right in. So work goes on – for a little while.

We finish a grid within walking distance of camp and now start a new one two miles west, canoeing to a spot on shore where we walk in a mile to the response site. We locate the conductor and return to camp, leaving the motor/generator unit, coil bag and magnetometer in the bush as usual, taking only the receiver coils, battery pack and headphones back to camp.

The weather has been beautiful lately, warm sunny fall days and with cool brisk nights to keep the mosquitoes and deer flies at bay (not the ubiquitous black fly, though. They can live through anything. I have even had black flies bite me after the first snowfall.)

106

That night the skies open up and it pours, and by 2 am we are getting wet. Every time a water drop hits that marvellous, treated, waterproof canvas, a fine spray of water continues on through. Some of it falls on us and little beads of water form on the inside of the roof to run down and drip on our beds.

We finally crawl out of our sleeping bags at daybreak, wet and miserable and call the airbase. "Tell Fred he forgot to send the tent fly."

The answer comes back. Fred says no fly is needed – the tent is guaranteed leakproof. Leakproof my ass!! OK for Fearless – he lives at the Monaco.

So we go out in the rain, dig out our 12x14 fly and rig it up. It solves the problem, sort of – it is just too small! We have to move all our stuff away from the perimeter of the tent, leaving a three-foot margin of unusable space on all sides. Our roomy tent has become much smaller.

It rains and rains, pouring all night and showering during the day, or vice versa. It never stops for long. Our eiderdowns and clothes are damp and everything is soggy. Even our Sifto salt stops pouring. (Remember? "Sifto Salt – when it rains, it pours.")

About day six the sun comes out. Whoopee! We cook a real breakfast and two hours later head down the lake, eager to do something other than huddle in the tent. A big black cloud is coming up – maybe it will miss us? No way! Half an hour later we are back in camp, soaked to the skin.

And still it rains. By day ten we are whining to Fred. "We want to go home."

"Stick it out," is the answer, "It will clear up." But it doesn't.

No longer do planes pass over, and it is two weeks before a meat order comes in on a showery but flyable day. We sadly watch the plane leave, wishing we were on board.

We are not short on grub. A student soil sampling camp had been disbanded September 1st and we were given three large boxes of canned goods from their camp, including a 48-can case of Beanie Weenies. (These will come in handy as cabin fever relievers.) Unfortunately the students had left some cans out in the rain and the labels have fallen off. No matter, by now the incessant rain and high humidity has removed some labels from our own cans. Let's face it – the air is so wet we are growing gills!

We are not doing much cooking – even bacon and eggs seems such a chore. We usually just play "What Have You Got?" We each take a can, sit on our bunks and open them with our axes.

Sidebar: Your axe is your friend, and you keep it sharp. Everyone has an axe preference as to manufacturer and weight. Some like the Iltis Ox-head. I prefer a Walters 2 1/4 pound. A six-inch file is always in the lunch bag. Some, myself included, carry a small whetstone to touch up the edge from time to time. Walking in the bush without an axe in one hand just seems unnatural. It can even serve as a balance pole when crossing a creek on a fallen log or on a narrow beaver dam, so you can understand why the three of us sat around opening cans with our axes instead of using a sissy can opener.

One guy says, "What have you got?"
"I've got peaches."
"I've got fruit cocktail."

107

"I've got string beans, dammit!" and we laugh like crazy.

We are getting goofier by the day. You can imagine three young, fairly normal adults in a space of no more than 150 square feet with rain pounding on the tent and a 24" tin heater stoked with firewood that is never really dry.

We never argue. What is there to argue about? When firewood needs to be cut, we cut firewood. If someone feels like cooking, he cooks enough for three. The days may be getting shorter outside, but they sure are getting longer inside.

Even a pot of soup has an element of discovery. We can tell the soup cans by their size, but what kind of soup? Answer: just open three cans, pour them in the pot and add a little water. The soup of the day: Chicken noodle-minestrone-tomato. What a taste delight!

Fun and Games II

Caution: The next part of this story may cause you to doubt our sanity – or perhaps my veracity – but I assure you every word is true.

First I will describe our airtight. It is made entirely of tin and although they come in different sizes, our standard summer camp issue is 24 inches front-to-back and 18 inches high. It is oblong shaped with four short, stamped-tin legs. The stovepipe fits into the top at the rear and the stove is loaded through a one-foot diameter hole in the front of the stovetop. A tin lid with a strap bolted to the top swivels aside to stoke the stove and an adjustable draft at the bottom front controls the fire. They usually last only one season, but the little rascals throw a lot of heat. (See photo, back cover.)

When the two prospectors down the lake broke camp before the rains, they left us a present. We were in the bush when we heard their plane leave, and on our return to camp we found a 50lb sample bag full of pocket books – westerns, detective novels – lots of reading. We will thank them often in the weeks to come, bless their hearts!

So we start our own version of Frotet roulette, using the Beanie Weenies instead of bullets.

Here is the drill – I kid you not. We have 50 lbs of paperback books, right? We also have 48 cans of Beanie Weenies, right? (Beanie Weenies are little half-sized cans of beans and wiener pieces. The soil sample students must have loved them, but we won't eat them.) And we have a stove that needs a firewood transfusion from time to time.

One of us starts a pocketbook. After reading about one quarter to one third, depending on the size of the book, he rips off the finished pages and passes them on to the second guy, who reads them and passes them on to the third man. When the third man finishes, he stacks the pages beside the stove to be used as a boost to the damp wood the next time the fire needs stoking. When the first guy finishes a book, we vote on the next selection and westerns usually win.

Now the fun really starts. Each of us takes a turn putting wood on the fire. The thing is, when you fill the stove with wood you also add a can of Beanie Weenies. As the fire gets hotter the can explodes. The lid bounces up a few inches and weenies spray all over the stove's inside walls. The trick is to wait as long as possible before your turn to stoke the fire comes up. Finally, amidst calls of "chicken!" or "cluck-cluck," you add wood, keeping your face averted. (Hal even wears safety goggles.) About every fourth or fifth time you catch the loaded chamber and the can explodes, spraying your face and

108

the tent ceiling with hot Beanie Weenies. Then we all laugh so hard the tears roll down our cheeks!

Now that's cabin fever reliever!

Fire on the Water

Twenty-one days later the skies clear, and off to the bush we go. We have two more weeks of work to do, but ice is already evident on the lakeshore some mornings, and we want to do what we can before the camp is pulled.

Ready to give her tar-paper, we head out to the gear, set up and go to work but nothing works – we have no signal! Although we had made sure that the headphones and battery pack were kept high and dry hanging from the ridgepole over the stove, we still have a problem. The constant humidity has screwed up the electronic components in either the battery pack or the transmitter, or possibly both. Back to camp we go and back onto the radio.

"Tell Fred we can't work – we have no signal."

The answer comes back that Fred is bringing down some spare equipment. The Beaver will be there by 1 pm. The Beaver arrives and we are at the dock to unload it. What is our spare equipment? Two batteries for our hip pocket battery packs, for Chrissakes! That is Fred for you.

But we are getting smarter. "Hold it right there," we tell the pilot.

It is easy to check things out. There is always a faint hissing in the earphones when things are working right – no hiss, no signal. We try the batteries in two different packs but it is a no-go. Back on the radio and there is no asking this time. "Tell Fred we're coming in. Meet us in two hours."

(Fred must have had a premonition, because he had sent out two batteries weighing less than two pounds in an otherwise empty plane, and not on a split charter, either. The Beaver would have gone back empty. It didn't make any sense.)

We are going home! Our adrenaline levels jump at least ten notches. But what to do? Our EM gear and mag are two miles down the lake and half a mile in the bush. No problem. While Jack and Hal tear down the camp, I go to retrieve the gear. It is pedal to the metal down the lake and it is Clear the Track Jack on the way in. I hang one strap of the motor/generator over one shoulder, one strap of the coil bag over the other, grab the mag and, carrying over 150 pounds of gear, boogie back to the canoe.

Heading back to camp I hit a rock hard enough to cave in the cedar strips next to the keel near the stern. Fortunately, it misses the prop. The thing is – I knew that rock was there. We had passed it several times, but the steady rain has raised the lake level a good 12 inches and the rock is now submerged. I keep the throttle full on with an eye on the water rising over my boots. It is a half-mile to camp and I make it, but Mr. Canoe is done for the year, headed for dry-dock.

I pull up on to the shore, run to where the tent used to be, roll up my sleeping bag, and stuff my possessions in my packsack. Meanwhile Hal has already emptied the canoe, passing the survey gear up to the pilot. He removes the motor, drags the canoe up beside a tree and overturns it. (While I was busy Hal had also dumped the gas from the cruise-a day.

We are leaving on the first trip with our personal gear, survey equipment and kitchen paraphernalia. No way is one of us going to wait for the second trip and chance being stranded by the capricious weather. Canoe, tent, motor and remaining odds and ends will wait a few days to be picked up on a back haul. The heater and stovepipes that had served us so well will be left to rust quietly on the shores of Frotet Lake. (Nowadays, 50 years later, you would never get away with that.)

So I carry my stuff down to the dock and toss it up to the pilot while Mike and Hal pile the remaining miscellany around the overturned canoe. I walk to the end of the dock, lean against the tail of the Beaver, look out over Frotet Lake and have one last cigarette to be enjoyed while I contemplate on the soon-to-end summer season. I throw my match into the water – and the whole bay goes up in flames! They must have been 40 feet high!

The pilot is screaming, "Cut the ropes! Cut the ropes!" With what? Our axes are already on the plane!

Mike and Hal are scrambling to untie the shore ropes and I am standing up to my butt in flames trying to undo the rear rope. The four gallons of gas from the cruise-a-day had been dumped in the water and had had time to spread out over the whole bay.

The flames die down as quickly as they flared up, leaving one still very excited pilot and three guys whose last drop of adrenaline has been wrung out of them. We are a pretty subdued group as we lift off from Frotet.

Damage is slight. There is some soot on the underside of the wings and belly, and one wingtip navigation light is a bit melted. The fire, although short-lived must have been pretty hot. The Beaver has belly tanks, but is of all metal construction. Had the aircraft been a fabric covered Norseman, also with belly tanks, the outcome would have been different – and no doubt tragic with the pilot being in the plane. I will never smoke on a dock again.

We overnight in Chibougamau and board the "nevair" bus the next morning for our return trip to Sudbury, glad to be on our way. (We were never allowed to spend much time in town in those days – we were like a bunch of little pups, full of energy, yappy, and barely paper-trained.)

Table Scrap - Pass Me da Hax

Chibougamau was Fred's territory. He had spent a year or three there, and one of his anomaly chasing crews was led by Henry Levac, the same Henry who had tossed Herb S. into the lake a few years earlier.

The summer had started off rather poorly for Henry. His EM gear wouldn't work. The motor ran fine and the generator kicked in and went into transmitting mode, but he couldn't get a signal. He tried every possible combination of batteries, pocket receiver packs, earphones and receiver coils. Nothing worked. He radioed Fred, who sent him out new earphones, receiver coils, etc. No go, so Fred sent him out a complete new motor/generator set and transmitter coil – still no signal!

"You are doing something wrong," said Fred, and Fecteau brought out the last spare unit.

Henry told the pilot to wait and the crew set up the unit right there in camp. Same old story.

Henry turned to one of his men and said, "Pass me da hax," and he beat the crap out of every piece of that equipment. By the time he was done, the motor/gen was a pile of twisted aluminium, bent shaft and broken wires. He turned his attention to the transmitter coil and chopped it into pieces as well. He threw everything into the aircraft, climbed in, and back to Chibougamau they went. Fred met them at the base and Henry threw the mangled mess on the dock. "I tol' you, she don' work!"

Fred wasn't even upset. That was Fred: he would drive you nuts, but always saw your point of view.

And Henry was right. There <u>was</u> a problem, and it had been back in Frood at the lab. A phone call to Frood elicited a sheepish apology. It seems that during the spring break they had gone over Fred's equipment, tuning up the motors, changing the couplings, checking everything including recalibration of the generator output. It turned out their own calibration equipment was faulty, and every one of Fred's units was off frequency. No 1000 cycles per second – no signal. The story should end happily right there, you'd think – but no.

A few days later the new improved on-frequency gear arrived. Henry had stayed in town, and Fred said that they would make sure the stuff worked before flying it into the bush. There was a wide back lane behind the Hotel Monaco and they set up the whole deal just as you would on the job, fired up the motor/gen and Henry went down the lane a hundred feet or so with the receiver.

LEVAC, Henry Paul — The family announces with sorrow his death April 11, 1999 at the Sudbury Regional Hospital (St. Joseph's Health Centre), at the age of 75 years. Husband of the late Germaine Levac (nee Jodoin), predeceased in 1983. Son of the late Domina Levac and of the late Emily Levac (nee McMurray). Loving father of Claudette (husband Richard Charron) of Sudbury, May (husband Denis Proulx) of Val Caron, Jim Levac (wife Tessie) of Hanmer, Gilles Levac of Sudbury. Predeceased by two sons Gerry and Marcel. Survived by 10 grandchildren and six great-grandchildren. Dear brother of Isadore, Auguste, Fernand, Leo, Lucien, Ernest, Roger, Marie and Blanche. Predeceased by his brothers Domine, Alcide, Andre and by his sister Florence. Also survived by one sister-in-law Jackie Roy. Mr. Levac retired after working for 30 years with INCO in Field Exploration. The family will receive friends at the Co-Operative Funeral Home, 222 Lasalle Blvd. East, (corner of Notre Dame), Sudbury, today from 2 to 5 and 7 to 9:30 p.m. Funeral services will be held Wednesday, April 14, 1999 at 1:30 p.m. in the Bishop Roger Despatie Chapel at the Co-Operative Funeral Home, Sudbury.

"OK," he said, and trotted back to tear down and pack up. Just then a two-ton delivery van came around the corner and ran over the whole works! Such is life.

The story never really ends. "Pass me da hax" has become part of our vocabulary. To this day, when things just plain refuse to work right, I will say to myself, "Pass me da Hax," and I remember Henry Levac.

Sidebar: This story, as with any true legend, has floated around the Shield for more than 50 years, and will continue to be passed on into infinity. Others have slightly different slants to the tale, and some have had their versions published. I can only say that my version came "from the horse's mouth," so to speak – actually, I got it from both horses on the team – Henry and Fred.

It is back to Copper Cliff to check in and scope out what the fall program will be and take three or four days off, which I spend with my aunt Thelma. She is my mother's older sister and lives in a neat little house near Sudbury's centre. She is a very nice lady, but a little wary of the world around her. Life has not always been kind to Aunt Thelma. She is a trained classical pianist who prefers to hammer out boogie-woogie and ragtime, and is a big Elvis fan. She earns her living teaching piano five hours every evening and all day Saturday, and her only social life is a weekly shopping excursion and the occasional recital with her students. Since I sleep on a sofa in a two-piano living room, I spend my evenings at a bushwhacker bar and visit with Auntie in the daytime and on Sunday.

She had married an underground miner back in the twenties and they had hit a lot of mining towns. There was always a better mine, or a safer mine, or a better paying mine just down the road. Her husband loved to work, loved his beer and was always kind and good hearted – too good hearted Thelma said. In fact he would often come home from the bar in his undershirt, having literally given the shirt off his back to someone less fortunate than himself.

They had lived in Bisset, Manitoba in the depression years. The San Antonio Gold Mine was in full swing, and she said that no one really felt the depression. Everyone in town had a job and no one was hungry. But he was quite a guy. One payday, instead of bringing home a paycheck, he brought home a grand piano – ordered from Winnipeg and trucked in on the turkey-track from Pine Falls!

When they moved to Sudbury he worked underground at Falconbridge. One day in the early '50s he didn't come home. He called to say goodbye and just plain skedaddled. Turned out he was kind enough to share more than his clothes.

Aunt Thelma decided to get a divorce and sue for support. But it turned out that she couldn't divorce him because they were never really married. The bugger already had at least one wife when he married her. Thelma was devastated. According to her own moral compass she had been "living in sin" all those years. She started teaching piano, living in fear that her hoity-toity customers would find out her terrible secret. I would always see Aunt Thelma as the seasons changed, but when I left Inco in '64 I lost touch – I was too young and self-centred to make the effort. Aunt Thelma died a few years later virtually alone and I still feel that I abandoned her.

So I spend a few nights barhopping with the boys and catching up on what the others have been up to. I find out that one crew who were camped 60 miles west of Chibougamau had undergone 30 straight days of rain that September, so I quit whining about our puny 21 days.

I run into Alec, my old packsack buddy from last May. It seems like ages ago. Later I go to his room with Dick Anderson and Gordy Sage for a couple of beers. Dick and Gordy get into an argument. Gordy gets up, pokes Dick in the nose and dives out the second story window! We look out but don't see any bodies in the parking lot. We figure Gordy was late for a date. We laugh like crazy.

112

Everyone says that Gordy is a loose cannon. I figure that he just makes quick decisions, some good, some not so good.

Table Scrap: Sage Decisions

That summer Gordy had been chasing anomalies north of Kenora. They were setting up the first fly camp of the year, and towards dark he asked a summer student to fill the Coleman lantern with gas and light it. The Coleman has a gas tank as a base and built into the filler camp is a small pump. The procedure is to fill the tank with naphtha (white) gas, screw the lid on tight and use the pump to pressurize the tank. Then you crack open the valve and light the raw gas which passes through a stem (called a carburetor) and into a cotton mantle. You light the mantle with a match, and throttle the flow until the carburetor heats up enough to vaporize the naphtha. This takes a minute or so. When you have a nice white-hot glowing mantle, you open up the valve fully and hang the lantern in the tent.

The student got things a bit backward. He filled the tank, opened the valve and started pumping. Raw gas was pouring out of the mantle, but he lit it anyway, and the whole lamp burst into flames!

Gordy was standing about fifty feet away. "Get the hell out of there!" he hollered, but the student froze, transfixed by the burning lamp on the ground at his feet.

Gordy, who could move through the bush like a moose, covered that 50 feet in record time, ran past the kid, grabbing the lamp handle on the way by, and in one smooth motion threw the burning lantern out over the lake. Before it hit the water it exploded twenty feet out from the shore. Snap decision – good result.

The following winter (62) would find Gordy at a semi-permanent winter camp in the same area. They even had a light plant, housed in a plywood structure with a two-hole facility (If you get my drift.) Gordy was making use of the facility one night and decided to check the fuel level using a match to do so. Snap decision – bad result - no more light plant, no more two-holer, and no eyebrows for Gordy for a while.

Copper Cliff sends us out to the old Victoria Mine area ten miles north of the little town of Whitefish and 30 miles west of Copper Cliff. We throw up two 14x16 wood frame tents and go to work cutting grid lines in the sparse underbrush and second-growth poplars.

The brass always has a camp for us out-of-towners. They won't pay for a hotel, and groceries don't cost much – we are more into barley sandwiches. We have no company truck, as most of us have a personal vehicle.

The local boys live at home and work out of the field office, sometimes with the permanent land survey crew. There are always survey lines to be cut, claims to be re-staked, and a couple of guys work at Frood where the equipment is tuned up for the coming winter. The whole work force is pretty much fluid in the fall anyway. Some leave to start a more civilized career or to continue their education, and some transfer into the mines or the smelter. The company does take good care of you in some ways. After proving yourself in the bush you have the option of a transfer to a basin job. You are classified as "designated personnel." This meant you can jump the Union queue and be fast-tracked to shift boss, either underground or in the smelter.

113

Sidebar: I always thought that the fall program was "make work." The Victoria area, for instance, was one of the earliest producing mines back in the day, when the nickel was first discovered during railway construction. In fact, our camp was smack dab in the middle of Mond, an abandoned townsite. Numerous partially filled-in basement dugouts were all around us and the old shaft was not far away, surrounded by rubble, which could easily be climbed to look down into the water-filled shaft: Eerie! The water was within 20 feet of the surface, with no hope of climbing out of you fell in. There was no fence, but within a few years Skidoos would become popular and a protective fence would be erected. Even later, a new shaft will be sunk, and the Victoria will be put back in production, so perhaps our work was valuable after all.

In early November we move the camp to the Vermilion River on the old Whistle Mine road north of Capreol. Capreol is situated on the CNR line on the north rim of the Sudbury Basin - nice town.

The camp is pretty boring. We cross the Vermilion every day using a small boat attached to a cable, and we pull ourselves back and forth with ropes tied to each bank. It is a mile and a half of good walking to the grid we are cutting beside a small lake, which is starting to freeze over. We even have a cook – Ted Jodoin from Sturgeon Falls.

He brings a dog with him. He had been gassing up his car at a service station and had seen this bedraggled, half-starved German Shepherd lying beside the building. She looked pitiful, ribs showing and weak as a kitten. The owner told Ted that she had just weaned a litter of pups and since she was of no more use to him, she would be shot. We all thought that Ted was mostly BS, but you know, he took that dog (had to carry her to the car,) brought her to camp, nurtured her back to health and gave her a loving home. That sure changed my perception of him.

And what a dog! Ted is her idol, no doubt about that, but she loves to jump into the skiff and go to the bush with us, happily checking out everything her new life has to offer. She loves to chase rocks – and rocks only – sticks do not interest her. If you throw a rock out on the new ice of the lake she will charge out after it, put on the brakes, slide into the open water, climb back out and come trotting back wagging her tail. "Throw another one, Okay?" We take pity on her and stop throwing towards open water. It is getting darned cold by now.

One day I am taking a smoke break on the lake shore. Dog is looking up at me. "Throw a rock, c'mon, throw one!"

There are no rocks around, but the ice is thin at the shoreline. She starts jumping up and down with her forepaws, breaks the ice, sticks her head in, rummages around and finds a rock. I am amazed! A problem-solving dog! What a girl!

One day Ted and I are the only ones in camp. The rule is that you never go into the bush alone, and I am feeling poorly anyway. I go to the cook tent to whine to Ted. "I'll fix you up." he says, and gives me a cupful of a nasty looking concoction. "My Mom's recipe… one cupful does the trick."

He is right about that. One cupful of whatever it was has lasted me the rest of my life. By that afternoon I am at the Copper Cliff clinic. The doctor tells me I have acute

114

gastritis, gives me some thick white stuff to drink and tells me to lay off the beer and cigarettes for a while. And I do – for at least 24 hours. How easy it is to bounce back when you are young.

It is December the 7th – time to head home for Christmas. I stop at the field office to tell Herb I am no longer interested in a transfer to air crew. I know he will never allow it anyway. This way I can have the satisfaction of cutting him off at the pass, so to speak. Herb gets a shot in, though. Where will I be heading on Jan 2? Maybe one of those cushy motel jobs? Uh-uh!

I am going back to Frotet Lake with Fred – such is life.

Chapter IV   Winter of '62: Chibougamau

I leave home in New Years Day. It is a twelve-hour drive, down to Duluth, across Northern Wisconsin and Michigan to cross back into Ontario at Sault Ste Marie, thence onwards 200 miles to Sudbury. Leaving my car at a friend's, I go to the field office in Copper Cliff to pick up my train tickets and join two crews on the round-about trip back to Chibougamau. One crew is led by Gus Levac, nephew of the legendary Henry. Kent and Bill will work with Gus. My party chief is Jack of Beanie Weenie days. He and Fred are the only familiar faces. Our third man is Grant McPhail – just a great kid. He is in his late teens and always happy – one of those "up" people.

We check into the Monaco (who have their liquor license back) and begin flying the next day. The whole camp will be flown in by Fecteau Air, mostly with single Otters. They can handle 4x8 sheets of plywood, 12' long 2x4's and lots of freight. They are workhorses – solid, rugged and oh-so-slow – 90 mph tops.

The camp is to be built on an already cleared section of shoreline on the southeast corner of Frotet Lake, exactly where I had beached the canoe when we did our trek last September. Three of us, myself included, fly in by Beaver the first trip with our snowshoes. The ice thickness is sure to be adequate but there is always the possibility of slush hiding under the snow. We unload the Beaver and immediately start packing a runway for the much heavier Otters. The rest of the crew is organizing lumber, etc. at the air base - the airlift will be underway tomorrow. There is no slush so we spend the day packing and marking a pretty decent airfield out on the lake, a mile long with a nice off-ramp and turn-around area near shore. The Beaver will return with another load later in the afternoon and take us back to town, where we will spend two more nights in the Monaco before the camp is habitable.

The camp will house ten men in three 14x16 tents, one for each crew and a third for Fred, who will forsake the Monaco and tough it out with the boys. Fred will share his tent with the helicopter pilot and mechanic who will arrive once the camp is ready. The cook will bunk in the kitchen.

I will now ask you to sit down while I conduct a camp-building tutorial: take notes if you wish – smoke 'em if you got 'em.

14x16 Tent- Canvas over wood frame:

Materials List per tent:
4 – log sills, minimum 18' long, cut from nearby trees.
1 – Log ridgepole, also cut locally, to support the fly.
7 – 2x4x14' floor joists (2' extensions to be scabbed on one end.)
7 – 4x8x5/8" construction grade plywood sheets (one ripped in half lengthways) for the floor.
8 – 4x8x1/2" plywood for the pony walls.
9 – 2x4x12' to be cut in half for wall studs.
11 – 2x4x12' for upper and lower wall plates (scabbed as needed for proper length.)
2 – 2x4x12' to be scabbed together to make inner 16' ridge pole.
18 – 2x4x8' for rafters.

116

10 – 2x4x8' for framing doors and miscellaneous.
1 – bundle – 4' laths (end use – see tutorial B.)
Lots of nails – 6", 4", 2", 1"
TOTAL: 36 pieces of dimension lumber, 16 sheets of plywood, laths and nails.

Equipment needed per tent:

1 – Coleman oil fired space heater 2'x2'x4' high
14 – sections of stovepipe and two elbows
2 – galvanized steel water pails (one always on the stove for hot water for washing up, the other for cold water for drinking. It sits on the washstand by the door.)
1 – tin wash basin.
1 – broom and dustpan.
3 – canvas tubes – we build our own beds – a simple frame at each end and two 2x4s stretch the canvas tubes.

Multiply this by three, and you know that Fecteau's two Otters will be busy as beavers (pun intended) for quite a few days. Allow one hour for loading, two hours flying (round trip) and a half hour unloading (just toss the stuff out - we'll haul it away

Welcoming committee at Frotet City Airport.

before the next load arrives.)If both Otters are available we can get four loads per flyable day. But we are not the only game in town. Falconbridge is setting up camp some distance west of Frotet, and Fecteau have their bread and butter work to do. The village of Mistassini is 40 miles northeast of us, and the east coast of James Bay and Hudson's Bay must surely be on Fecteau's radar.

Sidebar: In 1900 the Canadian Geographical Dept took the bay away from Henry Hudson by dropping the "apostrophe s." As natural-born traditionalists, we sill still call it Hudson's Bay.

Sidebar: I recently watched a documentary about Fecteau Airways – very interesting to me, and what an eye-opener! It dealt mostly with the base at Senneterre (west of Chibougamau) and there were no familiar faces.
Here's the deal. During that decade of 1955 to 1965 there were no topo maps of the James Bay lowlands in Northern Quebec. Maps ran out a hundred miles or so north of Frotet, so when Fecteau flew farther north, each pilot jotted down details. Back at base the info was shared and Fecteau generated their own maps! What a story!

117

Of course none of our pilots mentioned this to us – it was kept in-house, if you catch my drift. I am sure that Chibougamau pilots were doing the same thing, but knowing that the pilot was mapping as he flew would hardly generate confidence in a passenger – but it worked until government mapping caught up to the bush pilots.

Additional freight to be hauled includes:

3 – sheets good-one-side ¾" plywood – one will be our kitchen table. The others are to be cut to size for map tables, one for each tent.

(I have probably short-changed the lumber list. More will be needed for benches and shelves for the kitchen, stools for the map tables, wash stands and tent doors. If there are any scraps left over they become chairs, door stops, perhaps even a table for card playing or letter writing.)

- A propane cook stove and a stash of 100 lb propane bottles.
- Initial grub order to stock the kitchen shelves and meat locker – to be replenished weekly.
- 3 complete sets of work equipment (EM, mag and spares.)
- 8 men and their personal gear.
- Stove oil, aviation gas for the chopper, white gas for the Coleman lanterns – a fuel dump will be established, and will be continually replenished throughout the winter.
- Miscellaneous stuff - hammers, axes, snowshoes, gas cans, shovels, handsaws, etc, etc, etc.

We have no "How to" books on tent building. Jack, Gus and I have lived in framed tents, and we had built a couple at Whistle the year before, but those were pretty rough. Old-timers have an apt phrase for shoddy construction – "jack-knife carpenters." This is a winter camp, and more care must be taken. Mother Nature does not cut any slack for fools.

Fred, no doubt, did have some winter camp smarts, as he had ordered the appropriate materials and tools. To his credit he always accepts input from us semi-pros, and with the steady stream of aircraft coming in daily, it is easy to add to the lumber pile.

By the time the camp is finished there will be three more experienced winter tent builders in the north bush (no Ikea instructions necessary.)

When I said the construction site was already cleared, I meant cleared of trees. There is snow, lots of snow – five flippin' feet of snow. I now have Frotet Lake figured out, meteorologically speaking. In the western Canadian Shield the weather systems will generally move southeasterly, more or less paralleling the west shore of Hudson's Bay. The systems of the eastern Shield do likewise, only southwest around the east shore. Both systems chug along picking up moisture on their trip, until they collide – directly over Frotet! Then they have a tussle and dump all that moisture. The fall rain we had experienced had continued as snow after October freeze-up and had obviously never stopped. Out of curiosity I dig a five-foot hole, straight down through layers of visible strata showing at least three definite indications of freezing rain or perhaps chinook-like days of melting. Obviously <u>over</u> five feet of snow had fallen before our arrival. The firm, deep snow will actually be of great benefit in the months to come. The snowshoeing will be fantastic – just like walking on a packed trail.

But we have tents to build, so we go to work. Normally snow would be cleared to make a space large enough to lay the tent sills, but due to the deep snow we have to go to "Plan B." We also have the sloping terrain to deal with. The lakeshore rises up an incline of about 10 feet in 150 feet, not much, but enough to be a problem. The tents have to be level so we decide to dig out a terrace for each tent, starting with a slight cut at the front of the building site and removing increasing amounts of snow until we reach the back, if you get my drift – (pun intended.) Then jackpine sub-sills are placed every five feet crosswise under the dimension lumber floor sills, to provide more support on the snow. Our tents will essentially be floating, but as I said before, the snow is quite rigid. The pines are cut from the bush behind camp – no power saws, just swede saws and axes and six energetic young bucks. After the eight pine sills are in place the seven 14' 2x4 joists are laid and the 5/8' plywood floor is nailed down.

Now we frame the two sidewalls, 6 foot uprights with 2x4 upper and lower plates. These are framed on the tent floor and the ½' plywood pony walls are nailed in place. The walls are erected on both sides and held with temporary supports. Next the end walls are built and sheeted with pony walls. A door frame of two feet by six feet is built into the centre of the front wall (forget about 32"x68" – all cuts are made in full feet only – no plywood is wasted.) The 16' inner ridgepole is now scabbed together and rafters (7 for each side) are angle cut so as to fit flush on the ridge pole and top plates. We don't want them protruding and possibly tearing the canvas. No carpenter's squares for us – just eyeball it, draw a line and cut. While building the first tent we always make a real good rafter and this is kept to serve as a template.

With a little help from our friends the ridgepole and rafters are up. Now we throw the tent over and pull it down into place. The laths are used to fix the tent snugly along the bottom (which reaches half-way down the ½' plywood pony walls) and to all uprights, top plates and door frame. We don't want the winter winds working at the tent and finding ways to get in. At the door frame we simply fold the tent flaps back against the frame and fix them in place with laths. The flaps of course intrude into the door opening at the upper corners. The plywood door is now reinforced with 2x4s, and hinges are attached.

Helpful Hint from the Old Prospector: Door Latch.
Make a triangle dohicky from materials at hand. Nail it on the door frame so that it forms a right angle with the upper side parallel to the floor. Cut a strip of leather from and old mitt, belt or what-have-you. Tack the leather on top of the triangle. Fix a short piece of 2x4 to the door protruding 2 or three inches past the edge. Make sure a little upward pressure is necessary to slide the 2x4 over the leather-covered triangle when closing the door.

We now have a snug door latch able to withstand high winds. A rope loop and spike on the outside gives added security when we are away, but we don't work in a gale anyway. If the wind is really hammering at us as we wait out the storm, we just set the

119

water pail in front of the door. (We never did get that kind of wind in the months to come, but we were prepared.)

Now for the tent fly, always necessary even in winter. The snow cannot be allowed to accumulate on the tent roof. Heat escaping through the canvas will melt the snow with problems to follow. The fly catches the snow before it reaches the tent, and we make roof rakes to keep snow from building up on the fly.

To hang the fly, four pieces of 2x4 four feet long are nailed horizontally at the upper plates on each corner protruding one or two feet beyond the tent sides. Two more short pieces are criss-crossed at the peak. Three lighter poles are cut from our timber stand and serve as the second ridgepole and side poles. Now we stretch the fly over and tie it off.

The oil stove is placed dead centre in the middle of the tent. Stove pipes are installed exiting the tent at the asbestos ring near the tent peak at the rear, making sure that the outside pipe extends far enough out to prevent contact with the tent, and then enough sections are added above an elbow to clear the fly and allow for a good draw. Lighter poles are used to make a pipe support.

A simple log crib is built beside the tent to hold a 45 gallon drum of heating oil laid on its side. (For you newbies and third world residents, substitute 205 litres or 55 US gallons.)

Here is the bonus – we have a kitchen/dining room on site. The previous diamond drillers had left their core shack and a core rack near the beach. We never knew who the company was or how far back they had stayed here, but it couldn't have been more than a couple of years because the shack is in real good shape. It is a plywood enclosed building 12' x 18' with a sloped roof covered with roll roofing. The interior walls are open studs but it doesn't leak water or wind and has a solid floor, a proper door and even a window facing the lake. Claude the cook has the only window in camp.

While we were building tents, Claude has been installing shelves, a work counter, propane range, oil stove, kitchen tables and benches, and setting up his bed and night table in a rear corner. When he needs a hand he gets it, and vice versa.

During construction stuff is always arriving, so it is necessary to drop what we are doing to unload the plane and clear the load away from our parking apron to make room for the next load. The 4x8 sheets of plywood come in every trip. A few sheets are put in the Otter laying flat on the floor, then other stuff is piled on top. This is common sense – heavy stuff on the bottom, lighter gear on top for maximum use of space and payload.

Table Scrap

One day a load comes in with its usual mix of freight and Bill is carrying a 4x8 sheet of plywood off the lake. Plywood is awkward for one man to tote. The best way is to stand it on end, back up to it, grab both sides, lean over a little and carry it on your back.

Now picture this: every time the plane comes in to the offloading apron they pour on the power just before stopping and swing the tail around to end up facing the runway.

The plane is now leaving. The engine has fired up and the pilot is sitting there on partial throttle, bringing the engine up to operating temperature before taking off. Bill is standing on the wrong side of the plane, waiting for it to leave. His arms are getting tired and he finally says "To hell with it," and walks around the tail of the plane just as the pilot pours the coal to her. Bill is one of the most stubborn men I ever met. He won't let go of the plywood and the prop wash picks him right up in the air. He para-sails about 40 feet and gets dumped in a snowbank – and we all laugh like crazy.

On day six we build a chopper pad in the morning as the chopper is to arrive this afternoon. We make the pad just north of Fred's tent and closer to the lake. We tramp the snow down with our snowshoes and lay down two layers of poles. The first layer is spaced every five feet and the second is tightly fitted together across the first. We are careful to make the deck as smooth as possible with no branch stubs to puncture the rubber pontoons. The pad is roughly 14' x 20' and slightly higher than the surrounding snow level. Its proximity to the lakeshore plus the lack of tall trees nearby results in an excellent take-off and landing area.

We also move the av/gas drums. It is the mechanic's job to refuel the chopper but he is not expected to manhandle the 45 gallon drums up from the lake. Each drum weighs 350 pounds and is brutal to handle in the deep snow. We make a toboggan out of a piece of plywood and other material at hand, then with two guys pulling on ropes at the front and a third pushing at the rear the job is done – lots of bull-work.

With the camp virtually complete, we spend the rest of the day getting our gear ready for tomorrow. Gus and Jack spend a few hour going over the maps with Fred. The winter's work is about to start, and we are eager to get at it. Anomaly chasing will be a breeze compared to our last few days. The drill camp starts to arrive that same day, but that's another story.

N.B: It took us six days to complete the camp and we did <u>not</u> rest on the seventh. God was lucky that He was not on Inco's payroll.

Table Scrap
The core rack that had been left for us becomes our meat locker. Claude covers the sides with cardboard and banks it with snow. A mink had obviously laid claim to the rack before our arrival. Claude gives him the occasional donation and he leaves the rest of the meat alone. No bologna for Mr. Mink – he prefers steak.

Mink are not cuddly, friendly animals. They are predators and a poor judge of size differential. They will go toe-to-toe with anyone – humans included.

He has decided that Claude is semi-okay – Claude supplies fresh meat. If we call him he will come to his hole-in-the-snow door and all we can see is a pair of beady eyes.

One day Claude gives us a demonstration. He has a strip of succulent steak in his hand and calls the mink. The mink sticks his head out – no further – and waits for the offering. When he takes the meat Claude pretends that he wants it back. Nosiree! Once Mr. Mink has the meat in his teeth it is his! He growls, snarls and Claude lets go. The mink goes back to his own kitchen to chow down.

Our chopper pilot is an Englishman named Jacques (if you can believe it – guess he could as easily have been a Frenchman named Trevor.) At supper the night before we start work the discussion centres on safety. Three of us are chopper-wise and we and Jacques (pronounced "Jakes") share our vast knowledge with the rest of the boys.

Never approach the chopper without being in the pilot's line-of-sight. Always crouch to protect your noggin. (That was for five of us. Gus could've jumped two feet in the air and still been under the rotor blades.)

JACQUES IS OFF TO DELIVER ONE OF US TO THE BUSH

Never climb in or jump out unless signalled to do so by the pilot.

Never carry the bundle of unipod poles on your shoulder near the spinning rotor blades – bad things will happen.

We discuss what Jaques wants if a landing area needs to be cut and also explain anomaly-chasing rules – two men in to each work area, followed by one-and-one, never leaving one man in the bush alone.

Oh yeah – never fart in the bubble.

Jacques is a multi-hour pilot who admits to one problem. He can't navigate in the North Country worth a darn. The mechanic had read the map on their way in from Chibougamau, and earlier legs of the flight had roads and power lines and such to follow. We think it is hilarious, but never let on to Jacques. Don't piss off the cook OR the pilot – they are both much too important to your well-being. So we develop a system (plan B of course.)

The party leader always carries an aluminium clipboard with a hand-drawn map of the anomaly to be checked and an air photo of the same area. Each party chief and one man fly in to their own anomaly, guiding Jacques to the spot. They jump out and point the way back to camp. (Did I also mention that Jacques has a piss-poor sense of direction?) Then while Jacques is making his first trip for the other crew, the first crew dresses up the landing spot and builds a fire with spruce boughs on top to make smoke for Jacques to home in on when he brings in the third man. Of course the second crew does the same. When it is time to go home each crew builds another signal fire. Jacques climbs to 2000 feet, gets a good bead on us and homes in. The days are now clear, cold and still, and going back to camp is a snap. Jacques just climbs until we see the smoke rising from the oil stoves.

In Jacques' defense I have to point out that the expanse of deep snow really alters the topography. Sometimes a lakeshore morphs into swamp with no discernible shoreline, and the swamp now looks like a bay. Likewise, what appears to be a pond on the air photo is now two or three times as big because the deep snow covers the lower growth around the pond. Tree lines, creeks and rivers are our main reference points and imagination does the rest. But we have lots of sunny days, and by the time the skies do become overcast, Jacques is getting pretty good at it. He is a competent pilot and always tries to save us as much walking as possible.

Just to show that we are not always as smart as we think we are, get this. Gus' first job is to relocate the conductors that Mike, Larry and I had found on our long walk the previous August. The purpose is to spot the first bore-hole for the diamond drill. Gus extends our grid, and guess what? He picks up the response we were supposed to hit last summer! We were off by more than half a mile – and here we are laughing at Jacques. Again it's no problem. The drillers have enough work to keep them busy for a long time.

Sidebar: The drill program was concentrated in the area of that long walk and must have been very, very interesting. Inco never shared sensitive info with us peons, so I never did learn what they were finding. Research tells me that a zinc mine is operating near Frotet, but it appears to be north of our 1962 camp. Perhaps a deposit still lurks beneath the swamp southeast of the lake.

But never say never. The mining exploration game is a continual gamble. As they say, "There's many a slip 'twixt the cup and the lip."

Take the Geco mine in Manitouwadge, for instance: The main showing had been common knowledge for years, but drilling had failed to prove continuity. Then a couple of prospectors from Geraldton staked it and optioned it to Geco, a Noranda subsidiary. The company put a drill on the site under the direction of a green geologist. He drilled in the wrong place and in the wrong direction, and found a mine (so legend has it.)

The drillers come in and every evening we watch their progress next door. Their tents go up in a hurry. They are 14x16' like ours but use no dimension lumber. The only plywood used is for the floors, and the tent frame is made by interlocking a gazillion aluminium pipes. The tents are then stretched over the frame and secured to the floor at the bottom. They have five-foot sidewalls so no pony walls are used. The door is also part of the package with provisions for fixing the tent snugly around the frame. No fly is needed as the tents are made of heavy orange canvas and have a double roof. It looks complicated, but these boys come in in the morning and by supper-time they have two tents up, the kitchen running, and move in.

One tent houses the four drillers, the other is the cook tent, where the cook and the drill foreman bunk. The drill camp is definitely a no-frills affair. The tents are gloomy inside even on sunny days. These boys work 12 hours a day, seven days a week, two men to a shift, so between working, eating and sleeping, there is little time for socializing. We get to know Clarence the drill foreman, but have no contact with the others.

An Overhead View of Frotet City.

The core shack/kitchen is near the lakeshore. Our three tents are 75 feet upslope from the kitchen. First in line is our tent, then Gus' 50 feet north, and Fred's is another 50 feet farther on. This is Main Street. By the time our camp is completed the connecting paths, hardened by the bitter cold, are suitable for foot traffic.

123

Just north of our kitchen is the fuel dump, with the chopper pad not too far north of that. It is about 75 feet from Fred's tent down to the chopper pad. A hundred feet beyond the pad to the north is the drill camp.

Frotet City from our doorstep looking north - Gus' tent, with shadows of Fred's tent behind.
The drill camp tents are in the background.
The cook tent is offstage left.

We have a garbage dump 75 feet south of our tent with a one-hole facility nearby. Neither one has a proper pit – a hole dug in the deep snow will do for the winter. The garbage dump is used for tin cans and food scraps – plastic containers are rare in 1962. Claude burns the dump once a week and the dump is secure – we are not on the raven circuit. Next summer the ravens will clean up bones and such and all that will remain is the tin cans.

The facility is used only by the town folk – we six bushwhackers are like bears, leaving our digested extras in the forest. When spring comes the little shack out back will fall over and by the end of summer the rains will return everything but the plywood to nature.

We have no public works department. From time-to-time Claude or someone will shovel the sidewalks if we get a heavy snowfall. They seldom drift in – the prevailing winds here are from the east and northeast and the forest behind us is a good windbreak.

Taxes are low in Frotet City and we never complain to the Mayor and Council.

The drillers now have some finger-freezing work to do. Their caterpillar tractor and the diamond drill are on the ice in pieces. Every fly-in drill camp in those days used a small John Deere Model "M" cat to move the drill and haul in casing, drill rods and fuel to the drill site. They were relatively light, but being too large to fit into a Single Otter, they were dismantled at the airbase. The tracks and the track assembly were unbolted from the frame and differential. The gas tank, seat and instrument cowling came off. The two-cylinder engine, transmission and final drive were split and all the nuts and bolts went in a five-gallon oil pail. The pieces were then flown to Frotet and thrown off the plane into a snow bank, ready for reassembly. (And we thought their tent frame was complicated!)

Of course, every time the cat is disassembled and rebuilt stuff goes missing. No problem – these are rugged, simple machines. That part had four bolts? Three will do, and so on. Big spools of haywire are a part of any drill camp inventory, and by the time a cat has been moved more than twice it is pretty much a haywire outfit.

124

The drill is a BBS-1(Boyles Bros. Series 1) good for a depth of 1000 feet or more, but usually limited to four or five hundred feet when drilling anomalies. Powered by an aircooled Wisconsin V4, it can also be torn down and reassembled.

It takes the crew a day to put the cat together and when we return at the end of our day they are firing it up. The plan is to break a trail running between the drill camp and the chopper pad, up the hill and then on southeast six miles to the drill targets.

But things don't always work out as planned. The cat, although light, just doesn't have the flotation to stay on top of the snow. There are just too many pounds per square inch of track surface and it digs itself down five feet. After two hours of back-and-forth, back-and-forth, they progress less than a hundred feet. All we can see from our tent is the exhaust pipe sticking up. Plan "B" now swings into place, and John Deere will sit there waiting for camp tear-down.

Our Bell G2-A had arrived in camp with a full load. The pilot and mechanic had their own luggage and eiderdowns and the mechanic (for some reason they are always

CHOPPER SLINGING A FUEL BARREL.(FILE PHOTO)

called engineers) had his tools and spare parts. There was also a shroud onboard – made of light canvas, it covers the bubble and engine every night.

They also have a Herman-Nelson. This is a self-contained unit that supplies warm air ducted into the shroud. Its job is dual purpose. It warms the engine and keeps the bubble frost-free. One never scrapes frost from the bubble – plexiglass is easily scratched and we want Jacques to be able to see our signal fires. (The Herman-Nelson is quite heavy and has been flown in separately.)

Last, but not least of the gear brought in by the chopper are two sets of cargo slings. Always part of their basic equipment, they become Plan B.

Jacques now has extra flying to do. The drill, which had been put together to be hauled in by the little John Deere is now re-dismantled. Jacques slings the whole deal in to the first hole, and he is a busy boy for a couple of days, and remember – he has to do his scheduled bus route for us guys.

Clarence, the drill foreman, also makes the occasional flight and on the second day he has one of the rules of chopper etiquette literally pounded into his noggin. All the drillers wear hard hats with liners as their normal headgear. Clarence fails to duck and the rotor clips his hard hat and knocks him on his ass. No harm done, just a stiff neck for a couple of days, but it was kind of funny to us yahoos, of course. Added to our enjoyment is the sight of Clarence approaching and leaving the chopper on his hand and knees from that day forward.

The drillers also have a plan C – they bring in a Skidoo! This is the first time we have seen one and we are green with envy!

125

They were still in their puberty stage in 1962. The first models had a tin hood and an eight horse engine. It looks like a toy to us, but that little skidoo punches <u>way</u> above its weight class. It is soon doing most of the work, although Jacques still handles the heavy lifting. That little sled will run non-stop until spring and they will wear out two motors.

We are making progress. The weather is generally good and we can really move in the bush. The big snows are over and for the rest of the winter it snows in a normal, civilized pattern. It is strange, although the snow sets up quite firm in the bush, the lakes remain knee-deep and fluffy. (This will lead to a Beaver-related incident later on.)

We finish a grid and move on – and on, and on. It is still very cold and the few warm spells bring fresh snow – not much, but enough to cause some problems.

We have a bit of a staking rush. One day our crew goes to a new target 30 miles west of camp. Jack and I are on the first trip and while Jacques goes back to take Gus' first trip out we poke around and find fresh blazes! A quick reconnoitre soon turns up freshly cut claim posts, so when the chopper returns with Grant, Jack climbs in and goes back to camp to consult with Fred. They quickly and correctly surmise that our area of interest obviously overlaps that of Falconbridge, who are camped farther west.

Fred has claim tags, so Jack returns with them and we start staking. It seems that the land we want parallels their area of interest. When we finish staking 48 claims we are satisfied. It is hard work but kind of exciting, actually. We never did run into the other guys, but we knew they were around and they obviously knew we were in the area.

Competitive staking protocol works like this: say you are running a line west, your competition is running a line east and you meet in the bush. You pull out your map and they pull out theirs. You both want the same real estate, so you negotiate on the spot. There's no sense in overstaking each other and getting into a pissing match. "OK, we will run a line north, you get one side and we get the other." It usually works if the area is not a hot issue.

Table Scrap:

A few years later I heard a story connected to the Timmins Texas Gulf rush, which <u>was</u> a hot one. The deal is, you stake your block and record it later. If you are over-staked and it comes to an inspection scenario, the earlier staking date gets the ground.

One group of notorious claim jumpers intentionally over-staked a block. The lead stakers were followed by a guy with a can of turpentine and a brush. He painted their fresh blazes and fresh claim posts, which were predated. The turpentine artificially aged the appearance of the fresh-cut wood.

During the claim inspection process that followed, it looked like their work was obviously older. Because of their reputation and because the claim inspector knew where the bear crapped in the woods, the jumpers' claima were thrown out. Why? Because the inspector could smell the turpentine, that's why.

By day four we have all the ground we want and are waiting for the chopper. It starts to snow – a light snow at first, but getting heavier by the minute. Jacques finds us (he is improving), but with the weather deteriorating he doubts he will be able to make two 45 minute round trips. The Falconbridge camp is only ten minutes away, so Jack volunteers to be dropped off there. Jacques returns for Grant and me and we go back to camp. We snuggle in our beds that night, secure in our minds that Jack is comfy - not so much, it turned out. Falconbridge didn't have an extra bed (neither did our camp) and they were not very hospitable. Jack didn't even get any supper – pretty harsh, but not unexpected. Inco tends to be arrogant – not us peons so much, but the reputation is there, and rivals don't much care for us.

In the morning Fred comes to our tent and says that Grant and I will probably be working alone that day. Here's how he puts it: "If I know Jack, he'll want to sleep all day," (with sort of a sneer.) He can be such a jerk.

It turns out that Jack had spent a fitful night dozing on a hard bench in the kitchen. He said that every time he tried a more comfortable position he fell off.

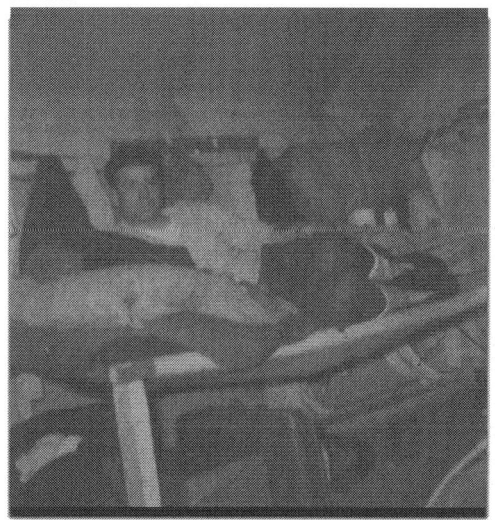

JACK MARTIN RELAXING ON HIS THREE SLEEPING BAGS.

We are getting into early February and each day gets colder. Grant is prone to frostbite on his cheeks, and every time we meet on the snowshoe trail both Jack and I check him out. If his cheeks show tell-tale white spots we have to tell him, he can't feel it. He then blows on his hands, warms up his face and pulls his parka hood closer. His cheeks have a semi-permanent rash from being frozen and thawed so often. We all carry snotcicles – little icicles of you-know-what. We could sure use Mr. Wolverine's help these day. (Wolverine fur never really frosts.)

One night the temperature really plummets. Trees behind camp have been cracking the last few nights as frozen sap expands and splits the tree trunks with loud "snaps," but this night it sounds like an armed insurrection back in the bush. Jack and Grant are already in their eiderdowns and I am finishing off a letter home. I seal the envelope and just start to strip down to my long johns when I hear a clicking sound! I know immediately what's happening. The clicking is caused by contracting metal joints. The oil stove is starving for fuel and as the fire dies the metal cools off quickly. I also know what the problem is. The extreme cold is causing the fuel to gel in the short section of flexible line between the oil drum and the outside tent wall.

I get back into my clothes and my parka and hustle outside and down to our equipment stash. I dig out the blowtorch and fire it up. No sissy little butane tank stuff for us – this is a real blowtorch, just like plumbers used back in the day. Someone, bless their heart, has put it away full. I play the flame on the flex line until inspection shows I have established oil flow and I re-light the stove. The tent gradually warms up and I jump into my sleeping bag, just in time to hear the ominous clicking again.

Out of the sack and into my clothes to do the blowtorch dealie again. This time I get the oil flowing before the stove is completely out – so far so good.

By now I am aware that sleep is not in my immediate future, so I light the gas lamp and prepare for a long night. Shucks! The stove is clicking again! (Even on high heat the stove just barely gets the job done. In this weather the heat radiated by the stove ends inches short of the inside wall of the tent, and when the stove goes out the cold moves in quickly.)

I haven't seen Grant since 11 pm, but I am sure he is under those goose feathers somewhere. Jack is just plain cruel. The baggage handlers had misplaced his sleeping bag on our way in from Copper Cliff. A quick SOS call to his sister and also to his mom resulted in two bags arriving post-haste – then his missing bag was found and sent in. As it gets colder he just crawls into the second bag and then the third, and he now has five layers of eiderdown to keep him warm. I call him a big suck, but he just giggles at me.

By the time I have finished giving our oil line the third treatment I see the smoke dwindling at Gus' tent, so I give his fuel line a torch job. Now our fire is once more fading, and I realize I am fighting a losing battle. Plan "B" swings into action.

I head for the garbage dump and dig out a couple of 32 oz juice cans. (For you young readers, who don't know an ounce from a hole in the ground, it's the big one on the grocery shelf.) Using my axe, I cut 2 strips of gunny sack eight inches wide by two feet in length. Then I find two wrenches and remove the fuel line at the shut-off valve, fill each can half full of stove oil and reconnect the line. Now I roll up the gunny sack and push it into the juice can with an inch of sacking sticking up out of the oil. This is the wick. I light the wick and place the can under the oil barrel right at the end where the shutoff valve is located. I make sure the can is well supported and that the flame just barely touches the bottom of the steel drum. I now thaw out the line and relight the stove. I repeat this process at Gus' tent and soon have two good uninterrupted fires going. Woohoo!

I have never heard of it being done before, but I am sure it will work, and it does - but it is too early to pat myself on the back just yet. By the time I have Gus' stove working I can tell that the fire is going out in the cookhouse. I can't find any more large cans at the dump, so I go into the kitchen intending to get a full can, dump it out and make another oil barrel heater.

I find Claude up and about. He has all four propane stove burners going, has the oven on high with the oven door open, and he is warm as toast. "I'm OK," says Claude, "Go get some sleep."

On my way back to our tent I look over and see that Fred's fire is failing. "Fearless can fix his own stove," I mutter.

128

It is 4:30 am as I crawl into my sleeping bag with my clothes on, cold and tired. It is kind of mean to leave the pilot and mechanic heatless, but I am still pissed at Fred's snide remark about Jack. (You may wonder how I did all this in the dark of night. Well, the big old moon on a clear, cold night, combined with the pure white snow made it seem like daylight. I'm sure I could have read a newspaper outside.)

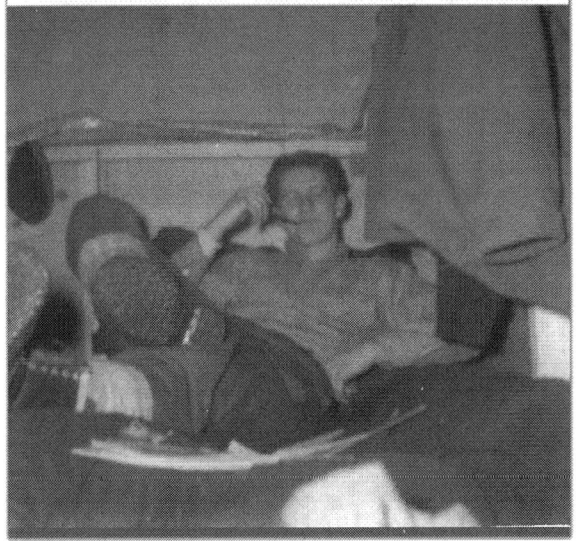

COOL DUDE GRANT MCPHAIL IN HIS BUNK. NON-SMOKER GRANT BORROWED MY PIPE. HE ONLY PRESSED IT TO HIS LIPS – I TOLD HIM IF HE GOT SPIT ON IT I WOULD THROW IT AWAY.

At 6:30 am I crawl out of the sack and join the others for breakfast, and we are a pretty subdued bunch. Everyone has checked the thermometer hanging under the eave of the kitchen roof. It is a circular dial type, marked to -60F. You can extrapolate where the -65 mark should be, and just beyond that is a pin to stop the needle. Well, the needle is right up tight against that pin! Radial thermometers are notoriously inaccurate at the extremes of their range and although I always say it was -60 Fahrenheit, it could have been -70. Any way you cut it, it is doggone cold. At that temperature you avoid taking deep breaths – the air seems to cut into your lungs.

(I stopped telling this story a long time ago. I noticed that people tend to break eye contact and start humming. Well, stop it! I will swear on a stack of mining acts that everything I write is absolutely true. I was there, and I done that!)

So we sit at the table, chowing down and waiting for Fred to come in. When he shows up he is sort of grumpy. "Pretty cold, eh, Fred?"

No answer... we are waiting for him to scrub the day's work, but not a word!

During one of our evening coffee gabfests, the mechanic had told us that it is inadvisable to fly the G-series chopper in temperatures colder than -40F. Extreme cold can hamper lubrication of the swash plates and cause excessive wear on the fittings, so we wonder if the pilot will talk some sense into Fred. The six of us gather in our tent and talk it over – surely the chopper will be grounded. I make it clear that I am not going out. Not that I am worried about crashing and burning, after all, we have already flown in temperatures below –40F. (Yesterday, in fact, it had been -52.) I am whipped after tending fires all night and I also point out the futility of going to work only to huddle around a campfire. I don't even put on my bush gear – and the other guys are as reluctant as me.

Sure enough, the darn chopper fires up and begins its warm-up. We have the routine down pat by now. We know how long the warm-up period is and the first two men out always get to the pad just in time to load up. Our crew is scheduled to be the first out that morning, and none of us move. We have a wobble going, baby!

Gus goes back to his own tent. He would never scab out on us, but he wants to distance himself a little. Gus tends to be a little nervous.

129

Table Scrap

Besides being a nervous sort, Gus was also deathly afraid of water. Jack told me that Gus had worked for him one summer. They had a leaky canoe, and one day while returning to camp with Jack on the motor, the water started to build up in the stern. Jack could have been bailing it out with the tin can in the bottom of the boat, but he decided to yank Gus' chain a little. There were three of them in the canoe with Gus in the bow.

"Hey Gus," Jack said, "Check this out."

Gus turned around, saw the water almost up to Jack's boot tops, screamed and jumped clear out of the canoe! Didn't even tip it. Why he thought he was better off in the lake God only knows. He just lost it, I guess. The guy in the middle grabbed Gus by the collar and they towed him the short distance to where he could wade to shore. As per usual, they all laughed like crazy.

And if there was the slightest chance that he might hit thin ice, even on the smallest creek, Gus would cut himself a 20 foot pole to carry. Like his Uncle Henry, he was a good bushman, but he hated it. This coming summer he will transfer to underground and I was later told that he and another Levac brother, working together, became an elite stoping team pulling in huge bonus cheques.

The chopper shuts down and it's very quiet in camp. We are getting a little nervous ourselves. Will we all be fired? Soon there is the crunch-crunch of boots on frozen snow. Fred comes in, hot as a firecracker. "You're not working today?"

"Too cold, Fred," says I, echoed by Jack. At least Jack backed me up – nobody else says a word, they just look at the floor.

"OK," says Fred, "But you're all a bunch of sissies."

He turns to the door, turns back and centres me out. (Fred always centred me out - part of the curse of my being tall, I am easy to home in on.) "You!" he says, "You kept your own fire going and let the rest of us freeze. "As Fred stomps out, Jack gives me a thumbs up.

Well, Fred is partially right. This is not the first time we have locked horns, and it won't be the last – sort of a mutual unadmiration society.

I hit the bunk for some sleep catch-up. The rest of the guys spend the morning modifying our stove fuel delivery systems. Small stands are built behind the oil stoves for ten-gallon drums which are plumbed directly to the stoves. Our fuel is now in a heated area and will never gel again. The down side is that the ten gallon drum has to be filled daily and our tents smell like stove oil. (In cold weather the stoves use almost eight gallons a day.) The kitchen oil stove is not modified as Claude has the propane range for backup. Of course Mother Nature does it to us again. The cold snap breaks and we never see -40 for the rest of the winter - the 45 gallon drums could have stayed.

At lunch time I am up and about and Fred has some good news. A Beaver will arrive shortly with three barrels of fuel, and Fred, Jack and I will be on the back haul to Chibougamau. Our licenses had been used for the staking rush, and in Quebec the licensees are required to be present when their claims are recorded. After staking you have 30 days to record each claim, so Fred has done the math and realizes that the trip will soon be necessary anyway. He has cooled down by now and is downright pleasant. Jack and I spiff up a little, put on our cleanest dirty shirts and climb aboard. We are excited. Why? Is it the prospect of pretty girls and good, cold beer? Nosiree, bub! We are going to have a hot shower!

As we taxi to the ice strip Fred turns to me and says, "Gus and Claude told me what you did last night. I apologize for what I said."

Go figure. Just when you think you have him scoped, Fred pulls a 180.

So we record our claims, have our shower, spend a pleasant evening with Fred in the bar, and return to Frotet early the next morning.

Sidebar: Keeping clean was not always easy in winter camp. It took forever to heat enough water and the only tub available was a round washtub used by Claude to wash dishtowels, so we had sponge baths. One night Grant hauled the tub up, managed to heat up some water and climbed in. It was sort of comical to see this grown man sitting in the tub with his legs hanging over.

We washed our socks and underwear, but for the rest of our clothes we adhered to the first rule of bushwhacker physics – heat plus pressure equals clean clothes. As we changed our shirts and pants the dirty clothes would go into our packsacks. Around midwinter we would start digging them up from the bottom. See? Heat + pressure = clean clothes.

About Fred – and About Us

About Fred: He was a loner, no doubt about it. He wasn't married, and never talked about his experiences or private life – he was a closed book. Post-Inco I worked with some different EM equipment for a couple of years before I found out that Fred was integral in its design and development. He was just sort of below the radar. When he was not in the bush or in Copper Cliff, he lived at home in Ottawa with his mom.

Fred had weird eating habits. We all sat cheek to cheek in the small cook shack and I would never sit near him if I could help it. His spot was at the southeast corner of the table and if I arrived early enough I would sit at the northwest corner. That suited us both, I guess. I couldn't stand to watch him eat – it drove me nuts. This man was in his forties, but you know how some kids are, the different foods are not allowed to touch each other on the plate. Say supper was mashed potatoes, peas and pork chops: First he would put the mashed potatoes and gravy on his plate and eat them, the same with the peas, Then the chops would be eaten one at a time. I'm sure he counted the chews – 20 chews and then swallow – 20 and swallow. I wanted to jump up and holler at him, "Eat your effen dinner!" Who was the nutty one, him or me? My dislike for Fred was becoming an obsession.

He rarely joined in on the mealtime banter. He would just giggle once in a while. I was glad it was Jack who paid the usual evening visit to his tent/office. I never once set foot in that tent. On the rare occasion that Fred wanted to talk to me, he came to my tent. I have never, ever been like that before or since, but he and I were like oil and water.

(During the spring season of '63 I was trained on the new fluxgate mag and had my plotting and contouring skills tuned up. I was essentially being fast-tracked for a huge mag job that summer north of Yellowknife. Why was I picked? Years later I came to

131

the conclusion that Fred must have recommended me for the job. Was it because of my skills as a mag man? Or was it because he wanted to make sure we were 3000 miles apart that summer? I'll never be quite sure.)

But Fred is gone now and those rough waters have passed under the bridge: Reste-tranquille, Fearless.

About Us: Gus is from a large family of mainly underground miners. Kent's father is a doctor, thus classifying Kent as middle class, although he never plays that card. Jack's father is a mining promoter and Jack will eventually join his dad in the business. Grant grew up in Copper Cliff – his father is a supervisor at the smelter. Bill is from Manitoulin Island, a spawning ground for Major League pitchers. (If Bill throws a snowball at you you'd better duck. You can actually her it go by, trailing a comet tail.) Both Bill and Grant will later become career policemen with the O.P.P.

As for me – I am a farm boy with no defined career path. I like the bush, I don't mind the semi-isolation, and I always welcome the opportunity to see new country and to try new things. To me, life is a game of snakes and ladders – roll the dice and take your chances.

Sidebar: As I look back to that winter so far away, I have to be proud of every one of us. Three men to a tent – 224 square feet of living space – 74.7 square feet each. It could have, and should have, led to the occasional flare-up, but we all got along.

Maybe it was because we never discussed politics or religion. We talked about girls. If the conversation started on cars - girls appeared. If the subject was sports – girls appeared. We were all experts on girls.

If we did have a discouraging word, it was generally about Inco. Mother Inco fed us, housed us and paid us the going wage – of course we had to complain about her.

Even Fred got a free ride from the rest, and although I have said I disliked him, I have to admit that I respected his management practices on the most part. He just irritated me, I guess.

So we all got along in that winter of 62, and I relish the memory.

STO (short take-off *with ground crew assistance.)
We have an isolated response on our to-do list. Normally the two crews will be out in the same general direction, being sure to be separated by four or five miles to avoid signal overlap, but this day Gus is 15 miles southeast, and our anomaly is 25 miles north. This makes for a long flight for the normal chopper shuttle system, so Jack, Grant and I will hit this target by Beaver.

Sidebar: Inco supplies fuel for the drill contractor in addition to our own fuel needs, so with stove oil for five tents plus the kitchen, gas for the drill, and av-gas for the chopper, we are conspicuous consumers of petroleum products. Barrels come in almost daily, and both Fred and Fecteau work together to economize. This is workable by daily radio contact between Fred and the airbase.

There may be a trip in to Mistassini. Because the aircraft is going in empty a slight detour can bring us barrels of life sustaining hydrocarbons. The charter will be split –

132

Inco pays the first leg – Mistassini picks up the rest. It is an all-around symbiotic money-saver.

Today the charter will be split in-house, so to speak. The Beaver will bring in fuel and take us to the target in the morning. In the afternoon another load will come in and we will be picked up before the pilot heads home. The days are longer now and we will leave and return as per normal. The pilot will have a longer day than us, but he is not on snowshoes, so we don't feel a bit sorry for him.

The Beaver comes in with three barrels of fuel. We load up our gear and off we go. Jack has the air photo and shows the pilot our destination, a fairly large lake about two and a half miles from our target. The pilot is Doug Fisher, the very same guy I had tried to immolate last September.

"How about that lake?" he says, pointing out a little pothole no more than a quarter mile from our anomaly. We had not even considered it for a landing spot, but Doug is a superb pilot. If he says he can put us in there that is more than fine with us.

"I can't pull you out of there though," says Doug. "You'll have to walk out to the big lake tonight, and in fact, you are also going to help me get out of this one."

Well, you learn something every day. Doug plunks us down in that little damp spot, keeps partial power on and swings the tail around right at the shoreline. "Now you guys grab that tail rope and hang on as long as you can."

So we toss out our gear and grab the rope. Doug firewalls the throttle, and 550 pounds of dumb bush-whackers with snow bullets bouncing off our cheeks hang on to that rope until the tail starts to lift! No shit – that tail ski almost clears the snow before we let go. The Beaver just plain pops into the air. We pace off his takeoff run and it is less than 150 feet! When Doug returns to pick us up at the big lake we tell him the distance.

"I figured as much," he says. "For a while there, I thought you crazy bastards were never going to let me go."

We give her tar paper for two days. The chopper has spoiled us and we don't like long walks. But the second day in we have a new pilot, a Parisian Frenchman we have never flown with before. He asks us where we want to go and we show him the pothole with the ski tracks still evident.

"Oh, no!" he cries, "Who put you in there? Feeshair? Non, non, non! Feeshair ees one crazee man! I do not land there!"

So be it: never argue with the pilot. We bite the bullet and do our two-and-a-half-mile walk, in and out.

(You know, over the years people have asked me, with all the flying I have done, why have I never gotten a pilot's licence? My answer is simply this: I am so lazy I'd probably kill myself and a few other people just to save a two-and-a-half-mile-walk.)

We Decapitate a Mag.

A week or so later we do another fixed-wing job. The chopper has to go into Fecteau's hanger for repairs. The official reason, fed to us know-nothings, was that a G-2A has had an accident in the Mideast caused by abnormal wear in the swash plate fittings, so every G-2A in the world is grounded to replace these fittings. But Hmmm...

133

Jacques has sent copies of his log books to head office for billing purposes, and I think some honcho picked up on the cold-weather flying angle and decided that those fittings should be replaced on spec. I also think Jacques may have gotten his tit caught in the wringer, because within ten days he is gone, supposedly to a better job offer. Uh-huh!

Once again our pilot is Doug Fisher. Poor Doug. I'll always think of him as proof of the Bushwhacker's Second Axiom: "The only people who never make mistakes are those who never do anything."

So here's the deal: two crews plus gear are going out. It's a little heavy for the Beaver perhaps, but Doug has already burned off some avgas, and besides, we don't see any D.O.T. Inspectors hiding in the bush.

We are to be dropped off first and Gus is to continue on to his target. The lake where we are headed is a mile long in total, certainly adequate for a Beaver's capabilities, but it has a dog-leg. Not much, but enough to require a curve in the takeoff run, so in the pm Doug will pick us up first and continue on to get Gus. This will make for a lighter load to lift off from our dogleg. (There is no slush, and packing a runway is uneconomical as it would take all day just for one anomaly.)

We land at Dogleg Lake and Doug drops us off at a little point just where the lake takes a right-hand turn. We are a hundred feet from shore, and do we pick up our gear and move it to the shore? Not us – that would require a few brain cells. We stand around the gear to watch the Beaver take off. (Bushwhackers' First Axiom.)

Doug motors off down to the end of the lake, pours on the power and swings the tail around. As he comes around under full power he is heading straight for us. Remember the deep snow? He has no one holding the tail today, and he needs some ground speed to get the tail up. Just as the tail clears the deep snow we panic! Two of us run one way and one runs the other. Doug has nowhere else to go and runs right over the gear. The mag is sitting on its tripod covered by the ever-present sample bag, and one wing strut clips it at the base and shears the head right off – scratch one magnetometer off the inventory. The right hand ski hits the motor/generator unit, smashing it all to hell. Scratch that one off the list also. Doug swings the Beaver around, taxis back and jumps out to count bodies. We aren't hurt, just laughing so hard we can't stand straight. Once we have wiped the tears from our eyes we apologize to Doug and appraise the situation. Plan "B" swings into place. We decide we will join Gus for the day. Doug figures he can get us out of Dogleg, as he has broken some trail already, and get us out he does with room to spare.

We get to Gus' lake and unload. Bill looks like he is ill. His face is kind of grey and he isn't moving much, he just stands there. He looks like he is in shock, which he is. He had been sitting on the floor in the cargo area and his only view outside was through the window of the rear cargo door. He told us later that when the mag head flew past he was sure the prop had decapitated one of us.

We send him back to camp with our trashed equipment. This is a good deal for our crew. Bill will tell Fred the story and by the time we all return to camp in the pm, Fred may have cooled down. (Fred may be our area manager, but sometimes we have to manage Fred.)

134

Well, we never do return to camp that day. The five of us double-time it with no break, and by 3:30 the whole grid is done and we are on the lake waiting for the plane. It had started snowing at noon and it is getting worse, but we listen and hope. "I hear it," someone says – dead silence – and there is no silence quite as complete as a still, snowy day, standing on a lonely lake, waiting for a plane you know will never come.

By 4:30 we know we are licked. Daylight is waning. It is two hours 'til sunset, but already we can't see the shoreline 200 feet away, it is snowing that hard. Darkness will come early today.

When we landed that morning we had seen that the bay we were in became a large swamp a half mile west of us, with lots of water-killed, standing dry jackpine, which is excellent firewood. (Bushwhackers' axiom #3 - always scope things out when moving into unfamiliar territory.)

We grab our axes and the ever-present emergency box and head to the swamp. We pick a spot near a good supply of firewood and set up a rough camp – more rough than camp. The emergency box contains a small 6x8 sailsilk tent, a two-burner Coleman stove, soup, coffee and a minimum of utensils. It is meant to serve three men, but is pretty slim pickings for five.

The first order of business is to clear the snow away, which we soon find is impossible. We only have snowshoes to shovel with, and less than a foot below the surface is a tangle of swamp grass and willows that can't be cleared. I have a good idea! We will fabricate a double layer platform of green logs and build our fire on top. It may have been a good idea, but we soon find out it was not a great idea.

While three of us build the platform and cut firewood, the other two throw up the tent and make a floor of spruce boughs on the snow. We soon have a roaring fire going and settle down to have a cup of soup followed by a good cup of black coffee. There are only three cups so we take turns. We finish our rations and prepare for a semi-comfortable night.

The main flaw in my good idea soon becomes evident. Our roaring fire burns through the platform and heads Chernobyl-style for the centre of the earth. The only thing being heated is the air above. More firewood! By ten o'clock the other guys say "To hell with it" and crawl into the tent and tie the flap, to sleep with only their body heat for warmth.

I elect myself fire-tender. After all it was my good idea, and there is barely room for four in the tent anyway. So I cut wood, and I cut wood. My theory is that if I keep the fire hot it will melt a circle large enough to sit in and we can actually be warmed by said fire. It's working! By 2 am I have a big circle melted out and even make a sort of ramp up to the tent. I tie the flaps back to let the sleeping beauties get some heat from the fire.

Let me tell you something about cutting firewood in the middle of the night. It isn't hard to find a dead tree – there are lots of them. It must have been a good stand of pine before they were flooded out years ago by a long-gone beaver dam.

The thing is, you have to be careful. For some reason standing dry pine will sometimes get really rotten a few feet from the top. When you start chopping at the tree it shakes, and suddenly, whunk! A five-foot spear penetrates the snow beside you. It is impossible to see the tops of the trees at night, so you can't tell which ones are

135

dangerous, so you give the tree a good whack with the back of the axe and bail out. Do this twice and then start cutting the tree down, pretty confident that you won't be speared.

(We are used to ducking by now anyhow. The alternate freezing rain in the fall has left big clumps of snow hanging on the boughs of the evergreens. When we blaze lines we discover that if you don't "whack and jump", you have a good chance of getting one of these clumps on the head. Even if you don't get your bell rung, you at least get a bunch of snow down your neck. We never wear our hoods up while working in the shelter of the bush. The trees keep the wind at bay and we have to let our body heat escape or we'd be swimming in sweat.)

At 4 am the boys join me in the fire ring and are now cutting their share of firewood. The skies have cleared, the temperature is dropping and my feet are getting cold. Alternate warming and freezing during the night has turned my moosehide moccasins into balls of ice.

We break camp just after daylight. It is a pretty sad looking line of guys heading out to the lake. No one is laughing now, we just want a hot breakfast and a warm sleeping bag. As we near the bay the plane comes in.

There is that beautiful, beautiful Beaver. I can still see it, grey with red trim shining in the morning sun, the prop flashing with each revolution of the engine ticking over at slow idle. What a welcome sight!

I am first in line and when I get to the plane and open the door, there is Fred! He pushes a cardboard box at me. Inside are two loaves of bread, a tin of butter, a container of mustard and a hunk of bung bologna. For a moment I think, "Nice gesture. He knows we'll be hungry. Then it strikes me. "The silly prick wants us to stay in the bush!"

Without a word I push the box aside, throw in my snowshoes and climb in – the others follow.

"What's this? You're not working today?"

"Job's done," someone says.

With everyone and the gear aboard, Fred is forced back into the co-pilot seat. Feeshair is flying. He looks back at us and looks over at Fred. Fred nods and home we go. Again we are a little overloaded, but as far as I am concerned we could have gone out two hundred pounds lighter, leaving Fred on the ice behind us.

Sidebar: Seven years later I am working for Noranda in a position similar to Fred's. We are flying out of Red Lake with Ontario Central Airlines (OCA.) I am sending out a three-man camp today and I am standing on the dock with Dave Harvey, the base manager. We watch as the dock boy, using an old feed-store style balance scale, weighs the load – first the three men, then the other stuff until the proper weight is attained. I can't remember exactly, but I think the DOT allows about 1400 lbs payload for a Beaver on floats.

"You know," Dave says, "It's just goofy. On the way in we cut it right to the pound, but when we pull them out we will stuff that plane to the roof."

Wrapping up the Season

In the second half of March the weather starts to turn towards spring. The days are long and sunny with never a cloud in sight.

It is too warm! By mid-afternoon the snow starts to ball up under our snowshoes and day by day afternoon walking gets more difficult. The usual rigidity of –40 snow is gone and at every second step we sink to our hips.

We adjust – it's not plan "B" – it is common to every crew in early spring weather everywhere.

We arise before sunup, hit the bush at 6 am, and travel like the wind over frozen snow. We shut down at 1 pm and eat lunch. We have no tea fire – it would be an insult to Mother Nature.

After lunch we make a spruce bough bed on a south facing slope and lay on our parkas, stripped to the waist, working on our suntans.

At 2:30 the chopper comes in. Our new pilot is Luke Monroe. He is a great guy and is an excellent pilot. We don't need a signal fire for Luke.

(You might be wondering why we didn't go back to camp at 1 pm. Well – we don't want to insult Fred either. Besides – it is a lot nicer on a snow-white beach than in a semi-gloomy tent.)

The drill camp moves out leaving nothing behind but empty oil drums. We say goodbye to Clarence the drill foreman, and I'll always remember him. When we met he grabbed my hand, gave it a good shake and said, "I'm Clarence Burkhold from Killarney, Manitoba. I'm the inventor of the Burkhold Beam."

That's how he introduced himself to everyone. We never did know what the Burkhold Beam was or did, but one thing for sure, Clarence Burkhold was always beamingly pleasant. I never ran into Clarence again, to my regret, so Clarence – if you're reading this, call me up and tell me what the hell that Burkhold Beam thing is.

(Before they pulled camp, Clarence brought out the tent and Coleman stove we'd left in the swamp last September – he found them in the tree where we had tied them, still tightly secured and unmolested.)

By now the warm weather is causing our floating tents to sink. Ours now has a definite list towards the lake and a stump is pushing the floorboards up at the centre of the tent. No problem – we are finished!

On April 8th we start to tear down. Personal gear is packed, and empty fuel drums are piled up. Only a couple of full and partially full ones remain to be flown out. Tents are pulled off the frames; cots, mattresses, cook stove, oil stoves and a plethora of dishes, cookware, gas cans, lanterns, tools and you name it are piled on the ice to join the steady stream being loaded in Otters and Beavers. Open food containers go into the dump, full ones and fresh meat are given to Claude to take home. There will be a pile of junk in the Fecteau warehouse to greet Fred on his return this summer.

In the afternoon of April 9th the last of us climb into the last plane and say goodbye to the lonesome looking tent skeletons and the little core shack that had served us so well. But we don't get too choked up. Wine, women and song await us in Chibougamau (after a hot shower.)

Planes, Trains, and Automobiles
We sleep in on the tenth of April and then stroll down to the dining room with Fred to plan our escape. On the way in so long ago, we had all left our suitcases containing

137

our town clothes in the custody of the Monaco. We are a little hung over and kind of shaggy around the ears, but we are at least presentable (and we smell better.)

Transportation connections to Montreal are convoluted to say the least. The train between Roberval and St. Felicien has been discontinued. We now have to bus to Roberval and overnight, then on to Montreal by train, and then another transfer with who knows how long a layover. We can charter a Cessna 185 on wheels, but there are seven of us. A twin engine charter is available, but too expensive.

So we call a taxi! No Shit! A bit of negotiation using Fred's bilingual skills (well, maybe not "bi", but adequate) and we soon have two cars lined up. Seven men, pack sacks, sleeping bags and suitcases are all off by mid-afternoon. I look at the tires on our taxi as I get in – bald as a baby's butt! No problem, I have made it through 66 below, and I'M sure not going to let treadless tires hold me up now. We head south and should arrive in Montreal at 5 am, barring a crash-and-burn.

By midnight it is pouring buckets. I'm sure our bald tires are riding a quarter inch above the pavement. But what's this? It is 2 am and our driver is stopping! Trouble? We rouse ourselves and look out the windshield. The lead taxi has stopped and the driver is out of the car and pointing to the sky. UFO?

We all pile out and look up, shielding our eyes from the pouring rain. We are directly under a high-voltage transmission line. It is raining so hard that electricity is arcing into the atmosphere! All along the line miniature blue lightning bolts flash continually! It's the first and only time I have ever seen a power line spring a leak.

We arrive in Montreal and begin to go our separate ways. Fred and the other five make their train connection. Fred will stop off in Ottawa for a few days while the others continue on to Sudbury. I take a taxi to Dorval and board a flight to Toronto to visit an old school friend.

My flight is an Air Canada DC-8. This is all new to me, as I have never flown in a commercial jet. What surprises me is how short the trip is and the roundabout way we get there. First we climb out for 20 minutes and I can't believe how steep it is. Then there is twenty minutes of level flight followed by an equally steep 20-minute descent. It is a breakfast flight and they don't start serving until the plane levels off. I am sitting right at the back and the galley is at the front, (no class distinction on these short hops, I guess.)

By the time the rear passengers get fed we are on the way down and I have to try to eat the grub with one hand while I keep the tray from sliding off the little fold-down dohickey with my other hand. I am told that by going to 20,000 feet they save fuel. I say BS to that. The shortest distance between two points is a straight line. They could have flown the whole flight at 400 feet and done an airborne survey at the same time.

Table Scraps: Stuff I forgot to mention

From time to time during that winter we would cross a grid, or extend one that had been done the summer before. The snow was so deep that the previous blaze marks were hard to find. You generally blaze a tree about chest height. My blazes would naturally be two feet higher than those of Gus, so the height of the men putting in the grid the summer before had a bearing on how easy it was to spot them in the winter. Some would be at snow level, some would be half covered, others were completely buried.

138

Another thing – claim posts: Here's how you make a claim post as prescribed by the Ontario, Manitoba and Quebec mining act of the time. First you chose an appropriate sized tree at the spot where the corner of your claim will be. A claim post is supposed to be four feet high and four inches square at the top with 18 inches of fresh face on each side. So you pick your tree – big enough to end up 4x4 but not so big as to require a half hour of hewing it down to size. You square it up to 4" by 4" and then cut the top off the tree. You now have a post four feet high, rooted to the ground. Now you nail your tag on the appropriate face and underneath it you write the staking info using a carpenter's pencil or maybe one of those new-fangled magic markers. (The info is for example, "Staked by Bob Bushwhacker, License H-9158 – Jan 5, 1962, 9:00 am.)

Now, how about the poor bugger the next summer who runs across your work, or perhaps wants to tie on to your claim block? The blazes on the trees are way up there! The claim posts would be nine feet tall! He would need a ladder to read the staking info. "Who the hell staked these claims, Sasquatch?"

Sidebar: They tell me that there is a claim post on display at a museum in Haileybury, Ontario. It was made from a tag alder and is 1" square. The deal is – if no trees are available within a reasonable distance you can use what is at hand.

Then and now

Regarding the various detritus left at Frotet – you can't do that these days, and rightly so. No building materials or empty oil drums are to be left in the bush. Do not leave full drums on the ice – they must be upright and secure on shore. Fuel spills must be reported. Tree cutting is frowned upon unless absolutely necessary. You can burn the garbage, but all tin cans and plastic containers must be flown out to be transported to an approved landfill. No more plywood huts with crescent moons on the doors. No more 2x4's nailed to a tree with a pit dug behind (commonly called a Johnson Bar.) You can establish a rock-filled grey water pit in an approved area by submitting a sketch to the Ministry of Lands and Forests, but personal waste is to be handled by chemical toilets or propane fired destroylets.

Drinking water is available at any lake or freshwater creek, but if you are an employer, you had better get the water lab-tested for potability or you will be fined heavily. Holy governmental interference, Batman! I have many times worked all day in hot, hot, weather with nary a lake or creek in the vicinity. To keep myself hydrated, I have pressed swamp moss down with my hands, and have drunk the cool water seeping out of the moss, filtering out frogs' eggs with my teeth (frogs won't lay eggs in bad water.) Fifty-some years later I am still hopping around – but let's face it – I may croak anytime.

Few self-respecting exploration companies will even attempt to build a camp these days. If you have road access, you bring in Atco-style self-contained buildings. For remote sites, it is now more common to leave the camp construction and operation in the hands of one of the many logistics companies.

Sidebar: A few years ago I was talking to a small diamond drilling contractor (he wasn't small - his company was.) He told me that if he did a job requiring four or five miles of cut road it was cheaper to hire a helicopter, even more so if a running creek

139

had to be crossed. An approved creek crossing would have to be built and then removed on their way out. He would also have to pay stumpage for any tree he cut down. Yet, he could cut a two-acre chopper pad on-site and nothing would be said. Go figure.

Chapter V   Summer of '62 - Back to Thompson

I am supposed to spend this summer at Cochrane with Barry Krause. Barry is moving his wife and baby to a rented house in Cochrane, and the rest of us will stay in a motel five miles west of town. Two anomaly crews, a chopper pilot and mechanic will sleep and eat at the motel. Barry has a room set aside as an office.

I pull out of Sudbury in mid-May with a company pick-up and rented trailer loaded to the gills with Barry's furniture and stuff. We arrive in Cochrane early in the afternoon, move Barry in, and my shotgun partner and I check in at the motel. No rest for us wicked guys – the chopper is already here and we'll hit the bush tomorrow morning.

After work the next day Barry asks me to go to Timmins with him to return the trailer. I hook the trailer up making sure the tongue is tightly secured to the ball. It's dark when we pull out and the highway is under construction, being prepared for new asphalt. There is a gravel ridge down the centre and we cruise along, not going really fast.

Barry is driving. I check my mirror occasionally, and halfway to Timmins I tell Barry the trailer is no longer behind us. He thinks I'm joking, but I tell him, "No kidding – the trailer is gone!"

Sure enough, we are trailer-less. We turn around, head back a couple of miles, and there it sits. It has come to a gentle stop, slowed by the tongue digging into the gravel ridge.

When I over-tightened the tongue on the ball, it worked the nut loose and the whole deal had come apart. We walk back a couple hundred feet with our flashlight, and there is the nut and washer lying on the road! We hook up, deliver the trailer and go home, thanking our guardian angels that we were the only vehicle on the road that night.

The next day Barry gets a call from Copper Cliff. Moak Lake is short one party leader – I am on my way back to Thompson with a battlefield promotion.

So once more I am rattling north on the Bay Line. I thought I would never see this old girl again, but you know how the saying goes, never say never.

She hasn't changed much since April '61. There is still the same mixture of humanity crowding the seats, and the rail bed may be bumpier. There are three main differences on this trip. Inco is paying the ticket, so I am in first class, which costs more, I guess. If there is a division it's an unguarded border – everyone sits and wanders wherever they wish.

The second difference is that I have cash in my jeans. A surprising third change is that they have a bar car.

OK – with money in his pocket, Suave Bob is going to have a drink befitting his upgraded status to first class.

I order a Manhattan – I don't know why, but I ordered a doggone Manhattan and I had never seen a Manhattan before.

Here's a hot tip – if you ever ride the Bay Line passenger train, (now called the Muskeg Express) don't order a you-know-what. It tastes like lighter fluid and a taste is all you get! The darn thing is served in a saucer-shaped glass, and almost all of it slops on my jeans as the car lurches over the permafrost. When I do get it to my trembling lips most of it runs down my chin. I catch the olive before it rolls off the train but even the olive tastes like a rabbit turd.

So take my advice – order anything but a Manhattan and have it served in a sippy cup.

P.S. I could have bought a 2-4 for the same price!

I get off the train to see good old Wray waiting. He will later transfer to an office job and I heard he made a decent career of it. He is a good, reliable man. His only drawback as far as I am concerned, is that he doesn't smoke. Why is this a drawback? Have patience – all will be revealed.

I don't hang out in town. I go straight to Cook Lake to be shuttled by 180 to Moak. We drive past the still expanding townsite, past the security gate and on past the new headframe. A large new office building is just inside the gate. I don't know if the mine is in production yet, but if not, it soon will be. The mill and smelter are almost complete. The H-huts are still there, but at end of construction they will be shut down. Permanent staff is already moving into the private homes and apartments with more being built each day. The airport is up and running, and soon Transair will be serving the community twice a day.

Sidebar: Carving a new mine, smelter and modern town/city out of the jackpine and spruce wilderness is always a big deal, but heads can get over-inflated.

Later on that summer, Inco's prez came up from Toronto for an inspection tour. As is befitting a royal visit, his itinerary was plotted out in advance with no detours allowed. The infrastructure painting crew was a little behind schedule, so a few days before the big event they were all put to work along the tour route. Steel support beams, overhead beams and various pieces of equipment were freshly painted on the sides facing the walkabout. The back sides were still in primer, or bare, unpainted steel. Weird, huh?

At Moak I renew acquaintances. The brass is still here, and some faces are new, but there are still some old-timers around.

I go over the maps regarding the summer's work with the boss, and in the afternoon my two crewmen fly in and we put our camp gear together. Norm Kearns is still here and is still flying Austin Airways Beaver CF-FHX. It had once been all black, and of course was known amongst us as the pirate ship. It has been repainted – still black, but now with red trim, red wings and silver wingtips. I like the old paint scheme better and so does Norm. Incidentally, this is Beaver serial number 60, if you can believe it, off the assembly line in 1948 and still looking brand new! It is Norm's baby.

The next morning Norm takes us into a two-mile long, nondescript, oblong shaped lake 50 miles west of Moak and about 35 miles northwest of the native community of Nelson House.

FLY CAMP AT JACKFISH

My right-hand man is Alfred Settee from Cross Lake. The third man is Dean, an 18-year-old kid from Saskatchewan. Dean is as green as grass, and although he is a willing worker, I will soon learn that he is totally deaf in one ear.

On a three-man anomaly chasing crew the jobs are delineated thusly: I am party leader, figuratively and literally. I direct the setting up of camp (Alf needs no directing,) I figure out the azimuth into our targets and I literally lead the way into the bush. (Once I have recognized Alf's intelligence and superior bush knowledge, we share the lead.) In camp I always make breakfast, Alf cooks supper and Dean does the dishes. (Camp rule #3 – If you don't like the meal shut up and eat, otherwise you'll be cooking the next day.) I run the radio Skeds, giving daily work and weather reports and making sure the next week's grocery order reaches Wray. If things are not going well I will be the one on the hot seat, as will happen a couple of months later.

Alf will be the magnetometer man. He knows the job well. In camp, I will help with the correction factors, he will make the corrections and I will plot the results and contour the map. I also plot my EM results on a separate map.

Dean, as junior man, will be pointing the transmitter coil during fieldwork, and we all share the job of pack mule. As Dean gains experience, he will be trained on the mag. It is always advantageous to change duties once in a while, just to get away from the same old, same old, every day.

ALFRED SETTEE ON THE DOCK AT JACKFISH

We set up camp back against some small poplars on a broad sloping outcrop 30 feet wide leading down to the lake 50 feet away. Good-sized clumps of willow flank our little point, ending 10 feet before the rock slopes into the lake. The rock slopes gently into the water and we build a strong dock of medium-size spruce poles.

There is no big timber in the area. There must have been a forest fire here years before. The bush is mostly new growth poplar and jackpine along with lots of chest high underbrush. We find poles suitable to hold up our 12 x14 sailsilk tent and fly, and

143

by nightfall we have a good camp set up, including a kitchen nook sheltered with plastic. We even have the old Lancaster radio up and running and do our 7 pm sked.

(Our call number is 52. One night in a fit of playfulness I do a little sing-song, "And it's QRU from 52." I sure get it from the other camps on the 9 pm party line. "Yahoo, 52!" "How are you, 52!" "52, are you blue?" Jocularity, jocularity.)

Table Scrap

Moak Lake has specific guidelines as to fly-in camp equipment. For example: 1 12" frying pan, cast iron; 3 dinner plates, Melmac; and so on – just like the army. Every camp gets the same stuff with no exceptions. One item is, 3 pots, nesting. The large pot is for potatoes or pasta, the small pot is for veggies or soup, the middle-sized pot is for brewing coffee or tea.

Another party leader is on loan from Copper Cliff. I don't know him well, but he is an OK guy. Two days after setting up his camp Dave gets on the 7 o'clock and orders a coffee pot and a pillow. Well, let me tell you, if I thought I had gotten the gears, you should have heard the 9 o'clock party line that night! "Do zem want a blankie, too?" "What's a coffee pot?" "Has anyone seen a coffee pot?" "Poor baby," et cetera. Once again laughter rolls out over the Northern Manitoba bush.

We go to work the next day and I soon realize Dean's limitations. For the first day Alf stays at the motor-transmitter with Dean, showing him the ropes. Simple: wait for my call, point the coil, pull the rope to start the motor, let it run for 30 seconds, shut it down and wait for me to holler again. I always orient myself using my receiver coil, so as to be facing directly toward the transmitter before taking my reading. I also have a good idea where the transmitter is located, and can tell if the coil is not pointed straight at me. During the first square I have to retake some readings. I simply holler again and Alf moves the coil to the right place. I then stand at the last corner and holler "Tear-down!" They put the coil in the coil bag, bundle up the tripod poles and home in on me while I wait there. (When there are only two of us on the EM, I go back to help with the move, of course.)

They show up and we set up for the next square. I ask Alf how it is going, and he just shrugs. He isn't about to rat the kid out. I read the second square, and I know we have a problem. I go back to the transmitter and ask the kid if he is having trouble zeroing in on my voice. He finally admits that he has only one good ear. His right ear drum was punctured when he was young and he is stone deaf on that side. It is impossible for him to locate me.

Now what? We sit down and I discuss the situation with Dean. He is downcast. He really wants the job and admits he neglected to mention his hearing loss when filling out his application. No one asked the question, and I can't blame him for not volunteering the info. Our only alternative now is to train Dean on the magnetometer, not ideal at the start of summer, as the new guy is usually worked into things more gradually. By fall he should have at least some knowledge of the whole deal and be an asset to any crew the following winter. If he is still around the next summer, he might even be groomed as a party leader himself.

144

The learning curve is a steep slope for a bush rookie. I know that very well, having been in Dean's boot exactly two years ago. But I have two good ears – Dean has one: I had Charlie McLeod – Dean has me.

It is not a good mix. Charlie was self-confident, even-tempered and kind. I have yet to gain those necessary attributes and I am as green in my position as Dean is in his. Had it not been for Alf and my determination to be at least a pale copy of Charlie the camp might have folded within a month.

We locate the conductor and start the grid. Dean is working with me and tends to be a little clumsy on his feet. I show him how to take a compass shot, blaze the trees and cut a rudimentary path through the thicker parts of the underbrush. I show him how to make sure his line is fairly straight by checking his back blazes, just in case the conductor has a strong associated mag high and is affecting the compass. (If you watch your back blazes the drift becomes apparent and you can make adjustments.) I teach him how to pace off the distances – a hundred feet is thirty-six paces for me, forty for him. I tell him that if he loses count he has to go back to the last station and start over. I teach him how to make a picket for the station if an appropriate tree is not at hand. I even have to show him how to use his axe without cutting himself. This kid is green! It is a slow process.

After we do our second set-up I leave the grid extension work to Alf while I get the kid started on the mag. He follows with the notebook as I do the readings. Every night at camp he helps me do the corrections and conversions to gamma values. I show him how to plot the readings on the map, draw countless sketches explaining the grid system, and on and on it goes. Alf doesn't have enough to do and I have too much. I am beginning to get frustrated and start to resent the kid. I should be kinder, but it has taken us seven days to do the first anomaly – a job that should take three or four at the most.

We locate our second anomaly, establish a base line and grid, and on the second day I turn Dean loose on his own to see how it will work out. He promptly gets lost while crossing from the end of the first line to pick up the second. (Reminder - the mag man traverses in loops.)

Dean panics, starts hollering and seems to be heading toward us. We answer, but he can't place us. He is now running! He passes us no more than 100 feet away – a moose on a mission. Alf cuts him off and brings him in. Dean is panting and sweating profusely.

So now we have a man who can't crew on the EM, is not great in the bush, and is prone to panic. I send a note to Moak suggesting they may have something more suitable for Dean. I need a new guy, but they refuse to replace him. I am beginning to suspect a bit of nepotism is going on (and Dean has thrown out a few hints.)

We go to plan "B." Dean will not leave a blazed line without one of us present. When reading a mag on his own he will backtrack to the baseline before moving over. Alf and I will make sure we are always working somewhere near and will check on him hourly. This really slows us up and adds to our work load.

On top of all this we are getting frequent showers, mostly at night or early morning, meaning at least a two hour delay to the start of our day. We are taking five days to do a three day job. Every night I call in my pitiful daily report and I can picture the raised eyebrows at the other camps. I even start to skip the Sked occasionally, partly to have a little more work to report, and sometimes because we stayed out late in an attempt to make up for lost ground. We no longer join in on the 9 pm party line. I'm not interested in discussing problems on a party line, and I am certainly not going to embarrass Dean in public, even though I am barely civil to him in camp. Alf is kinder than I. He talks to Dean and does some evening fishing with him. We all have fishing rods, of course, and can borrow Alf's tackle. Life goes on, and my stress level rises.

We always check out the fishing in any new camp. It appears that this new lake is a dud. The only thing we can catch are scrawny little northern pikes less than eight inches long. Maybe the big ones are hiding on us, or maybe the seagulls have not yet begun to stock the fishery. We name our lake after the little jackfishes. How original – there are now 100,001 Jackfish Lakes in Canada.

Sidebar: Some provinces, Manitoba in particular, do not allow the use of live minnows for bait. Biologists claim this can transfer undesirable species between lakes, but I have been told, and have witnessed myself, that seagulls do their own fish stocking. They feed on nearby lakes and when returning with a bellyful to regurgitate for the chicks, they will sometimes drop undigested fingerlings into the water, to swim away and reproduce. I have never heard of a seagull being fined or having his fishing privileges pulled.

King Oscar

Across the lake a little more than a half mile away is a seagull colony. There are two rock reefs quite close to each other a short distance from the north shore. We know it is a seagull colony for two reasons. Tthere are seagulls flying around the large reef, and the small reef is white with seagull poop. Seagulls are like pigs –they never poop where they live.

Two weeks after we arrive, Alf, Dean and I take the canoe over to Gull Island. There amongst the swirling, screaming gulls we find four nests of chicks. One nest of three is obviously an earlier hatch than the others. We choose the largest chick and take him home. Mommy follows us for a ways, and though we can't speak Seagull, I'm pretty sure she is saying bad words. She soon gives up. She can't count, and after all, we have left her two kids and the little guy cuddled in Dean's jacket is too scared to speak up. (I know right now some of you are shaking your heads and being a bit judgemental, but honestly, things couldn't have worked out better, as you will see.)

We put the little guy in a box with sides high enough to prevent his escape and cuddle him up in a nice clean towel. We gently stroke his feathers, talk soothingly to him, and keep the box beside our beds. We are also careful not to make any sudden movements around him or talk loudly. We are cool, calm and collected – model foster parents.

The next morning we are up early to check on him. He is already talking to us. "Where's my breakfast? C'mon, guys, Mommy always gives me breakfast in bed."

What to feed him? We didn't think of that when we brought him home. We don't have a minnow trap, and besides, none of us feels like regurgitating at the moment.

146

How about sardines? So we open a can of King Oscar sardines packed in natural sardine oil. He loves them! He eats almost the whole can, burps and goes back to sleep. We now have a name for him - King Oscar. He may be Queen Oscarina, but we dummies can't tell the difference, so King Oscar he becomes, shortened to Oscar when we get to know him better.

That morning we put Oscar back in his box and go to work earlier than usual. We don't know it yet, but Oscar will soon prove to be quite a slave driver. We return early to check on the little guy, and is he ever hungry! "Let's go, fellas, my belly-button is saying hello to my backside!"

He is served another can of sardines with a cut up wiener chaser, followed by more sardines an hour later. He is already bonding. We turn him loose with someone watching him at all times. He sits and watches us cook supper, always letting us know when another food transfusion is necessary.

By day three we are no longer hand feeding him. He is picking them right out of the can and even cleans up the remaining oil. We started with twelve cans, but supplies are dwindling fast. I phone in an additional order on the Sked. "Send in a case of King Oscar sardines." I hang that one on Dean."No Brunswick or any other brand, he only eats King Oscar." Dean is cool with that.

We are substituting other foods by now. Oscar likes the cut up wieners and once in a while we snag a little jackfish for him. We cut off the head and tail and dice the rest for him, guts and all. The case of sardines comes in on the next Beaver flight and Oscar settles into a more or less variable diet suitable to his taste. He isn't shy about letting us know his preferences, believe me, not shy at all. He has little interest in our food other than the sardines and wieners. Seagulls are notorious garbage pickers, but not Oscar. I'm sure that later on in his adult life he will be called a snob by the rest of the seagull community.

Oscar is the perfect house guest. He has the run of the tent and never makes a mess. We never did find out where he went for a crap.

By the end of his first week with us Oscar has total freedom. We throw the box and tea towel in the garbage pit and Oscar never sets foot in the tent again – never! If we put him inside he goes straight to the mosquito netting and pushes his way through the opening. He is growing like a weed, is smart as a whip, and he sure takes my mind off my troubles.

Oscar walks us down to the canoe every morning, squawks goodbye, and meets us every afternoon to say hello. We start leaving open sardine cans out for him and still give him the odd diced jackfish.

One evening while Alf cooks supper, I do a little shoreline casting and catch a seven-inch jack for Oscar. He of course is helping me. He knows what is going on, and as usual, is being critical of my stance and follow through. I reel in the fish and lay it on a rock while I go to the kitchen for a knife. I come back, and to my horror, the little shit has swallowed the whole fish head first! The thing is, the fish is as long as Oscar and three inches of tail sticks out of his beak,a classic case of one's eyes being bigger than one's stomach.

I call Alf and Dean and we study the situation. Should we perform a fishectomy? Inadvisable as the dorsal fin will never come out backwards. Oscar doesn't seem to be

147

in any discomfort. He is just sitting there calm and quiet - he seems to be wondering why his family is in such a tizzy. We decide to just keep an eye on him while we eat supper. That little puke has an amazing digestive system. As we eat our supper the fish just keeps slowly disappearing, and by the time we are finished eating, so has Oscar. We have satisfied his appetite today, for sure.

Oscar is starting to swim. We can't believe the speed of his growth and skills development – every day brings something new. He starts out with tentative dips into the water, does a circle and walks back up the gently sloping rock, then back into the water and out again. Before nightfall he is paddling out of sight around the corner, only to reappear to paddle around the bushes on the other side of our rock.

And what a show-off! Every time he goes past us he checks us out to see if we are watching and does a dipsy-doodle. "Look at me! Look at me!" and another little flourish. This guy has personality plus!

We return one afternoon to a real calamity. The tackle box had been left out and Oscar, that bad boy, had tipped it over. Maybe because the lures smell of fish, or maybe, as all youngsters are, he is just curious and mischievous, but there sits Oscar, hooked on a bass plug, looking sad as all get out. It is a large lure with a treble hook at each end and two barbs have penetrated Oscar's skin, one in his neck and the other well embedded just behind his rectum.

Oscar is obviously in pain. If he moves he will deepen the barb in his neck. The one at the back has already gone right through the skin with the barb protruding. It is hard to say how long he has been impaled, but luckily he is no longer trying to get loose on his own. He just looks at us imploringly. "Help me, please."

There isn't much blood and it seems nothing important has been punctured – at least we hope not. I manage to pull the front hook out of his neck without much damage, just a little "Eep!" from Oscar. But what to do with the other one? We don't have side cutters.

Dean holds Oscar gently while I get a clean razor blade from my shaving kit and a bottle of iodine from the first aid box. Dr. Bob is going to operate. While Dean holds Oscar upside down I carefully slit the half-inch of skin between barb and shaft. Success! The hook has not passed through any organs and it seems that Oscar will be OK!

I douse the wounds with iodine and Dean turns Oscar loose. He heads right for the lake and jumps into the water. What a sight! We roar with laughter as this little seagull paddles around, first dipping his head in the water and then rearing up to plunge his bum in, back and forth, back and forth. The iodine must sting something fierce! After about fifteen minutes of this Oscar walks back onto the rock, avoids eye contact and goes into a corner under a tree branch to sit and sulk. He is always hurt when we laugh at him. Like Bugs Bunny's Aladdin Lamp genie, Oscar is very sans-si-teeve.

A couple of days later we come back to camp to another, happier development. Our boy is learning to fly! He might have been at it for a few days, as far as we know. Oscar prefers to practise alone so we won't laugh and embarrass him but he just cannot help showing off.

He is a student pilot from the ground up, you might say. As foster parents we have been a little lax and Oscar has had to design his own training plan. For days he has been working on his pecs by flapping his wings, and maybe on his leg work too. To take off he starts fifteen feet from shore and runs down the sloping rock. Just as he gets to the end of the runway he lifts off. He's Flying!

The three of us stand there, amazed and as proud as punch! I can still recall that special feeling, watching that young seagull take flight. Five years later I will remember that day as my daughter takes her first steps.

Oscar is pretty proud of himself, too. He does a few swirls and banks, making sure we can admire him. After a few minutes he decides to take a break and comes in for a landing – not so good – the kid isn't ready for his float endorsement yet. He doesn't flare out at all. He comes straight in, low over the water, sticks his feet out, his toes dig in, and he goes ass over teakettle! He bobs up to find his family laughing at him. I laugh so hard that a tear runs down my leg. Oscar gives it two more tries but the landings aren't getting any better. After the third attempt he gives up on us jerks and paddles around the corner to sulk for a half hour. When he comes back he eats his supper quietly and doesn't look at us or chat the rest of the evening. We all think he is doing quite well, actually. He just needs to work on his VTOL. (Vertical takeoff and landing.)

Oscar is a hard worker, and by the third day he has learned to keep his toes up and his landings are more graceful, although still rather un-seagull-like.

He has also been working on stamina. After supper we watch him do his rock run again, and this time Oscar heads straight north across the lake. Either he has retained some memory of his early life, or more likely, his superior hearing and eyesight has alerted him to the fact that other gulls live on the lake.

We wait anxiously for Oscar's return. By 8 pm we are having evening coffee and no one is very talkative. We are pretty sure our boy has flown the coop. We tell each other how it is the best for Oscar, but we are pretty sad. Alf and Dean wander into the tent, perhaps to read, more likely to think of the good old days. I get my binoculars and look out over the water towards Seagull Rock. The gulls are less active, no doubt settling in for the night and I can imagine Oscar sitting with his family, telling stories of his wild younger days with the hairless bears on the south side.

But what is this? There is something bobbing on the water this side of the island. I can barely make it out in the gathering dusk. Is it Oscar? I call the others and we take turns on the glasses checking the object out. It seems to be getting closer. We quickly slide the canoe off its poplar pole cradle and pile in, crank up the three-horse kicker and head out.

There is Oscar, in the middle of the lake paddling home, and are we ever glad to see him! Likewise for Oscar! He calls out to us and we scoop him up. He sits in the bottom of the canoe, slightly water-logged and tells us "Thank you!" as we head home. We are a family again!

(We liked to think that Oscar felt the stronger pull of his bonded human connections and decided to come home, and no doubt this was true to a certain extent. But the fact that he was swimming home indicates that he was probably rejected and not even allowed on the reef for his take-off run. We will never again make fun of him.)

149

Two more days and Oscar has the whole flying thing down pat, and even starts doing his own fishing. We open a can of sardines for him and he turns his beak up at it. "If you like those smelly things so much, why don't you eat them?" he says.

Maybe the girls over on Seagull rock told him he had sardine breath, but from that day on, Oscar does his own grocery shopping.

(You may think all this is just a lot of B.S. I admit I may have gone overboard in my interpretation of Oscar's human-like interaction with us, but the facts are true. Oscar had become an important part of our crew. I say that Oscar had bonded with us, but it would be more accurate to say Oscar bonded all of us together as a bush family.)

Oscar is a hard taskmaster. Ever since he moved out of the tent he has taken on the job of alarm clock. Every day, right at daybreak, he stands outside the mosquito netting, rain or shine, and tells us with little chirps to get up. We try to ignore him, but he won't give up, and by 6 am we are tired of his nagging and crawl out of bed to feed him and start our day. Oscar continues this even after he takes on his own feeding duties. We grow used to it and actually appreciate his input. It's a tough job to get the boys going in the morning, but someone has to do it.

Smokeless in Seattle

Neither Alf nor Dean smoke. I smoke roll-your-owns, Players Fine Cut rolled in Vogue papers. The tobacco comes in pocket-sized plasticized pouches, six to a carton. One pouch typically lasts me a little less than two days, so a carton will keep me puffing for about ten days, more or less. I order one carton per week, which means I always have three or four days' supply on hand when the new carton comes in.

Did I mention that Wray, our expediter, is a non-smoker? Well, the Beaver comes in with our weekly order, and when we unpack the goodies I find one pouch of tobacco – not one carton – one pouch! I guess Wray had a brain cramp, bless his heart and pink, unpolluted lungs.

I have a pouch and a half left, meaning I have less than three to last me a week. No problem, right? I will just have to cut back for a week to conserve my supplies, right? Wrong! Any smoker can tell you that when faced with a cigarette shortage you always smoke more. "How much do I have left?" You open the pouch to check and since the pack is already open, you have another cigarette, naturellement - non? Oui!

Added to the mix is the seagull chick. We all take turns sitting with him and talking to him. He really likes to watch the smoke curling up from my cigarette and is especially interested in smoke rings. Big help, Oscar.

So it comes to pass that I run out the day before our next order comes in. I begin to pick up butts. Curses, why do I insist on camp cleanliness? (Sidebar A)

I carefully open the butts, discard the burnt ends and amass enough tobacco for ten cigarettes. Phew!

The next day is plane day, and wouldn't you know it, as dawn arrives so does the pitter-patter of rain on the tent fly. Soon it is pouring. The sky is low unbroken overcast and visibility is less than half a mile – one of those days when you know you are in for an all-dayer, if not longer. There will be no grub run today.

I have saved my ten butts, which produce two more cigarettes. Then those two butts are rolled into one skinny. Like the dad said to his boy sucking on his straw at the soda fountain, "Let's face it, kid, it's empty."

Now, what does one do on a rainy day in camp? One drinks coffee and has a smoke. I only have half the equation and am going into withdrawal. Alf digs into his tribal lore memory bank and remembers an elder telling him about smoking Labrador tea in times of tobacco famine. I really don't have time to pick Labrador tea leaves, spread them on a rock to dry and grind them into tobacco. "C'mon Alf, you can do better."

He then remembers that some of the old folks used to smoke real tea, so out comes the Red Rose tea bags. I open a couple and roll a fatty. Hmm, might work. I light it up, take a drag, lean over to put the match out on the dirt floor and all the damn tea runs out of the end of the cigarette. I roll another, this time being careful to keep it horizontal. It doesn't taste too good, and I have to take care to lean ahead so that the constant shower of sparks doesn't set fire to any important body parts. I try one more and give it up. The tea isn't tasting any better, and besides, Alf and Dean are getting far too much enjoyment out of my efforts. Even I have to admit that I am having more fun trying to smoke than smoking.

The next day it stops raining before noon and the plane comes in. Guess which box I open first? Let me tell you, the rest of the summer I make darn sure that I always have an extra carton on deposit in the tobacco bank. (Sidebar B)

I couldn't possibly screw up again, you think? Think again. I forgot to order matches, for crying out loud! I have a Zippo lighter, but no lighter fluid. I carry wooden strike-anywhere (Sidebar C) matches in the bush, and there are just a few left in the match box in camp.

We also use matches to light the cook stove. Moak has upgraded from the Coleman two-burner naphtha stove, and now each camp has a nifty three-burner propane Jiffy Range. No peizo-electric spark lighter – you just turn the valve and light the burner with a match. Do I smoke or do we eat? I decide we can do both.

So once again we go to Plan B. I scour my packsack and office box (Sidebar D) for matches. Every smoker I have ever known has a stash of penny match folders. They get tossed into your gear when you leave town and can be found in a shirt or jacket pocket. I come up with two full folders and two partials. Some of them don't light so good – they are too old, or have too often been soaked in sweat and dried out again. I carry the penny matches to work and during breaks I never have more than one cigarette. At lunch I have three but only use one match, lighting the next one from the butt of the previous smoke. The wooden matches stay in camp to light our cook stove. At camp I risk eyebrows and moustache, lighting one off the propane stove after supper.

The penny matches last three and a half days. We now have four wooden matches left. After supper on day four I leave one round on the propane stove burning. Our butter comes in tin cans roughly twice the size of canned tuna or flakes of ham. With the top completely removed, the upside-down can fits over a burner with a little room to spare. I turn the flame down as low as possible, cover the round with the can to protect the flame from the wind, and go to bed. It works – in the morning the flame is still

151

going! The stove is kept burning in this way until after breakfast on day six. Our grub order comes in and we are no longer matchless. (Sidebars E)

Sidebar A: A clean camp means no bear trouble. After every meal leftovers, eggshells, bones and such are taken to the garbage pit, which is burned periodically. Meat and bread are always left high in a tree large enough to support a rope pulley arrangement, but with not enough girth for a bear to climb. Dishes are washed after every meal. On our last night a bear showed up, sniffed around and left. A clean camp had paid off.

Sidebar B: Two years later I am drilling in the Smoky Mountains in North Carolina. We stop at a country store and I think I might roll my own today, so I buy some rolling papers and Prince Albert in a can - no kidding - a steel flip-top can a little larger than an American cigarette pack. "Prince Albert. Guaranteed smoking satisfaction. For your pipe or cigarette."

Well, now, Prince Albert turns out to have all the qualities of Red Rose tea, although it does taste a bit better. For an added attraction, the papers have no glue strip.

So I ask Pat, our official hillbilly liaison, what the drill is. "Just use your spit," says Pat.

So I do, and no luck. So I spit some more, and then again more. By the time I get the cigarette to hold together and fire it up, it is so goopy it won't burn. When I do get it lit, the spit dries out and the cigarette paper opens like a blossoming flower and Prince Albert is "Gone With the Wind." Oh, well, back to tailor-mades.

Sidebar C: In 1962 "Flick your Bics" have only recently appeared on the market and are unreliable. The flint falls out or the butane bleeds off. We smokers carry Zippo lighters. The works are in sturdy metal cases. To refill you just slip the innards out and soak the cotton batting with lighter fluid from a reclosable can, and reassemble. To replace the flint you unscrew a little spring-loaded dohickey and put in a new one. You light it the same way as a Bic, but if you are cool, you hold it with your thumb on the base and your first two fingers on the lid. You press the base, and let your fingers slide off. The lid snaps open and with an appropriate flourish you spin the striker wheel and light up your smoke. This supposedly shows the ladies that you are one cool dude, but in reality it only impresses kids up to age four – and Oscar.

We don't carry Zippos in the bush. When running a mag, anything steel in your pants pockets might affect the magnetic sensor. We even wear our belt buckles at the back. You can carry the Zippo in your shirt pocket, as any steel above the instrument's head doesn't affect it, but you already have a tobacco pouch in one shirt pocket and a notebook and pencil in the other. A prospector's life is sure complicated, isn't it?

You hardly ever see Zippos anymore. They lie forgotten in the back of Grandpa's sock drawer, to appear on the yard sale table after the old boy bites the bullet.

Sidebar D: Office box: We are supplied with a semi-waterproof, letter sized steel box. It has a hinged, latched top with a carry handle, and in it are our pens, extra pencils, paper clips, rubber bands, tracing paper, note paper, etc. etc. It also serves as a junk drawer (you know what I mean.) Our larger maps and blank map paper are rolled up in a cardboard map tube.

Sidebar E: Wooden matches come in two styles, safety matches and strike anywhere matches. Safety matches will only light if you rub them on the sandpaper strip on the side of the box they are packed in. Strike anywhere matches can be lit by rubbing on virtually any dry surface – table top, ass of your pants – anywhere. (Cool dude trick: hold the match in your clenched fingers with the match head high to avoid burning yourself, strike the match with your thumbnail, and of course with a flourish, light your smoke Once again you have only impressed little kids and Oscar.)

Strike anywhere matches have one big drawback. When carried loose in your pocket they can rub against each other and the whole bunch will go up in flames. This is tough on pants and thighs – especially thighs. You get singed by the fire, and also end up black and blue from pounding on the hot spot. Best remedy? Drop your pants and poke your pocket inside out. Added risk? Those two yahoos on the trail behind you may get herniated from laughing so hard. Best all-around remedy? Get a waterproof screw cap container from a sporting goods store. They are generally made of steel though, so are unsuitable for a mag man. Plastic pill containers work well. (Incidentally, in fire season, tailor made cigarettes and matches are outlawed in logging camps.)

Sidebar Lebenty-leben: Tinned butter was great in camp - easy to pack and ship and requiring no refrigeration. Small print on the can extolled "Product of Denmark." Cool – imported butter!

Years later I learned that surplus Canadian butter was shipped to Denmark, canned, and shipped back to Canada. The butter made the round trip – only the can was imported.

Even now, mushrooms grown in Winnipeg are sent to China and return in cans, more correctly labelled "Packaged in China." What a strange country we live in!!

Zippo handy hint #1 If you see one at a yard sale, scoop it up – they're becoming scarce.

Zippo handy hint #2 (not so pertinent these days) This is a real handy hint, as anyone who has leaned over at the end of the dock to untie the canoe, only to watch helplessly as his Zippo slides out of his shirt to disappear into the murky depths will tell you. This is easy to prevent. Find a strong rubber band and wrap it tightly around the case several times – now it will stay in your shirt pocket.

Zippo handy hint #3 When you hit town next fall, remove the rubber band as it may cause an accident. I'm pretty sure mine did. I sat down next to a good-looking chick at the bar, she took out a cigarette and I suavely whipped out my Zippo, my inner-tube-wrapped, jackpine-sap-encrusted, mosquito-dope-soaked Zippo. She took one look at it, fell off the bar stool and hurt herself. I know she must have hurt herself, because she ran off real fast to get medical attention, and never did return to finish her drink.

OK, you obviously carry your Zippo in the bush, since I see you appreciate handy hint #3. So here is – wait for it –

Zippo Handy Hint #4. Never throw the can of lighter fluid in your packsack when you move camp. Sure as hell, the can will leak and your underwear, shirts, and jeans will smell like lighter fluid. Upside: you will always find a seat on the bus o in fact, you will have two seats to stretch out in. You can avoid crowds (maybe the crowds avoid you,) and you always get to sit in your own pew in church.

Mid-Summer Break: Nelson House/Footprint River

A couple of days after our match shortage was rectified I am told on the sked that a pack-sack drill crew is coming in to Jackfish. The next day we hear the Beaver make two trips in and on our return to camp we check the shoreline and find the driller's camp about a half mile east of us. There are no good camp-sites nearer to us, and besides, their camp is closer to their first drill target. I liked the results of our first four grids, and obviously so did Moak. They are wasting no time following up.

The three-man crew is led by Alan Paupanekis from Cross Lake. I ask him how Charlie is making out and he tells me that Charlie has opened his pool-room/restaurant and is doing OK. Allan's two helpers are Alec Linklater from Nelson House and a burly young flat-lander from Saskatchewan whose name I can't remember. We'll call him Pete.

Alan is a tall, rangy kind of guy and sharp as a tack. He has done a tour in the Canadian Armed Forces and I am told he is the current record holder for the five-mile run (Army-Navy-Air Force competition.) I will run across his trail 26 years later near Bisset, Manitoba.

Alec is very reserved, taciturn, but friendly. He was born in Nelson House and recently returned home with his family after spending 15 years working on a section gang on the Flin Flon/The Pas CNR line. I will soon be amazed at his talents in the wilderness – a true north-woodsman, as I'm sure you will agree.

Pete is OK. This is his first season in the bush. Alan tells me that he is a hard worker and as strong as an ox.

Guess what? They are all smokers! Where were they two weeks ago? Now that I have people to bum matches and tobacco from, I have my own. Sometimes life is just plain unfair.

So we now have neighbours to visit and vice versa. We soon find out that they treat Oscar unkindly. I don't know why. They tease him, but in an unfriendly sort of way. Oscar is still in his early development stage, but he is already wise beyond his years (weeks, actually.) He makes himself scarce when he hears their canoe coming down the lake.

One evening while trading BS over a cup of coffee, Alan tells me that Alec's wife is coming to visit. When Alec heard about Moak's plan to move the drill into Jackfish, he wrote to his wife. She knew the lake and wrote back that she would bring their son up from Nelson House by canoe in late July to visit. Cool. You will also find out, as did I, that she is one amazing woman. (I know that I am overusing that adjective, but I was amazed, I am amazed, and now, years later, I am still amazed.)

154

Work goes on. We are still slow and the drill is catching up to us. We should have been out of here by the end of June, but it is now near the end of July and we still have two more anomalies to hit. At four to five days each plus rain days, it could be two more weeks before we move and I am in need of a stress reliever.

I talk it over with Alan. Alec's wife plans to stay two days. Two extra people in a three-man camp is not much of a conjugal visit, even if they throw up another shelter. Why don't we all accompany her tomorrow on her return to Nelson House and overnight there? Alan agrees, and tells his two men of our plan. They are both cool with it, so the stage is set.

Alec's wife and son arrive and we prepare for our trip. Alan and I each have 17-foot freighter canoes, wide and stable. So does Alec's wife. Alan and I each have a three-horse outboard motor with integral gas tank. So does Alec's wife. Alan and I each have a one-gallon gas can, and so does Alec's wife. We will use our canoe as it is in a little better shape than Alan's. (The packsack drill and rods, etc, are hard on cedar strip canoes.) We plan to leave early tomorrow morning. The weather is great and Alf's instincts indicate continued sunshine and he is usually right (always much more accurate than any meteorologist.) Alf, Alan, Pete and I will be in our canoe, Alec and his wife and son will follow in theirs.

Dean wants to stay in camp, no doubt looking forward to almost two full days of solitude. He is no longer nervous in the bush, so I have no qualms about leaving him on his own. With no canoe, what trouble can he get into? None. I tell him to listen in on the sked, but not to join in. As far as Moak is concerned, we will just plain miss it that night. Dean will also charge the batteries. We don't use steel cots in fly camps, so there is no chance of another mattress fire.

We hang two motors on our canoe. It's easy. The v-stern is just wide enough to catch one clamp on each motor. With the two tillers so close together they are a snap to operate. The only danger is whiplash. When you crack the throttles on those two monster motors, look out! We should have a salt-water set-up.

Sidebar: In the early part of this century I am on Lake of the Woods in North-Western Ontario shuttling guests and groceries by boat to a fishing lodge forty miles from Kenora. The camp boat is a 21 foot Lund Baron Magnum, seating six people legally (and up to 14 usually.) With a 200 horse Mercury Optimax on the stern, and with a heavy load, it takes more than a little bit of power to pull it out of the "hole" and "up on the step." The torque twist is pretty strong – just crack her wide open and steer left.

A neighbouring camp (keeping ahead of the Joneses) has the same boat with <u>two</u> 200 horse Mercs hanging on the transom. I could never figure out how the driver could handle the torque until I was told that they use a salt-water kit – the two props rotate in opposite directions, cancelling out the torque.

Oscar is an accomplished pilot by now. He escorts us to the end of the lake, loses interest as we start the first portage, and goes back to Dean.

Here's the route: The first leg is fairly easy. A half-mile portage to a little loon-shit bog hole, not even worthy of a name, then a one-mile portage to a larger lake.

The portages are not well travelled but are in good shape. Our Jackfish Lake is obviously part of a well-used system, or has been in the past.

Our portage procedure is thus: We each don a life jacket. (Isn't this goofy though? We sit on them in the canoes, then wear them on the portages. I guess we are unclear on the concept.) Alan and I are about the same height, so we carry the canoe. We turn it over and sit the narrow bow and stern on our shoulders. Alan gives me the bow. I know why, and I silently thank him for it: I can set the pace and call for rest breaks. Alf and Pete each carry a motor and a gas can. We all carry one paddle each. (You never leave home without a paddle, just like you never go on a car trip without a spare tire.) At the end of the second portage I am doing OK. I am sweating like crazy, but I haven't asked for a break yet.

We quickly throw the canoe in the water, attach the kickers and shove off. It seems that Alec has told Alan that he will give us an hour head start and then he will catch up to us on the way but Alan isn't about to let that happen. Alec's theory is that although we have two motors to his one, we are more heavily loaded, will sit lower in the water and have more drag. I kind of doubt his theory as I scoped out their canoe the day before. It was in good shape, but years of use have naturally allowed the cedar strips to soak up water. Our canoe probably weighs about 100 pounds and I estimate his weighs twenty pounds more.

As it turns out Alec will overtake us. He doesn't gain much on the water – it is the portages that make the difference.

Our next leg takes us five miles down a lake, followed by an eight-mile portage, (wow) onwards another ten miles on a larger lake, then the final three-mile portage, thence westward four more miles by lake to pick up a good-sized creek for a mile or two, and then join the Footprint River three miles above Footprint Lake. There are twelve miles of portages – a tough row to hoe, but they are not rough and we are going down river, meaning every portage drops a little in elevation – we are going downhill.

But Alan is getting a little impatient. He asks if I want to trade with Alf. I shake my head. I have my pride to deal with and I'm stubborn. During a break I ask how the others, Alec et al, will split the load. Alan says that Alec will carry the canoe, his wife the motor and gas, and the boy will follow.

I make it through the eight-mile portage, stopping frequently to rest and change sides. My shoulders are getting sore and I am powering out. Alan is starting to push me a little bit. "Hastuum, neechoggin." (Hurry up, pardner.) I am looking forward to the upcoming ten-mile boat ride, but I am worried about the last three-mile portage.

I should point out that we have no map – Alan doesn't need one. Alec had described the route and location of each portage to Alan back in camp. You know, "Two miles past the orange coloured rock, around a little point, and on to the left of a big jackpine that looks like your mother-in-law."

Alan just filed it in his phenomenal memory bank. I'm not worried, and neither is Alf. This is my third season working among Northern Manitoba Cree, and believe me, every one of them knows where the bear craps in the blueberry patch.

We are one mile in on the last portage and that darn Alec catches up to us. We hear him coming a ways back and lay down our loads to clear the track. I ask Alan how he does it, and he tells me to watch as Alec passes. "He walks with his knees bent," Alan says.

I watch Alec go by, and sure enough, at every step he seems to flex his knees a bit, and the canoe sort of glides by, moving in rhythm with him. On the way by Alec gives Alan a little nod, and so does Mrs. Alec and the boy. That's it. I pack it in. I give up my end of the canoe to Alf, Pete takes two motors and I stagger along in the rear carrying two whole gallons of gas.

We motor on down the lake and creek to the Footprint River, down the three miles to Footprint Lake, turn left to go to the village, and there on our left are the footprints!

Alan shuts the motors off and we drift by in wonderment. There on a sheer, unfractured, vertical rock wall are two perfect footprints, 25 feet above the water. There is no lakeshore. The wall, about 60 feet high, rises straight up out of the lake, and almost in the middle of this dark, virtually smooth expanse, someone, sometime, had taken two steps – left, right – exactly as if he had walked through wet concrete.

These are no mere depressions resulting from a softer rock eroding from surrounding harder stone. The rock face elsewhere is unblemished and the footprints are deep, as if the moccasins, (it had to be moccasins) had sunk in two inches or so. The prints are about size nine, with a definite foot shape and flat, if you know what I mean. The heel, the arch, and the ball of a size nine moccasin-covered foot would fit right in.

Sidebar: A power dam has raised the levels of the Footprint watershed. The prints would have been underwater, so that piece of rock wall was removed, and now sits high and dry near the new village. (Google – Nelson House Footprints.)

A boulder just doesn't do it for me – the prints on that majestic wall now look bogus. I wonder why we ignore our past to satisfy our present.

I asked Alec what the footprint story was and he just shrugged. I put it down to shyness and his usual reluctance to converse in English, but I have recently changed my mind. Stories are still coming out about the terrible injustices done to children in the residential schools. They were beaten if they spoke Cree, and they were made to believe that their culture was worthless. Alec's age would indicate that he was among the unfortunates, and that might explain his reticence.

What really did me in when Alec and his family passed was that they each carried a full packsack, even the boy (Alec carried a canoe and a packsack.) They had caught up to us carrying at least an extra 60 to 75 pounds each! Astounding! Another wipe-out was that I knew what was in the packs – food. They had raided Alan's grub supply, obviously in collusion with Alan and I didn't like it at all. My opinion of the whole group plummeted, and I felt a little sick at heart. I didn't say a word, and I was later glad of that, because once we arrived in Nelson house, it didn't take me long to change my mind.

(Daily newspapers are not available in bush camps. The radio is tuned to country/western and we don't have that much interest in the outside world anyhow. When Marilyn Monroe died I heard about it a month later.)

157

Well – it turns out that Nelson House had been front and centre in the Winnipeg papers and on CBC-TV that summer. The village was near starvation due to a brainless change in policy at good old Indian Affairs, smugly complacent in their belief that they knew best (when in reality they only know how to cover their asses and cash their paychecks.) They had decided that Nelson House residents would not receive food vouchers that summer, instead they would get bullets – no shit! The Hudson's Bay store would hand out bullets and shotgun shells to the 75% of the village without jobs, but Indian Affairs had made no provisions for outboard gas and oil, or for staples such as sugar, flour and butter.

So what could the people do, stand on shore and call the moose over? "Here, moosey, moosey, c'mon over here, I need a good shot at you."

There are fish in the lake but 400 people fishing from a half mile of shoreline? You get the picture.

I was being pretty smug myself, thinking that Alec was ripping off Inco for his own benefit. Families like Alec's who had a paycheque coming in could barely support themselves due to the high HBC prices. Those heavy packs on Alec's family's back were not for themselves, nor was the visit for conjugal purposes. It was a relief mission, pure and simple. Had I been aware of anything but my own self-centred existence, I would have gladly carried a pack of our own camp food and my sore shoulders would have been an insignificant irritation.

I heard later on during a one-night stay at the Midwest camp that an anomaly chasing crew nearer to Nelson House had also been heavy on the grub that summer. The thing was, they traded the grub for certain favours, if you get my drift. Sometimes I cringe in embarrassment for my fellow men.

Table Scrap – A further item re: Indian Affairs.

Thirty-four years later I spent a lot of time with my old friend Sam Duggan before he died. He told me a lot of stories. Here is one of them.

In 1972 Sam is in the Savant Lake Hotel just after the New Year, mobilizing his crew for a diamond drill job. The six-man crew is drifting in and the afternoon train brings one of the runners, a native from Armstrong. Sam tells me he is a good man and a good driller. A while later Sam goes to the dining room for supper and soon the Indian guy joins him.

"How about an advance, Sam?" he says, "I'm all tapped out and I'd like to buy my share of the beer."

"Have a coffee while I finish eating," says Sam, "Then we'll go to my room and I'll write you a check, and the bar will cash it for you."

Just then a suit comes in and sits at a corner table,

"Hang on, Sam," says the driller, "I know that guy. I might not need an advance after all."

He goes over, shakes the suit's hand, they confab a little bit and they both go upstairs.

Sam finishes eating and re-joins the crew in the bar. The driller comes back downstairs, pulls out two hundreds, peels one off and buys a round.

Sam's eyebrows go up. "What's the story?"

"No big deal," says the driller. "He's from Indian Affairs and he's going up to Pickle to fly in to a few reserves. I told him I was going into a drill camp and needed a stake for some underwear and such. He pulled out a briefcase full of money, slipped out two hundred bucks and gave it to me. Happens all the time."

"Did you have to show him your status card?"

"Nope."

"Sign a receipt?"

"Nope, just took the money, said thanks, and that's all there was to it. Happens all the time."

Sam was astounded. "Can you just imagine the opportunity for graft in that set-up?" he asked me.

(Certain chiefs have recently been accused of running their reserves like banana republics, more to their benefit than their peoples' - but let's face it, they have been trained by pros.)

Table Scrap

In 1998 my wife and I were managing a lodge near a remote native community. The building had once served as a combined school/community centre and had been built by a Christian brotherhood (name withheld) in the early part of the century and the monks ran it until the early '60s. One day our friend Wesley, an elder with a huge store of memories, told us this story.

Saturday night was movie night and for a dime you got to watch a movie. One Saturday night a lady, who had obviously been into a batch of moose milk, was trying to get in to watch the show. A big burly brother (we'll call him Aloysius) was the doorman/bouncer and he wouldn't let her in.

"All right for you, all right for you, Brother Aloysius," she cried, "No more piece of tail for you!"

We pull the canoe out of the water in front of the Hudson's Bay Store and set off to see Downtown Nelson House – well, downtown is the HBC, that's it. Modest houses are spread along the shore in either direction, with others a little ways back – Front Street and First Street, if streets existed. The town is on a flat sort of mesa, about fifteen feet above the lake level and I estimate a population of 350 – 400 people. The shoreline is dotted with little private docks with a larger dock nearby for float planes.

(The old town, Like the footprint wall, is now underwater – the new Nelson House is now a town of 3500 with an up-to-date business district and all-weather road access.)

Alf and Alan renew some acquaintances – not many, most of the men with jobs are away from home. Pete and I check out the store, buy some snack food and have supper on the store verandah. This is definitely not going to be a wild night on the town, but we are only mildly disappointed. Night-life is not the real reason for the trip. We just need a break, and so we sit back and enjoy it.

We see the Catholic Mission on a point across the bay, so Pete and I round up Alf, who is at loose ends, and we motor over for a visit. The mission is comprised of a small white church, a small graveyard surrounded by a white picket fence, some small white

159

storage buildings, and two small white houses. It is staffed by a small priest and two small nuns in white habits

We tie up at the dock and walk up the path to the larger of the two neat houses and knock on the door. It is answered by a pleasant looking, albeit mildly surprised older priest who, after discovering the purpose of our visit to Nelson House, invites us in for a cup of tea. It is a very neat place, and very comfortable.

We have a nice chat, during which I ask him about what seems to me to be very high prices at the HBC store. He says he knows about that, but they don't shop there. They order from a wholesaler in The Pas. The order is shipped by train to Wabowden and on to Nelson House by float plane or winter road.

I ask why natives don't do the same thing with a shared meat order. He says that the day the meat came in would be the day that those involved in this obvious mutiny would find their credit curtailed, and perhaps refused entry to the HBC store. Brutal, eh?

Sidebar: However, the People were starting to wise up. Native run co-ops were starting to appear, slowly at first. By the 1980s after two hundred years of a not-so-benevolent, paternalistic and predatory business model, Hudson's Bay Stores/Northern Division, had lost their market share and shut 'er down.

At one time HBC had controlled the fur market from Montreal to the Pacific Coast, where they had posts as far south as the mouth of the Columbia River and into parts of what is now Alaska. They continually fought for territory with John Jacob Astor's American Fur Trading Company to the south and west, and with the Russians to the west and north.

An upstart calling themselves the North-West Trading Company was making strong inroads into HBC's Canadian sector, but as a result of strong political manoeuvring and even a few armed battles, they were defeated, to be assimilated by the HBC.

When HBC Northern went belly-up, a new company incorporated under the name of "The Northwest Company," bought HBC's assets. What beautiful irony! Graveyards throughout England must have ripped open as those greedy peers of the realm turned over.

Today, bright, modern, well-stocked and competitively priced Northern Stores are found in many communities, sharing the pie with private and co-operative competition.

Alf, Pete and I sleep on the floor of Alec's tidy, if spartan, self-built log bungalow. There are no extra beds, and we are not about to commandeer the kid's beds. Alan, the lucky guy, spends the night more comfortably (nudge-nudge, wink-wink.)

Early the next morning we gather at the HBC, and when it opens we buy some junk food, do a little shopping and prepare to disembark. In addition to the usual small toiletry items I make a couple of compulsive purchases – a flat-top guitar ($10) and a Buddy Holly album. ($4) How can these be so cheap, when fresh meat is so expensive? Elementary, my dear Watson; the meat is flown in, while The guitar and LP had come in by cat swing the winter before. (Later, I gave the guitar away but I still have Buddy Holly.)

160

You may wonder why I would want to haul a guitar back across twelve and a half miles of uphill portages. Well it looks like twelve of those miles can be circumvented. Before the store opened I was grumbling to Alan about the return trip and Alan said that Alec knows an alternate route. Years and years before, as a boy, he had spent many moons on the trap line with his father. Alec thinks we can go 30 miles or so up the Footprint River and pick up a creek leading into the little pothole next to Jackfish. His wife was not so familiar with the area, so had used the more overland route, and Alec had elected to stick to the same path on our trip into Nelson House. He had spent the evening thinking about the old path. He told Alan about it, and we agree to try it. (Alan has a lot of confidence in Alec – I am just lazy.)

So we load up, climb in, and wave goodbye. For some reason I can't recall, I am motorman. Alec, our guide, sits in the bow, Alf and Pete on the thwarts, and Alan builds himself a seat of life jackets in the middle. This is how you distribute the load in a canoe, anyway; light in the bow, most of the weight in the middle, and some in the stern.

With five if us in a 17-foot canoe we don't have much freeboard. We are bucking a bit of a current now as well, and our progress is notably slower. After five miles or so the current seems to get stronger. We pass the mouth of the little river we had used yesterday, round a bend, and see rapids dead ahead. They are not too long, perhaps two hundred feet and it doesn't look too rough, nor is it really fast water. In fact, we can see the placid water above, where it pours smoothly over a rock shelf as if a low dam spans the waterway. Below the imitation dam the water tumbles over hidden rocks. Alec points to the right, and I think he is indicating the best channel to take, so I head a bit toward that side and keep the motors on full power. Alec glances back at me and grabs a paddle, and so does Alan. Now I see the portage Alec had pointed to, but I am already past it, and heading into faster water. I briefly consider cutting power and turning back, but I now have enough bush smarts to resist the impulse. Had I turned our overloaded canoe, we would have been momentarily broadside to the current and most surely would have taken on water – lots of water.

Onward, onward, and slower and slower – we cross our fingers that a motor won't catch a rock and bust a shear pin. The bow is at the dam, and we are holding our own, with progress measured in inches. Finally, with more than half the canoe across, the bow settles into slower water, the brave little combined six horses give a gurgle and we are over, and we all laugh like crazy!

Now the river is broad and still as glass. We chug along through wider, lake-like stretches, and beautiful, tranquil scenery. We pass near thickly forested shores and large, grassy, reed-choked bays. Alec points out patches of wild rice, harvested by the whole fam-damily when he was a boy. We see a couple of moose, up to their bellies in the wild rice, absent-mindedly chewing on their dinner as they watch us troll by. Waterfowl with their second hatch ducklings in tow feed along the edges. They exhibit no fear of us. Wonderful!

We have stopped twice to refuel our motors, and our gas cans are empty. We left Nelson house with full motors and the cans half full, so we had used less than half of the gas on the trip south. These little three horse kickers are stingy, but now we are heavily loaded and travelling into the current, so our miles per gallon has decreased.

With our gas reserves exhausted we motor on, with Alec studying every bay on the east side of the river as we pass. It is after 4 pm now, and I am getting a little nervous. I am up to speed now on Alec's capabilities, but crikey! He's working on childhood memories! We pass a large bay and Alec put his hand up. "Hold it!" I cut the motors and we drift along as Alec checks things out. Nope, not this one, and we forge onward.

An hour later we round a river bend and a very large bay appears on our right. It is choked with reeds and cattails. Once again Alec puts up his hand and I stop the motors. Alec finally nods and points, but at what? I see no sign of a creek, just an unbroken expanse of water plants, but I dutifully head for the spot Alec indicates. What else can we do? We are low on gas, so backtracking is out of the question.

With silent, lifted motors, we paddle blindly eastward through reeds towering six feet above us. We can't see the shore. The bow parts the reeds, and we are not really paddling, we are poling with the paddles and it doesn't look promising at all.

Suddenly trees appear, and we break out of the tullabees and hit the creek dead on!

Words cannot do justice to this feat of memory. This is far beyond any hokey, over-dramatized TV plot. I could give you Google coordinates of that bay on the river, and I doubt you could find that little creek; yet Alec did it on a remembrance of times long past.

Sidebar: Mother Nature and Nanabush worked for thousands of years to give the People wild rice. Now we have built dams, flooded rivers and lakes, and have destroyed much of the rice habitat. Will it recover? And if so, how long will it take?

Google maps will not help now anyway. The bays no longer exist, and I cannot find Jackfish Lake. If I could return I would not – it would break my heart.

Higher water may eventually inundate Venice. Progress has already inundated the Footprint. Compensation has softened the blow, but money is just money, and we cannot buy back the Footprint River.

But we are literally not out of the woods yet. The motors are dangerously low on fuel. I shut one down, lift it and continue on with one motor at half throttle. The creek meanders left-to-right so we can't travel fast anyhow. The creek narrows and gets shallow, making it necessary to tip the motor up almost out of the water in some places. The motor runs dry, so I start the other. The water gets shallower so we pull that motor up and paddle, pushing aside the low hanging fir branches and tag alders. I am once again of little faith – we are five guys up the creek with four paddles. I calculate sixteen hours to paddle back to Nelson House unless the search party finds us first. If my ass isn't grass already, it sure will be if we don't make it back to camp.

But the creek is getting deeper! I fire up the motor and we soon pop into that little loonshit hole we left so long ago (only 32 hours, but it seems like 32 days.) We turn left, hit the portage, practically run the half mile to Jackfish, leaving the empty motor to be retrieved the next morning. We return to camp and to a relieved Dean and Oscar, just as darkness falls.

No problem. We dig out five T-bones and some spuds, and food never tasted so good! We are home!

(Just barely – at trips end there was only a thimble-full of gas in the last motor, maybe enough for one more mile.)

162

Alan and crew share our grub for a couple of days. Two days later the Beaver comes in, replenishes our larders and Alf, Dean and I get our marching orders. We are to finish up the anomaly we are on and prepare to move camp in two days. Life is back to normal.

So you might say that I've been there and done that. I've seen, done and heard about things that only a small percentage of our 400 million Canada/US population have experienced, but that two-day trip to Nelson House stands out above them all!

Sidebar: I've bragged up our trip into Nelson House, but what about Alec's wife? She carried that heavy canoe over the same portages, uphill, and with a 12-year-old boy as her helper. How about that?

The Last Camp

We have the camp packed and ready when Norm comes in. He gives me the new maps, and guess what? Our next camp is to be located on the east shore of a lake we had been on only a few days before. I don't tell Norm, of course, but he must be surprised at how quickly I select a campsite when we circle the lake. What he doesn't know is that we had passed within 200 feet of the spot on our Nelson House trip. It's just plain natural to scan the shoreline while travelling, and I have been around long enough to unconsciously compare potential campsites.

We have no rock point this time, just a sparse unbroken timber line with two-foot high banks and an almost underbrush-free plateau rising gently to outcrop about 200 feet inland.

The camp goes up lickety-split. Dean is now an old hand at it, and by nightfall we are comfortable. Our tent is only ten feet from shore, and we have built a simple, very simple dock with jackpine poles supported by two three-inch posts driven into the lake bed about six feet out. "Good enough for us." I think, but it will prove to be a little short of Norm's standards in a couple of weeks.

Naturally, Oscar comes along. Well, not quite naturally – he flies in the Beaver, although a Stinson Reliant gull-wing might have been more fitting for Oscar's maiden flight. Norm won't hear of one of us carrying the bird, so we pack him in a cardboard egg crate. Oscar is cool with that. He always seems to know that his adopted parents would never do him harm. The drillers stay behind to finish their Jackfish work and will join us in two weeks.

Three of our targets require a canoe ride. The rest are to the east of camp and we will walk in from our tent. I elect to do the lake work first. We are getting a little faster, but with rain delays it takes us twelve days to finish the canoe work.

Oscar escorts us every day. On the morning run he flies point, scanning the lake around us for periscopes. He circles as we off-load, calls goodbye as we head into the bush, and spends the rest of the day doing Oscar stuff. When we fire up the motor on the way home, Oscar appears within minutes, peeling around a point in a steep bank to settle on the bow. He chats happily for a moment or two, then turns around to do his hood ornament impersonation, balancing with spread wings. He is our winged angel, (with apologies to the old Packard Motor Car Co.)

163

A couple of days later Norm shows up unexpectedly, just as we return from the bush. We tie off the Beaver cross-wise at the end of the dock and Norm climbs out the passenger door, steps down on the float, and eyes our class "C" dock with distaste. He is kind of grumpy-like, which is not the norm for Norm. He announces that he has gas for the packsack crew who will be arriving the next day. I am mystified. Our gas always comes in ten gallon barrels, not so hard to handle. He opens the cargo door and I can see why he is so upset. Wray has shipped in a 45-gallon drum weighing almost 400 pounds! To load or unload a drum of that size from a good strong dock is not a problem. A camp of any size has such a dock and always has a ramp available, perhaps eight to ten feet long and strong enough to support the weight. At Cook Lake they loaded the drum by rolling it up a ramp – not easy, but do-able.

We make a short ramp out of a couple of poles and a piece of spare plywood, but I can see that we have a problem. If we let the drum roll down onto our flimsy dock, we might as well toss it into the lake, because that's where the whole shebang is going to end up anyway.

It turns out Norm has a Plan B. The Beaver has two recessed rings mounted in the floor of the cargo area near the fuselage wall opposite the cargo door. I knew those rings were there, but had always assumed that they were used only to tie down cargo. Norm pulls a long length of 5/8-inch rope out of the rear storage compartment, ties an end to each ring and lays the rope on the floor, leading out the door. He gives me the left-over rope to hold and lays the gas drum down on its side so that the ropes are about a foot from each end. I give him back the extra rope and he hangs a loop over each shoulder. He reaches back and brings the rope criss-crossed around his body under his arms, then holds the ends criss-crossed in front of him, with the extra rope at his feet. (Are you still with me?)

Then Norm pushes the barrel out the door and with each foot braced against a doorpost, eases that 400 lb drum gently down the ramp, playing out the rope as the drum descends. Alf and I stand in the water on each side to steady things if need be, but the old expert has it well under control. Alf and I roll the barrel along the narrow dock and upend it on shore. Mission accomplished!

Norm is not a young man – he is pushing 60 for sure, but let me tell you, strength plus experience is hard to beat. Norm is almost two axe handles wide at the shoulders, and I can imagine him 30 years younger. He wouldn't need a rope. He would have just hoisted that drum onto a shoulder and carried it ashore. Anyway, as I have said before, you learn something every day.

The next day we hear the drill camp arrive. When we get back to camp we find them set up next door. We are OK with that. We like the company, and besides they have rebuilt our dock. They also have a mascot now. In the spirit of competitiveness, they have upstaged us (they think) with an eagle. I don't entirely agree with stealing an eagle chick. Seagulls are plentiful, but the word on the street is that the Bald Eagle population is under pressure. Was the use of DDT blamed? I can't remember, but we have seen few eagles this summer.

I give the drillers credit for bravery though. Eagle parents don't take kindly to home invaders and those talons can do real damage. The boys staked out the nest which was at the top of an old bull pine. (Bull pines have lots of good sturdy branches.) When

164

mom and dad were both out hunting Alec scampered up the tree, snatched a chick and scampered down. They wrapped the chick in a jacket and the parents never caught on – they still had two in the nest. (Since then, thanks to TV, I've learned that if bald eagles have three chicks, one is likely to go short on grub and will not survive – so maybe the drillers didn't do such a bad thing after all.)

But the eagle is a dud as far as I am concerned – he has no personality and is basically an uncommunicative, surly teenager. I think eagles must be much more reliant on their parents than seagulls, and for longer. This guy is in full juvenile plumage, but is unwilling to fly. When tossed in the air he just flaps his way clumsily to the top of a nearby poplar and then sits there glowering, and refuses to budge. Food offerings are ignored, and finally the tree is chopped down. The eagle rides it almost to the ground before he jumps off, lands nearby and waits to be picked up.

Our boy Oscar, it turns out, has a mean streak. It must be instinctive, because he didn't get it from us, but he hates that eagle. He teases him unmercifully, bugging and chattering at him on the ground and taunting him when the eagle is in the treetops. I've never seen an eagle rob a seagull nest, and no wonder. Twenty Oscars would deter any robbers.

The story has a somewhat happy ending. The drillers know they can't return the eagle to the wild so they ask Norm to talk to Moak, and Moak talks to the Department of Lands and Forests, no doubt lying through their teeth – its a big no-no to mess with an eagle's nest. The eagle is put into a little box with a nice lid and plenty of ventilation, flown out to Cook Lake, and Wray puts him on a plane to Winnipeg. Mr. Eagle will live out his days in relative comfort in the Winnipeg Zoo.

With the canoe work finished we will now do the anomalies behind camp. There are many responses to cover and all are in the same general direction, so I decide to start with the longest walk. Thus, as we knock off targets, our walks will decrease and we can possibly use the same trail every day.

….We head in to our farthest target, and two things surprise me that first day:

#1 – I can't believe the bush, or rather lack of bush. We cross at least a mile of flat outcrop with patches of thin overburden, The widely spaced trees are picket-sized jackpine and poplar, struggling to survive with hardly any topsoil to support them. Underbrush is sparse, though occasionally we cross through stands of spruce and alders growing in slight depressions, and of course there is the Labrador tea and moss that are always present in wetter places. But these are small, isolated areas. We pass through them quickly, and are on outcrop again. "Piece of cake," I think. How wrong I am, am I.

#2 – I have never walked across such interesting rock, before or since. It is full of quartz stringers, criss-crossed with diabase dykes, and there is gossan everywhere. There is evidence of folding and changes in strike and contacts of rocks of different types.

(I can use these terms now, because I have over the years gained some skills in rock and structure recognition – not much to be sure, "some" being the operative word. In 1962, I only knew this: something had gone on here hundreds of thousands of years ago, and it looked good. It turned out to be too good for me to handle.)

165

So we blaze a straight line into our first target, set up and immediately find a nice conductor. We establish a base line and continue on with the grid for the next couple of days. By day four we have a 3200 foot long grid completed, with nice mag association and with two parallel conductors, one on each side of the first. I leave the first conductor open at each end and extend the grid to cover the other two. They also have good mag association. I cover the east conductor first, do enough on it to satisfy myself, work on the extended grid nearer to camp, and find another parallel conductor! Now I have four conductors on one grid, and I'm losing location control. Remember – this is a compass pace-and-blaze grid and as the grid gets larger, errors are magnified.

As I do my map work in camp, I realize I have covered my next target on the map. I am working my way back to camp and have not run out of conductors yet. I decide to move a little closer to camp to the third target. We find the conductor and establish a grid, but the baseline has a different azimuth than the first, no doubt caused by folding. Once again there are parallel conductors, and the east side of the extended grid encroaches on our first grid, but on a different angle. Added to my problems is the fact that the conductors sometimes change direction very sharply, requiring a change in baseline direction as we continue to trace the conductor.

I am becoming more frustrated daily. It isn't the crew's fault – Dean has become a competent mag man, the sparse bush means that he can work on his own without fear of getting lost, and Alf is his usual sturdy self. I blame myself – I feel I am in over my head. We have been days and days on this small area and I cannot explain the situation properly to Moak. Once again our production is suffering, and Moak is getting hostile. I go on the sked and ask to be replaced. I want to go home.

Moak refuses to pull me. The next grub order comes in and with it is a long letter from my supervisor. Among other things, he suggests that I might be better off in another line of work. Now I am really dejected, although I should have been irate! I should have held Norm hostage at axe point and forced him to fly me to Moak! I should have stormed into the supervisor's office and punched him in the face! But my confidence is shot all to hell. I literally give up and spend the next few weeks going through the motions – definitely not one of the high points of my life.

Sidebar: For fifty years I held myself responsible for the failure to do an adequate job in that multi-conductor zone. Then I met Bob Marvin, a meeting that was a mixture of happenstance, curiosity and good Karma.

When we moved back to the Rainy River Valley in 2005 I began hearing rumbles about a gold play in Richardson Twp. sixteen miles northeast of our new home. An old friend, Don MacEachern, was a partner in the initial find and I gave him a call. He came out from Fort Frances and we did a tour of Richardson.

We checked out a dead-end side road and at the end was a cluster of three Atco units and a sign – "BAYFIELD VENTURES CORP. GEOLOGY AND SAMPLING FIELD OFFICE."

Don dropped me off at my place, and I hit the internet.

Sidebar within a sidebar: Don seemed kind of fuzzy that day, and I called his wife soon after he left for home. Was Don OK? He was <u>NOT</u> – he was slipping into Alzheimer's.

For the next six years I visited Don and his wife often until he mercifully passed away in 2014. He had been out of Richardson and the prospecting game since 2002. He left this world quietly and left his family in good shape financially. Don did it right – and I'm sure he somehow knows that.

Sidebar again: So I get on the internet and find out that the Bayfield Prez is Don Huston! I know Don! (Be patient, stick with me and the story – the overlapping Olympic Rings will eventually reveal all.)

The next day I hustle my butt back to the Atco camp and introduce myself. Don is in Vancouver. The field geologist is Bob Marvin, and right off the bat we click.

For the next seven years I will visit the core shack on a weekly basis, so between Bayfield Bob and visits to Don MacEachern, "The Price is Right" moves to third place on my busy schedule.

I told Bob about the multi-conductor zone near Nelson House. Initially the conversation centred on the interesting possibilities of said zone. When I reluctantly brought up my feelings of failure we exhumed the cadaver and did a cold-case forensic investigation.

Bob felt that Moak was the prime suspect. They should have recognized that the area needed a cut, chained and picketed grid. I was found "not guilty" and received a pardon fifty years after being convicted.

Bob became, and still is, a close friend. He is a pretty good guy despite being one of those Goldurn Murricans.

Sometimes in the evenings, I sit near the lakeshore, smoking and contemplating. Oscar knows I am troubled, and he sits beside me, mumbling seagull words of sympathy as he grooms his feathers. Oscar is never judgemental with his family – he loves us unconditionally.

But we are losing Oscar. It is early September now, and the seagulls are heading south. We haven't seen any large flocks, but every morning there are groups of three or four heading south down the lake past camp. Oscar is getting edgy. He flies out to meet them, travels with them a little way, then returns to camp and hunkers down, mumbling and eyeing us quizzically. We know exactly what he is thinking. "C'mon, folks, the good parking spots will all be gone before you dummies get going."

We have been worrying about him. What if he hangs in with us? He has to leave. It's like raising a kid and saying goodbye as he starts his own life. You are happy for them, but you still feel an emptiness inside.

For about a week Oscar repeats the process, but goes a little farther every day. Then one day we return to an Oscarless camp. Our boy has gone, never to be seen again. We are a pretty quiet group, each with our own memories of a good friend and I have lost my confidant. It is an indication of my mental state at the time. My best friend is a seagull, for goodness sakes.

167

Sidebar: I have often pictured an older snowbird couple sitting on their beach blankets by the ocean, and Oscar walks up to have a chat. The old lady turns to the old man and says, "I swear, Reuben, that bird acts almost human."

Right on, lady. He was almost human.

I've also wondered if Oscar ever returned those camps in the following years to look for us. Nothing would remain but some weathered tent poles and a few tin cans. I hope the fishing was good down in the Gulf. Oscar always had too much class to pick through the garbage.

Goodbye, my old friend. I hope you lived well and prospered.

I may have given you the impression that this camp was a total downer, but you can't stick six young men in the remote Manitoba bush and expect them all to sit in the corner sucking their thumbs.

One evening in August the six of us are sitting around after an early supper drinking coffee. It is a lovely evening, the lake is smooth and inviting and the water is still warm. Someone says, "If we only had a boat it would be a nice night to go water skiing."

Well, what the heck? We have two honkin' three horses and a V - stern canoe – we swing into action. We hang the motors on one canoe and say, "Hmmm...Where's the skis?"

No problem – we have the piece of plywood we used for a barrel ramp – we will surf-ski.

We pull the nails and punch three holes in one end. With a short rope for the skier to hold on to and a longer rope tied the centre hole going to the canoe transom, we are ready. (It is a pretty poor camp that has no extra rope.)

Surboard Skiing
Alf Settee is doing a poor job of holding down the bow.
I am at the controls of the twin-thrust turbo 3-horses.
Allan Paupenakis is the graceful guy on the plywood.
That's a pretty good wake considering we are moving
at the pace of a walking dog.

We all take a crack at it. One man stands on the board with the rope in his hands. The canoe, with one man in the front to hold the bow down, takes up the slack and the motor-man pours on the power. It works! The board comes up and we have lift-off! Actually, it's more like slo-mo snow plowing. There isn't enough juice to really get her up on the step, but we don't sink, either. Alf, Dean and Alec are the lightest. They can even turn if we keep the circle wide enough and if they follow directly in the wake. If they go wide on the turn the canoe slows to a standstill. Alan is heavier –we can only pull him straight out, turn, and pull him straight back.

I am a boat anchor. We even try taking a run at it, but as soon as the rope snaps taut the canoe stops, almost swamping at the stern and I just sit there like a rooted stump. I quit after a couple of attempts, afraid we will pull the transom off the canoe.

So I drive. We burn up all the gas in those two motors and have a great time. We all laugh like crazy.

M y right-hand man in the summer of '62 was Alfred Settee from Cross Lake. I've given a lot of credit to Oscar for getting me through those tough times, but Alf was equally responsible for maintaining my equilibrium.

Alf, like many of the People of Northern Manitoba, was intelligent, hardworking, and not a heavy drinker. But also, like many of his compatriots with a steady paycheque, he was unable to deal with the do-nothing folks on his reserve. If Alf had money, everybody had money. If Alf had food, everybody had food. If Alf had a case of beer, everybody had a case of beer, and that was a problem. Alf told me that on one Christmas break he had rented the Bombardier taxi to take him into Cross Lake from Wabowden. A few others from Cross Lake were in Wabowden, and Alf paid the taxi and took them in with him along with Alf's groceries and Christmas presents.

Now 24 beer would have lasted two weeks in Alf's house, but by the time they had left Wabowden the others had talked Alf into filling up the empty space with more beer – 20 cases more. Alf said that the moccasin telegraph preceded them to Cross Lake, and he was greeted by a house full of "friends" who stayed until the beer and food were gone.

Finally, in an attempt to break the mold, Alf bought a surplus CNR section house, and at no small expense, had it hauled in to Cross Lake by cat swing on the winter road. Alf set it up three miles out of town, hoping to put some distance between him and those so-called friends. I don't think it worked out for long. In '88 I ran into some Cross Lakers in Bisset and asked them about Alf. They didn't know him well, but they said he still lived there and was now back on the reserve.

I always thought that Alf would have been a good, steadying influence on the Cross Lake or even provincial political scene, but knowing Alf, I surmise that he just wouldn't fit in. He had his own definite views of how his race should handle the negative aspects of the white man's world. He would not be a team player for sure. Because Alf was unique in his approach to northern life, I believe he became sort of a loner, if you know what I mean.

Alf and I traded letters for a couple of years, but to my regret we fell out of touch. My fault – wine, women and song and my broadening horizons got in the way. It was my loss, and one of my few regrets.

169

By the middle of September Moak has had enough, and so have I. The camp is shut down two weeks early and we are pulled out. Alf and Dean go to Moak, I go straight to the Midwest camp to wait for the next train south. Moak doesn't even wave goodbye.

The next morning Wray drives me to the station and I board the train to Winnipeg to stay a couple of days with my sister. I plan to bus it home, but an ad in the paper catches my eye. Transair is establishing a scheduled flight to International Falls, Minnesota.

At that time the Fort Frances airport was only a flying club field, unsuitable for commercial aircraft. International Falls, our twin city on the Minnesota side of the Rainy River, had a paved runway and a control tower (sort of). When a commercial flight arrived, the airport maintenance guy would climb onto the roof of the airport building, take the mike off it's hook on an air vent and talk them in. I'm not kidding! Elbonia Airlines!

I make a phone call, and sure enough, I can fly out the next morning. My sister drives me to the airport, I plunk down my 32 dollars, and board a modern DC-3. That's right, a DC-3! Everything's up to date in River City.

(Thirty-two dollars was pretty pricey – the bus was seven dollars. Nowadays 32 dollars is ante-up money, but then it was ten percent of my month's take-home pay.)

Actually, the passenger cabin is pretty nice – clean, comfortable and lots of leg room. I am surprised by the number of people boarding, ten or twelve well-dressed men and women chat with each other as they buckle up. It seems that things must be hopping in the border area.

After we get airborne, the suits (even the ladies wear business suits,) mingle, laughing and talking and ooh-ing and ah-ing as they shmooze and check out the scenery below.It takes a while – I'm not the sharpest post in the claim block, but I finally clue in. My fellow passengers are Transair execs, government hacks and media people. I am the only paying passenger! The stewardess, very young, very pretty and very nice, is leading people to the cockpit to chat with the pilots and even sit in the co-pilot's seat if they want to. When she gets back to me I politely refuse. I am an insufferable snob, the old prospector who's been there and done all that. I stare out the window, wrapped in my cloak of invisibility, wishing I'd taken the bus.

(Within a month Transair will abandon the run. I think I was their only paying passenger.)

We arrive at the Falls airport, mission control talks us in and we deplane, wave to thousands of screaming teenagers and my brother meets me and drives me home.

The Terrible Summer of '62 is over.

Tutorial: The Cat Swing
Back in the day, northern communities were supplied by a winter road system. Unlike Hector's Highways, these roads were for winter use only, and therefore made use of lakes as necessary.

Caterpillar trains, aka cat swings, hauled everything that you can imagine: Groceries, fuel, (either in bulk tanks or in barrels,) building materials, boats, motors - you name it.

170

The trains were exactly that – a string of heavy sleighs, one behind the other, pulled by a caterpillar tractor.

After freeze-up when enough ice had formed to carry a bit of weight, smaller cats would prepare the roads, plowing snow off the ice to speed up the freezing process and packing and plowing the connecting overland trails. The end result was a smooth, wide road avoiding any steep hills. River crossings were made where the current was slow and would be periodically flooded to maintain ice thickness.

The trains were pulled by a caterpillar tractor large enough to supply the needed horsepower, but not too large, as weight versus ice thickness was a critical factor. You would usually find a Cat D-6 or an International TD-9 at the head of the swing. A typical swing would be made up of three sleighs, maybe four, if lighter, bulkier items like fishing boats or furniture were included in the shipment. The sleighs had either wide flat decks or enclosed bodies for cases of canned goods (or guitars.)

On longer trips more than one community might be supplied. Two cats ran in tandem with the cat skinner (operator) on the second machine. He had a system of ropes to control the lead cat: one each for the steering levers, two apiece on the throttle and clutch engagement lever. It must have been quite an act to master. With two cats, seven or eight sleighs could be pulled. At the first unloading point one cat and the empty sleighs were dropped off, to be picked up on the return trip.

In Manitoba the winter roads started at the nearest rail point or all-weather road. Flin Flon or Cranberry Portage was the terminus for the western side of the province. The rail line from The Pas to Churchill serviced a large part of Central/Northern Manitoba from stations along the line. Wabowden and Ilford were busy stations. East of Lake Winnipeg was accessed from Riverton, and the North-east Triangle was serviced from Red Lake and Pickle Lake in Ontario.

Let's take a trip with the boys. We are going to join a swing from Wabowden to Nelson House. Our sleighs have been loaded from a rail car on the siding. Forty-five gallon drums are secured on the first sleigh (a couple of them are for our own supply of diesel for the cat.) The fuel load also includes ten 100lb tanks of propane. The third sleigh has two 16-foot aluminium boats, one canoe, a sofa and matching chair, a kitchen table and chairs and some other stuff. There are also some outboard motors, two ten-horse and one 25-horse. This load has been covered by tarpaulins. The second, enclosed unit, is loaded with cases of canned goods on the bottom with lighter miscellaneous articles piled on top, right to the ceiling. My guitar and Buddy Holly record are in there.

The more heavily loaded sleighs are always at the front. This makes it easier for the sleighs to follow the cat around the corners.

Last in line is the caboose. This is the bunk-house/kitchen. There are four beds, one for each cat skinner (this is a 24/7 operation, they trade off every four to six hours), one for the cook and one for a trainee or hitchhiker.

The cat is warmed up, the skinner is in his seat, and away we go across Setting Lake. We rattle along at 4-5 miles per hour (full throttle) if the ice thickness warrants it. Earlier in the season we may have throttled back, and we certainly slow down when we near the shore. Our weight creates an almost imperceptible wave in the water beneath the ice ahead of us, and like a Tsunami, it is amplified as the water gets shallower. If allowed to build, it has more than enough force to blow the ice up ahead of and under us – and Caterpillar tractors are wonderful machines, but they don't tread water very well.

Beyond Setting Lake we will be mostly on bush roads until we hit Three Point Lake, where we take a short hop across a neck of land to avoid fast running water, then swing back out onto Footprint Lake three miles from our destination. The trip is 60 miles, and what with rough sections and one stop to refuel, we arrive at Nelson House just before midnight. We hit the sack.

Just before 6 am we arise, have breakfast and unload at the HBC warehouse. Some items such as the sofa and chair were ordered by individuals, and are carted off by helpful youngsters on the spot.

Freight is charged by the pound/mile, and bulky, lighter items such as a boxed lampshade are cubed (weight times three). Boats and canoes have to pay for the square footage of space they use.

We are almost finished by lunch break, and by 2 pm the skinner is bundled up and we head back to Wabowden.

The cats have no cabs. If a cat goes through the ice, the skinner does not want to go down with it. He has some protection from the elements provided by a "heat houser" that covers both sides of the motor shroud and extends back beside his seat, directing warmed engine air past him. He has a small opening facing the road ahead, and if he has the XL option package, he might have a plastic windshield. We in the caboose are not worried about following the cat into the water. The sleigh tongues are designed to disengage when the angle of pull is 45 degrees below horizontal.

We are running light on the return trip, so we arrive back at Wabowden at 3 am, with only one unlucky/lucky incident. Around midnight the cook steps out on the rear balcony of the caboose to take a pee, and closes the door behind him. The sleigh hits a bump and he falls off. Dressed only in long johns and socks, he runs along after the swing, hollering like crazy. He is no spring chicken, and he can't catch up and we can't hear him. We are sound asleep, and the skinner is a hundred feet away, behind a roaring diesel engine.

But the cook is lucky. The driver has also had one cup of tea too many and stops for his own pee break. Tragedy is averted. (This actually happened once. My sister's husband's brother told me the story. He also told me than when the old, cold cook climbed into the caboose they all laughed like crazy.)

Winter road season in those days would run from January first to April 15th, weather permitting. The Wabowden swing would make at least two more trips into Nelson House, and several to Cross Lake, a larger community 35 miles south-west of Wabowden. Wabowden and Ilford also had Bombardier taxis available for personal transportation.

But change was in the air. Within a few short years natives were building the swing roads, and white swing operators had to submit a competitive bid to get the winter freight business. One outfit, in a hissy-fit after losing his run to aboriginals, refused to sell them his sleighs. Since he was the sole manufacturer and supplier of these units, it put the locals at a distinct disadvantage. He forced their hand, and of course, Plan B swung into place.

Why not use trucks? The Thompson Highway was finished by now, and other short access roads were being built.

Nowadays, winter roads are used by all sorts of vehicles, and Hissy-Fit's equipment is still rusting in the weeds.

Chapter VI Fall Season '62 - Redemption.

It is October and I am heading to Sudbury via Duluth in my '55 Monarch with my packsack in the trunk. I am not exactly eager to get to Copper Cliff. I know there will be music to face, and sure enough, when I check in with Herb, he gleefully lays the hammer down. I am demoted to mag man and will spend the fall at Cochrane with Barry Krause, who will be in charge of three 3-man ground crews. A large airborne survey has recently been completed covering an area west and south of Highway 17 (sort of a quadrant: Hwy 17 runs north from North Bay to Cochrane, then takes a sharp left to run west towards Port Arthur/Fort William (now Thunder Bay) Ontario.

Sidebar: Some explanation here: In the U.S. Ford had a one dealership system – Ford/Mercury/Lincoln/Ford Trucks. In Canada we had a two-dealership system – Ford/Monarch/Ford Trucks, and Mercury/Lincoln/Meteor/Mercury Trucks. Monarchs were manufactured from 1948 to '61, only cosmetically different from the Merc. The Meteor was a slightly upscale Ford, always with its own front end treatment and models, and even using some Mercury body components. In '49 and '50 they used a modified Mercury grill, to the chagrin of California customizers. The West Coast boys had to spend a lot of time working sheet metal to fit a Merc grill into a Ford front end. We Canadian boys could buy the same thing right off the lot! In '53 and '54 the top of the line Meteor Niagara even used the Mercury dashboard – cool! As a teenager I once owned (for a very, very short time, until my dad made me sell it,) a whipped '51 Monarch convertible, one of less than 800 produced. I would also soon trade my '55 Monarch for a '60 Meteor Skyliner, one of 2300 manufactured. I was born a Canadian and I will die a Canadian, but sadly I can no longer buy a uniquely Canadian vehicle (or much else made in Canada.)

(Incidentally, from '52 to '57 Meteor outsold Ford in Canada.)

I've won the lottery this time. Not only is Barry my boss, he is also a personal friend. He tells me right off that Moak had suggested returning me to the wild, but he (Barry) had told Herb that he had no problem taking me to Cochrane with him. I am not about to let him down. My second lottery win is that my party leader will be Henry "Pass Me Da Hax" Levac. I am pretty nervous about meeting the legend, and am just as nervous about working with him.

No problem. We like each other immediately and develop a mutual respect. Henry is in his mid-forties and has slowed somewhat, they tell me. Yeah, right – but by putting the pedal to the metal, I am able to hold my own with him.

(Henry never let on, but one night in the bar four weeks or so later, he tells me that Barry wanted him to keep an eye on me. After working with me for a week, Henry told Barry that watching me was no problem – I was usually ahead of him.)

We just flat-out tear up the bush. It takes about two weeks for the other two crews to realize they are out-gunned. Now, they also have to work harder and their resentment shows. We don't give a rat's ass, we are having fun, and so is Barry. (High production never hurts a supervisor's rep.) I can't remember who the third man on our crew was.

174

We use them up faster than a blonde typist uses white-out. I think the only one that stayed was Arnie, a native from the Georgian Bay area.

We are a chopper camp. There are very few lakes in the area, just bush – heavy bush and tall bush, short bush and wet bush. This requires a lot of pad-cutting, but we have a great pilot. He always finds an area of sparse vegetation within reasonable walking distance of our target. He takes care of us, and we do our best for him.

Inco has put us up in a motel on the banks of the Abitibi River five miles west of Cochrane, a clean, pleasant town of 1500 – 2000. (Cochrane is now a clean, pleasant town of 6000, and soon to be much larger due to the opening of the Detour Lake gold mine 40 miles east of town.)

Cochrane has a full service business district, a bowling alley, a nice bar and lots of pretty girls. We fill every room at the motel and eat all our meals in their dining room. It is quite a bonus for the owners – Inco will be here all winter.

I like motel/hotel living; a shower every day plus access to a laundromat is great, even more so in the fall. But within a couple of weeks I begin to think we are a little too close to Cochrane.

I have heard many times that Henry was a hard charger in the bush, and he certainly is, but Henry loves the bar scene, and Henry never drinks alone, and never drinks with anybody else from camp – just me. I thought I could hold my own at a bar stool, but I am no match for Henry. His philosophy is, "Drink all night and never get drunk, work all day and never get tired." I can handle one or the other, but not both - I am sure getting tired. Besides, I always try to send some money home to Mom, but Henry is keeping us broke. By the end of October, he cuts back a little, and we start to get some sleep.

The Cochrane project is put on hold on December 7th for the Christmas break. Most of the men, myself included, have enough days in the bank to last us until Jan 1/63. I stop in at Copper Cliff, find out that I am to be returned to party leader status, and toodle off west to spend Christmas with my folks.

Inco has given me a clean slate. Thank you, Barry Krause, and thank you, thank you, Henry Levac.

Bill Barilko

This is truly an epic saga spanning sixty years, and will touch us briefly at the Cochrane camp.

Bill Barilko was a rawboned, hard-hitting defenceman for the Toronto Maple Leafs. In his fourth year with the Leafs he scored the Stanley Cup winning goal in overtime, still the most widely viewed of all internet NHL photo images (even eclipsing Bobby Orr's own airborne after-goal flight.)

In August of '51 Barilko and Dr. Henry Hudson left South Porcupine (near Timmins) on a fishing trip to the James Bay area in Dr. Hudson's personal aircraft, a Fairchild on floats. Hudson was known as a competent pilot, but a bit risky at times regarding fuel. On August 25, 1951, Hudson and Barilko left on the return flight to South Porcupine, and disappeared. A massive air search, the largest and longest ever undertaken in Canada to that date, failed to turn up any sign of the Fairchild. Hudson and Barilko, a

175

shoo-in future hall-of-famer, were gone. The official story is that Ron Boyd, a helicopter pilot, discovered the crash site in '62.

Well, my apologies to the Timmins Centennial Committee, but I must disagree. Ron Boyd was our chopper pilot. He didn't find the crash site, but he did confirm it. A fixed wing pilot returning to Timmins from Mosonee that fall got a glimpse of colour in a stand of tall, dense timber 62 miles north of Cochrane. He circled the area and decided he had definitely spotted the wreckage of a downed plane. He marked it on his map and flew on to Timmins. As there were no recent reports of downed aircraft, could this be Barilko/Hudson? The area was inaccessible to fixed wing – there were no nearby lakes. They needed a chopper and we had one.

Phone calls were made. First to Barry: "Yes, we have a chopper." Barry called Copper Cliff. "Can we fly someone in?"

Copper Cliff called Dominion Helicopters. "What are the insurance liabilities?"

Dominion consulted with their lawyers. "Copper Cliff must sign a temporary waiver, the Timmins people must sign a short-term charter agreement and foot the bill."

Copper Cliff gave the thumbs up, but our work must not be disturbed, Timmins had to provide a couple of men.

This gave Boyd a window of six hours to fuel up, fly in 62 miles, locate the wreckage, find a landing spot (he put them down 2 1/2 miles away,) wait while they hustled in to collect evidence and hustle out, return to the motel, and refuel again. We went to work earlier that day and returned later.

Enough personal effects were brought back to confirm that it was the Hudson/Barilko crash site. There was no indication of fire. The Fairchild had run out of gas, and had gone straight in (explaining why the search eleven years before had failed to see the wreckage; you had to be directly above it to see it.)

In 2011 Timmins celebrated its centennial. A crew using a large VTOL chopper retrieved the wreckage, flying it out piece by piece to Cochrane to be loaded on a flatbed truck. The radial engine had to be dug out of five feet of muskeg, added evidence that the Fairchild had been in a vertical dive.

On the way back to Timmins, the truck made a short side trip to the float plane base at South Porcupine. Water from the lake was poured over the mangled floats.

It took them over 60 years, but Bill Barilko and Dr. Hudson finally made it home.

Sidebar: I may be leaving the impression that Inco was sort of heartless. Not so. When necessary, they could step up to the plate for sure.

The following winter Barry got a call from (company name withheld.)A worker in a logging camp near Lake Abitibi had been hit by a falling tree and had suffered serious head injuries. Their 25-mile winter road was passable, but very rough and slow. The man would not survive the trip. Would Inco chopper him out to Cochrane?

There was no hesitation. Barry told the pilot to warm 'er up while he called Copper Cliff to tell them he was going in.

"Of course you are," said Copper Cliff.

They flew in and delivered the injured, unconscious man to the Cochrane hospital. Barry cradled the man's head in his lap, supported by a couple of camp blankets. The

176

man's injuries were stabilized, he was transferred by air ambulance to Toronto, and he survived.

What thanks did Inco receive? Two weeks later Barry got a bill from (name withheld) for two blankets – two blood-soaked blankets that had been burned along with Barry's ruined shirt and pants!

177

Chapter VII   Winter of '63 - Bancroft – Bissett

This is one thing I can count on in life. It's New Year's Day and I am on the road again.

I have two passengers – a recently married couple in their late teens. Her sister was a classmate of mine and now lives in North Bay. I will see to it that the kids get there in one piece. We will do the usual drill around the south shore of Lake Superior.

The young lad is worried - he's been a bad boy and two years before had been barred from re-entering the US. I tell him they won't remember his face. Customs people deal with thousands of faces every year.

We cross the bridge at 6 am on a frosty New Year's morning. There is an old gent at the checkpoint – "Oh-oh," says the boy.

The officer takes one look at him, disappears for a minute or two and comes back. The girl and I can cross, but the boy cannot. I am amazed – two years, and the old fellow had recalled the face – no flies on this guy!

Now, I know a bit about the kid, and I know his family. This is an important trip for the young couple. He is apprenticing as a mechanic and will continue on to Toronto for a semester of classroom work. She will stay in North Bay with her sister.

The border was kinder and gentler in those days. I ask the officer if I can have a five minute one-on-one with him – he agrees. I go in and explain the situation. I tell him how important the schooling is for the young lad. I point out that we will go non-stop to the Soo and will be back in Canada before nightfall. I make a veiled reference about an addition to the family – probably a good guess on my part.

The sharp-eyed old gent is also a compassionate guy. We are allowed entry! I pat myself on the back – no muscle spasms now – I have the pat down pat.

With one hurdle cleared, we head for Duluth, and my '60 Meteor starts to act up. She had shown signs of a problem on my trip home before Christmas. I had crossed my fingers – hoping she would heal herself, but it is getting worse.

It is a bad head gasket. She can't help it, but she is losing anti-freeze. We stop to replenish the rad when we run out of heat, the stops becoming more frequent. We limp across the border back into Ontario with 200 miles left to get to Sudbury. With thirty-five miles to go the 332 drops a valve and chews up a piston.

We arrive in Sudbury on a hook. There is only room for one in the wrecker, so we freeze as a group. I call Chibougamau Grant and borrow his car to deliver the kids to North Bay. The Meteor goes to drydock in need of a transplant.

Sidebar: I don't know why I bothered with wheels in those days. For eight months in any given year the car would be parked in a friend's driveway or left at home. The other four months I would use it – if it was running. I should have been investing in International Nickel instead of Detroit Iron.

However, this is shaping up to be a good winter. I am taking two men to Bancroft. Our area supervisor will be Jim O'Neill, an Inco geologist who is nearing retirement age. The two men with me are Carl Haglund and Norm Armstrong. I had met Carl in Sudbury in April/May of '62 in the Bushwhacker's Benevolent Society club room at the Belvedere Hotel. Actually Wes Marsaw had met

178

Carl, introduced him to me, and the three of us had spent more than a few evenings solving the world's problems over barley sandwiches. Carl was a sheet metal union man at the time, working in construction. He was making a good dollar when he worked, but constant union bickering and wobbling were affecting his paycheque and peace of mind.

Carl had spent some time in the Gowganda with Hasaga Gold and had also worked one season with an exploration syndicate. So with Wes' and my recommendation (more likely despite,) Carl joined the Copper Cliff field exploration crew.

Carl and I became close friends that winter, a friendship that has stood the test of time. We are different in many ways – Carl is built like Deputy Sheriff Barney Fyfe, I am 6'5 and 245 lbs. Carl is smart, inventive, a problem solver and a good judge of character. I am none of these, but we both enjoy a good story, a good laugh, and an ongoing interest in the ever-evolving mineral exploration saga in Canada.

Norm is a new hire – a Sudbury boy. He had been a heck of a junior hockey player, but lack of size has kept him from turning pro. I hate Norm for two reasons: He is half my weight, and on snowshoes he hardly makes a dent in the snow, while I sink almost to the ground in that soft, sugary, Southern Ontario two-foot-deep white stuff – and Norm whips my butt at cribbage 19 times out of 20 and then laughs like crazy – everybody hates a poor winner. The three of us will be a team all winter.

Jim, our supervisor, is about 60 years of age. He has a bad stutter problem, especially if he is excited or under stress. He is a very kind and compassionate man, always on the side of the working stiff - probably because he has come up the hard way. Jim had been an elite soccer player as a young man in Ireland, and in the '20s had emigrated from Ireland to Canada under an Inco sponsorship. Jim went to work in the smelter and settled in Copper Cliff. A couple of years later he assessed the situation, decided that the smelter held no challenges, and went to university with Inco's blessing, earning a degree in geology. He joined the exploration division working out of the Copper Cliff field office and remained there throughout his career.

Jim has two problems that hindered his advancement at Inco. One is his speech impediment. The other is that Jim has no patience with petty upper-strata politics and stuffed shirts. He will never rise higher than Field Geologist/Area Supervisor, and that is fine with Jim. He never complains about it, but once in a while indicates to us his frustration with upper management.

Sidebar: The Sudbury mines needed men, and the men came from across the Atlantic. Soccer was big in the basin at that time, with lots of interdepartmental/mine rivalry. If you were going to import labour, you might just as well import soccer-playing labour.

Table Scrap

Jim tells us a story of coming into work one morning dressed for the office. He had been spending some weeks analyzing data on a project and so was wearing a suit and tie to work. (When Jim was not in the bush he was a pretty snazzy dude.) The district geologist told him that a showing in a hot area north of Sudbury had to be looked at

ASAP, and he (Jim) had better get down to Lake Ramsay (right handy in South Sudbury) and catch an Austin Airways Beaver, already warmed up and waiting.

Jim did so, flew north to a lake, hustled in to the showing with his prospector's hammer and packsack, mapped the showing, hammered off some samples and booted it back to the plane. It had started to rain after they left Sudbury, but that didn't deter Jim. Copper Cliff wanted that info, and Jim got it.

He returned to the plane soaked to the skin and in ruined street shoes, the pilot said, "Sorry Jim, I can't chance the weather, we'll have to spend the night." which they did, trying to get some sleep on the floor of the Beaver, wet and shivering.

The weather broke shortly after daybreak, and they returned to Ramsay Lake. Jim hustled back to Copper Cliff, walked into the geologist's office and handed over the samples.

"I'll wr-wr-write up a r-r-report for you r-r-right away." says Jim.

Oren P. (the district geologist) looked at Jim standing there in his ruined clothes with disgust. "You've got no business coming in to work dressed like that. Go home and come back looking half-decent."

About five years later Jim would exact some revenge at his own retirement party. I was not there. I had severed my ties with Inco and was 1200 miles away, but the story rippled out through the northern bush and hundreds of old bushwhackers gave Jim an "Attaboy!"

The party was well attended by management brass and older field personnel. Henry Levac and Freeman Marshall were there, among others. All had worked with Jim and/or for Jim.

The brass made a few speeches, making sure to blow their own horns as they congratulated Jim on his years of faithful service. Then it was Jim's turn to speak.

Jim stood up at the head table and said, "I want some of you men to stand up as I call your names."

He didn't need a list of these men. He knew who he wanted. "Henry Levac, Freeman Marshall, Al Beauchamp." …and on he went until 15 or 20 men were standing in the audience.

"How many of you have college degrees?" he asked. No hands went up. "I thought so."

Jim turned to look at the others at the head table, and then back to the crowd. "I want everybody to look at these men," said Jim. "These are the men who froze in the winter, sweated in the summer, swatted the mosquitoes and brought us the information to process. You are the men who made Inco, not us. Give yourselves a hand!"

And they did – even the head table honchos, although a few looked like they had eaten a lemon for dessert.

During the after-dinner press-the-flesh session, the district geologist shook Jim's hand. "How's the ulcer, Jim?" he asked.

Jim replied, "I'm g-g-glad you asked me Oren, because y-y-you're the s-s-s-sonofabitch who g-g-gave it to me."

That was J-J-Jim – and we all loved him!

B ancroft is a nice little town about 25 miles south of Algonquin Park and 100 miles north of Trenton, Ontario. First settled in 1853, it was a logging, trapping and mining community. This area on the south side of the Canadian Shield had hosted many small mines: gold, molybdenum, copper etc. It is known for excellent rock hounding and is gaining popularity as a summer cottage destination.

In the '50s and '60s uranium came into play and Bancroft experienced a bit of a boom. By 1962, only two mines were operating, and one of these was winding down. The last one, Madawaska, operated fitfully until it closed in 1982, and Bancroft dwindled to 2500 souls.

Now it has grown back to 4000 people as folks have begun to appreciate its quiet beauty. There is talk of a new mine, but I imagine it would have to be a darned good prospect, due to high surface rights values and environmental concerns.

We check into the Grand Hotel, one room for Jim, one for me and one for Carl and Norm to share. The Grand is a long two-story structure and quite old-fashioned. The rooms are all on the second floor with a long centre hallway and the only bathroom is unisex (more about this later) with a tub, biffy and sink. All the rooms are large, including the bathroom. Downstairs is a large lobby, a large dining room with good food, and some leased commercial space. To our dismay, there is no bar – Bancroft is a dry town!

Our dismay does not last for long. It is kind of nice to relax after work without having to prove what a hard-drinking bunch of yahoos we are. Bancroft has been there and done that. They are happy and so are we.

Inco had flown the area a few years ago and Jim had been running the follow-up groundwork. He has spent a lot of time here, and knows a lot of people. We wonder why we were chosen for this job, and soon discover the reason. Jim wants to build a fibreglassed plywood surfing sailboat and somehow he has found out how handy Carl is at any project. Carl is going to help him build it: Norm and I are just along for the ride.

This is a clean-up operation. There are some straggler anomalies to be checked and some claims to be staked. We work, of course, but Jim doesn't push us. Snowing heavily? Take the day off. Need to go back to Sudbury to pick up your car? Take two days off. Just stay away from the field office.

Carl never gets a day off. Jim has borrowed a corner of the local lumber yard's shop and he and Carl are busy. Carl works a lot of evenings with Jim, while Norm gives me cribbage lessons, and some days Norm and I work alone. The job could have been wrapped up by the end of January, but the boat isn't finished until February 15th, the same day work is completed. What a coincidence!

The four of us had driven to Bancroft in a company truck, an International Scout, International's supposed answer to the Jeep CJ. It is a nasty little rascal: underpowered, cold, noisy and unwieldy. It takes six turns on the steering wheel, lock to lock; and, since its wheelbase is only slightly longer than your kid's Hot Wheels, it is rough as hell. With Jim, Carl and I on the front bench seat and Norm stuffed in the back with the mag, EM gear and four packsacks, long distance travel is very uncomfortable.

Norm is a newlywed and wants to visit his new wife. My Meteor is sitting in Sudbury with a new motor, so Jim gives Norm the keys to the Scout and we head for Sudbury, a five or six-hour drive. When we return with both vehicles, we have more transportation options. (Jim and Carl spent the two days working on the boat.)

We do the trip. Norm gets to see his honey, and so do I. I had left the Meteor with the towing company three weeks before on the transplant waiting list. Two days later a young fellow piled up his 55 Ford and the towing company bought the wreckage.

The kid wasn't hurt anywhere but in his wallet. He had stuffed a 410 Edsel motor in the Ford and had taken it out on a shakedown cruise. Ten miles later he tried to widen the road at a rock cut – no plates, no insurance, no money. The towing company guy paid him enough to cover his fine and sold me the motor for $400 installed. The tow guy rents a lot of vehicles to our field office and is a pretty good joe. For an extra $200 I can have the kid's triple two-barrel carb setup, linkage and all, but I am unwilling to part with another half month's pay. What a dummy! I now have a balanced, blueprinted, solid lifter, roller cam 410 honker – with my 332 intake manifold and two-barrel Carter carburetor sitting on top – sort of like feeding a sumo wrestler crackers and cheese.

Sh-sh-shovel

One evening shortly after Norm and I return to Bancroft, we are all invited out to dinner. Jim had met Hans a few years before probably because Hans, a timber cruiser for Ontario Paper, was familiar with the area and would have been a good source of information for Jim and Inco. The land around Bancroft is a mish-mash of private ownership, Crown Lands and Ontario Paper leases, and a liaison with Hans was undoubtedly very valuable.

Here's how we were invited. Jim says, "Hey Hans, since your wife is gone for the weekend, why don't we all come down for supper?" – That's Jim.

Hans lives on a little lake along the highway 20 miles south of Bancroft. He is originally from Austria, and what a craftsman he is! He built the house himself of logs and stone. It is all varnish, with lots of glass facing the lake, and no two floors are on the same level. Most of the furniture is hand made. Light fixtures, fireplace surround and wall hangings are all in self-hammered brass – altogether a comfortable place on a small, almost private lake.

And what a meal for us hamburger-and-fries yahoos… smoked venison, smoked bear meat, smoked horse meat, smoked sausage, and on and on. We are awestruck!

After eating Jim produces a bottle of rum and we settle down for some drinks and conversation. Now, 26 ounces of rum doesn't go far with three men, (Norm is a teetotaller) but after I have a couple Jim starts to get nervous. "M-maybe Norm should drive," he says.

I ignore him and pour another, sip on it for half an hour and Jim says to Norm, "G-g-get his k-k-keys."

Norm and Carl are grinning at each other, and it tees me off! "I'm OK Jim," I say and pour myself another just to rattle his chain.

Finally Jim says, "We'd b-b-better g-g-go," so we don our parkas, thank our host and head out the door. Jim won't give up. I hear him whisper to Norm, "C-can't you g-g-get the k-k-keys?"

Well, four drinks over four hours after a big meal certainly didn't impair me, and besides, I am not all that impressed by Norm's driving skills. I am not about to trust my baby to anyone else but me. I get behind the wheel, Norm and Carl crawl into the back and Jim sits on the passenger's side, resigned to his fate.

But we have a problem, Houston. Hans' driveway leads off the highway approximately 300 feet with the last 100 on a slight downslope to his house. I knew enough to park at the top of the steeper part, but other factors come into play. We'd had about two or three inches of snow since the driveway was last plowed. It is just below freezing, and the warm, wet snow has a lot less traction than the cold, crispy variety. The Meteor has summer treads, and the back tires are pretty bare. Also, Hans and his wife drive a Land Rover and a BMW, both with a narrow track, and the Meteor does not match the packed ruts under the fresh snow.

So I back the car up about ten feet and slide off to the side.

Jim says, "G-g-got a sh-sh-sh-shovel?"

I ignore him, pull ahead a little and back up again, make a few more feet and spin out.

"L-l-let's sh-sh-shovel," says Jim.

Third attempt – not much progress and still the sh-shovel bit.

Little squeals of suppressed laughter are coming from the peanut gallery and I am getting cranky.

"Jim," says I, "There's a turn-off plowed out on your side a ways back. You watch out your side and tell me when we get there."

So I roll down my window and stick my head out, thinking Jim will do the same. I give her tar-paper, and we head backwards once again, the straight-pipes barking. But – what is this? The car is spinning out again, and we are still on the track! I look over to my right and lo and behold, Jim hadn't opened his window – he'd opened the door, and it has caught the snowbank, has wrapped itself around to the front fender, and there is Jim, still holding onto the door, almost out on the ground! As the Meteor slows to a stop, I can hear Jim saying, "H-h-h-hold it!"

Carl and Norm are obviously in some distress and will need an underwear change when we get home. I can see tears running down their cheeks.

I pull ahead, get out, reef Jim's door shut, get back in, and, mad as hell, give her the gun to the turn-off. We drive back to Bancroft in silence, except for the occasional "Peep!" from the back.

The next morning before we go to work Jim asks for my keys. I return from work to find the door in pretty good shape, with only the slightest evidence of a wrinkle in the front fender. Jim had taken the car to the local Ford dealer, and had paid for repairs.

And all is forgiven. Shit happens.

Table Scrap: F- F- F - Finish the job
A few days later we go to work in a light snowstorm. By 9 am it has turned to freezing rain and we beat it out of the bush back to the hotel. The bad weather gives us a slow day to spend mapping, snoozing and playing cards. We change into dry duds, and I am relaxing on my bed with the door closed. Carl and Norm are across the hall with their door open.

183

The Renfrew-to-Peterborough bus makes a daily lunch stop at the Grand. There is only one washroom downstairs, but enlightened passengers know that there is a bathroom on the second floor. I hear footsteps coming down the hall, but don't think anything of it. Carl and Norm see what is happening, and have already started to giggle. Suddenly a muffled scream comes from the bathroom, followed by the sound of someone running down the hall to the stairs. I whip open my door, and there are those two across the hall, laughing so hard they are crying. "What's happening?" I ask.

"Jim is having a bath and he never locks the door," they croak, and there comes Jim down the hall in his robe.

"What is all the commotion?" I ask.

Jim replies, "I was j-j-just lying in the t-t-tub having a b-b-bath, when a l-l-l-large lady came in, p-p-pulled down her drawers and s-s-sat on the throne at the end of the tub. Sh-she looked over at me, s-s-saw me in the tub, screamed, p-pulled up her p-p-panties and headed for the d-door. I told her, "G-g-go ahead and f-f-f-finish the job," b-but she r-ran off."

That night at supper we asked Jim why he didn't lock the door.

"I n-n-never lock the door at home," he said, "And right now th-th-this is home."

Apparently Jim had some doubts about Carl's ability in the bush. Norm and I know that Carl can out-pack anyone, but Jim has never worked with Carl. One day Jim drops Norm and I off at a job and he and Carl head off to supposedly check out some claims three miles in on an unplowed logging road. They have to snowshoe it. (The Scout is useless in snow. Putting it in four-wheel drive means that you got stuck within 30 feet instead of ten.)

I have already mentioned how difficult the soft snow is for snowshoeing. When you sink, the loose snow falls on top of the snowshoe, adding extra weight with every step. Jim is in pretty good shape. They take turns breaking trail, and at the two-mile mark Carl is played out, but no way is he going to holler "Uncle."

At two and a half miles, Carl leans against a pile of pulpwood, absolutely whipped. He waits for Jim, ready to pack it in.

"L-l-let's go back," says Jim.

Carl perks up. "I don't mind breaking trail, Jim," he says, crossing his fingers inside his mitts.

Jim shakes his head and turns around – Carl has called his bluff. Carl always thought that there were no claims to check – Jim was just testing him.

On February 15th we are on our way back to Copper Cliff. Norm goes with me, and spends two nights at home. Carl and Jim stop at Carl's house on the way. Jim knows that Carl's mom is in the hospital and he makes sure to dally a bit so that Carl can see his mom and dad. Once again that was Jim – stubborn, crafty, brusque, and kind.

The three of us head off to Lac du Bonnet, Manitoba, two days later. I will never see Jim again - but the legend lives on.

Bisset

We head west in my car. As usual we had been provided train tickets to Winnipeg, but we cash them in and share the gas. We could take highway 17 around the north side of Lake Superior, but I want to spend a night at the farm with Mom and Dad, and we take the shorter route, through Duluth, Minnesota and up to Fort Frances.

So we cross the border at Sault Ste. Marie and head west through Northern Michigan. Before we reach Marquette it starts to snow and I mean snow! Those winter storms just roll in off Superior and dump on you! Visibility is so poor we can only make 20 mph. It is getting dark, and we are the only vehicle on the road. Even the plows have been called off, it seems… we sure don't see any.

Our road map shows an intersection coming up with a cross road leading to another highway paralleling the road we are on, but 25 miles farther south. Maybe we should try that? And so we do, blindly, slowly zigzagging down the featureless, trackless road, with only the occasional glimpse of the high snow banks on each side as a guide. But the snowfall eases off slowly, and by the time we turn back westward toward Duluth, we are in the clear. We overnight at the family farm and continue on the next day to join Eric M. at Lac du Bonnet, 90 miles northeast of Winnipeg.

Eric has a crew doing some lake work. With our help the job is soon finished and we all move another 90 miles northeast to the old mining community of Bisset.

We check into the San Antonio Hotel, another old-timer – probably built around 1932. It shares the same basic design as the Grand Hotel in Bancroft, with one big difference, the San has a bar! Woohoo!

This is also clean-up work. Inco had flown the Lac Du Bonnet /Bisset area five years before, and the Manitoba Department of Mines regulations require Inco to turn over the maps to them after five years. We are crossing the I's and dotting the T's before the airborne results become public knowledge.

We fly out daily in a leased Cessna 180. Merv, the other party leader, works from a Bombardier, which had been shipped down from Moak Lake, complete with driver attached. Their driver is Howard Malcolm, From Lundar, Manitoba, a predominantly Icelandic community in the Interlake district. Our family had farmed six miles north of Lundar in the '40s and my sister is married to a Lundarite, so Howard and I have some common connections. Also, remember "Charging the Batteries?" Howard was the other driver on that outing two years before. That party must have been rough on Howard - he no longer touches the stuff. Howard and I talk a lot about Moak Lake, Lundar and things in general. One day he tells me about something that is bothering him, and it bothers me, also.

Howard sometimes spends the day at Merv's drop off point, if they are a long drive from Bisset. Howard waits for them, perhaps fixing something on the machine or reading a book, firing up the engine from time-to-time to keep it warm.

One day Howard takes a little walk to stretch his legs and smells wood smoke. Curious, he pokes around a little until, through a gap in the trees, he spots Merv and his crew dozing around a campfire a hundred feet in the bush. They are supposed to be a mile in, working on a grid. These boys are doing stumpwork!

Howard now realizes this is Merv's habit, more often than not – perhaps even on every target. Merv is pretty thick with Eric, and Howard has to tell someone. I share the story with Carl and Norm. We surmise that, given Eric's own disinterest in results, both Eric and Merv have already written off the area. We keep our mouths shut. It is sort of deflating to us, but we continue to do our best. I will run into Merv two years later during the Timmins/Texas Gulf rush – more on that later.

Sidebar: Stumpwork is a self-explanatory term. If you are lazy, or just don't give a damn, you sit on a stump and work up a fantasy grid. To me, it always seemed to be more trouble than it was worth.

One day we fly into a job on the south shore of Long Lake, 20 miles east of Bisset. Long Lake is at least ten miles in length, but no more than three miles wide. The area had hosted a few small gold mines in the 30's, and as we land we can see the old abandoned townsite of Long Lake on the north shore, directly opposite our target. "Nobody lives there now," the pilot tell us.

In the afternoon, the weather closes in – there will be no return trip today. No problem – we are not upset at all. We have our survival box, it's not too cold, and there is plenty of firewood amongst the sheltering evergreens. As we prepare our bivouac we joke that we will gladly trade a comfy night for a day off tomorrow.

We build a lean-to out of branches, collect a stash of firewood, have a bite and settle into a night of telling lies and trying not to stare into the fire. (It's just natural to stare at the fire – what else is there to look at? The problem is, you'll end up with the same symptoms as snow blindness. The next day your eyes will be red and itchy, and it can be quite painful.)

Two hours after sundown the skies clear and the stars come out – but what is this? It looks like a light shining in one of the buildings across the lake. This is supposed to be a ghost town! As usual, we discuss it a bit and then say, "What the heck," strap on our snowshoes and head across the moonlit lake to solve the mystery.

Sure enough, there is a dim light in the windows of a comfortable looking bungalow, with smoke curling up from the chimney. We knock on the door, and are greeted by a pleasant older couple who are surprised, but not nearly as surprised as us! We are invited right in and the teapot is put on the stove.

We explain our presence, and they tell us that they are staying the winter, well supplied and revelling in the isolation and their enjoyment of the surroundings. They had heard us fly in and assumed we were stuck, and in fact were even a bit worried about us.

We spend some time chatting in this friendly couple's comfy cabin in the light of an old kerosene lamp. They tell us they had bought the house a few years before as a summer getaway. After retirement they started spending their whole summer here. This is their first shot at winter in Long Lake, and they like it. "How fortunate they are," I'm thinking.

"Well," we say, "We should be getting back to our campfire. Too bad we can't call Bisset."

"Why not call Bisset?" they reply. "Perhaps they could come out to pick you up. The road hasn't been plowed since the last snowfall, but we are sure they can get through."

"You have a phone?"

"Right outside," says they, "Just pick up the receiver and give it a crank and the operator will answer."

186

I go outside and there is the phone on a hydro pole! I call the hotel, Eric picks us up and we spend the night in bed after all.

The only thing missing was Arnold the Pig.

(You gen-Xers will have to research the old sit-com "Green Acres," starring Eddie Albert and Eva Gabor to catch my drift.)

Sidebar: In fact – the old couple told us that the whole town was for sale.

In 1988 I did a small mag contract near Wallace Lake, halfway between Long Lake and Bisset. The gravel road had been upgraded and one evening I drove down to check on Green Acres. The hydro pole was still there sans telephone. Children were playing in the yard – perhaps the old couple's grandchildren. All the old houses were neat, and yards and hedges were trimmed. Long Lake lives again!

The Old Diana Mine - Requiem for a Lady

Another day and another target takes us to a lake farther east, only a few miles from the Ontario border. As we fly in that morning, we spot an old wooden headframe at the west end of a lake just north of ours.

It is a long, sunny, mild day in March, warming up to a couple of degrees above freezing by the afternoon. It had been warm enough during the previous days to settle the snow on the ice so that it is a virtual packed runway, and there had been a two-inch fall of snow the night before.

At 4 pm the 180 comes in and the pilot tells us he cannot take us out. The warm sun on the fresh snow has made it so sticky that the skis do not want to glide. He can get off the ice by himself, but with the added 700 lbs of our three-man crew it is hopeless. He has thoughtfully brought our sleeping bags and says that he will drop them off at the old Diana mine where there is a cabin of sorts near the lake. He also drops off our survival box and then flies back to Bisset.

We snowshoe across a short portage to the other lake and thence about a mile west to the old cabin. We pack our sleeping bags and survival goodies up to the cabin and check it out. Not bad – it is kind of a shack, about 16 x 20 with a couple of busted windows, but it has a big old barrel stove and even a supply of dry firewood under the porch. It is unlocked, as per the rule of the north – never lock your cabin.

We fire up the wood stove, set up our Coleman on the table (the only furniture) and cook up some soup, coffee, and Klik sandwiches. Then we cover the windows with some cardboard boxes and settle in for the night.

The Diana Mine had been opened in the early thirties and had closed in 1937. It was a pretty remote place, and all the infrastructure had been freighted in by horse and sleigh on winter roads from Long Lake, 30 miles or so to the southwest. Other supplies would have been flown in, winter and summer, and there must have been a number of buildings when the mine was running, but now only the lonesome old headframe and our cabin remain.

In the gathering gloom of dusk the old head frame looks rather eerie, sitting about a half-mile upgrade from us, outlined by the last light rays of sunset in the west.

187

Carl and Norm decide it would be great fun to burn the headframe, but I talk them out of it for two main reasons. It is now a dark, moonless night, and I have a vivid picture in my mind of one or all of us falling into an old shaft or such. It also seems an inglorious fate for the old structure to die like that.

Carl and Norm settle into their sleeping bags for the night. I keep the fire hot for a while, set up a candle (always part of our survival gear) and read a 1929 J.C.Whitney catalogue. (J.C.Whitney was an automobile after-market supplier in Chicago.)

Two items stand out in my memory: a genuine leather crank hanger and a chrome plated battery box, "Guaranteed to enhance your running board."

When the plane comes for us at the crack of dawn we are in such a hurry to get back to a hot breakfast and a warm bed that I neglect to take the catalogue. I wish I had it now.

Sidebar: I heard a few years later that the Department of Lands and Forests had gone in the following summer ('64) and had burned the headframe. Carl and Norm might as well have had their fun.

Perhaps it would have been more fitting for the old girl to have died that night. Had we watched it disappearing in flames we would have had our thoughts of the old miners and their hopes and dreams. I hope the young fellows that burned it in '64 had the same feelings.

Spring is here. The last job we do is near a small river, which is starting to open. We have our last winter tea fire on the riverbank, soaking up the warm sun.

Three young otters come by, unaware of our silent presence – probably last year's litter, striking out on their own. We hear them coming around a bend in the river and they are having a great time, talking and chirping as they dive into an open stretch, swim underwater to the next hole, then over the snow to another. They alternate over and under as they make their way upriver. What a joy to watch. Priceless!

The next day we pack up and I go home to the for three weeks. Merv's two men don't have far to go – they are from Lac Du Bonnet. The rest of the crew goes back to Copper Cliff.

It had been a pretty darn good winter.

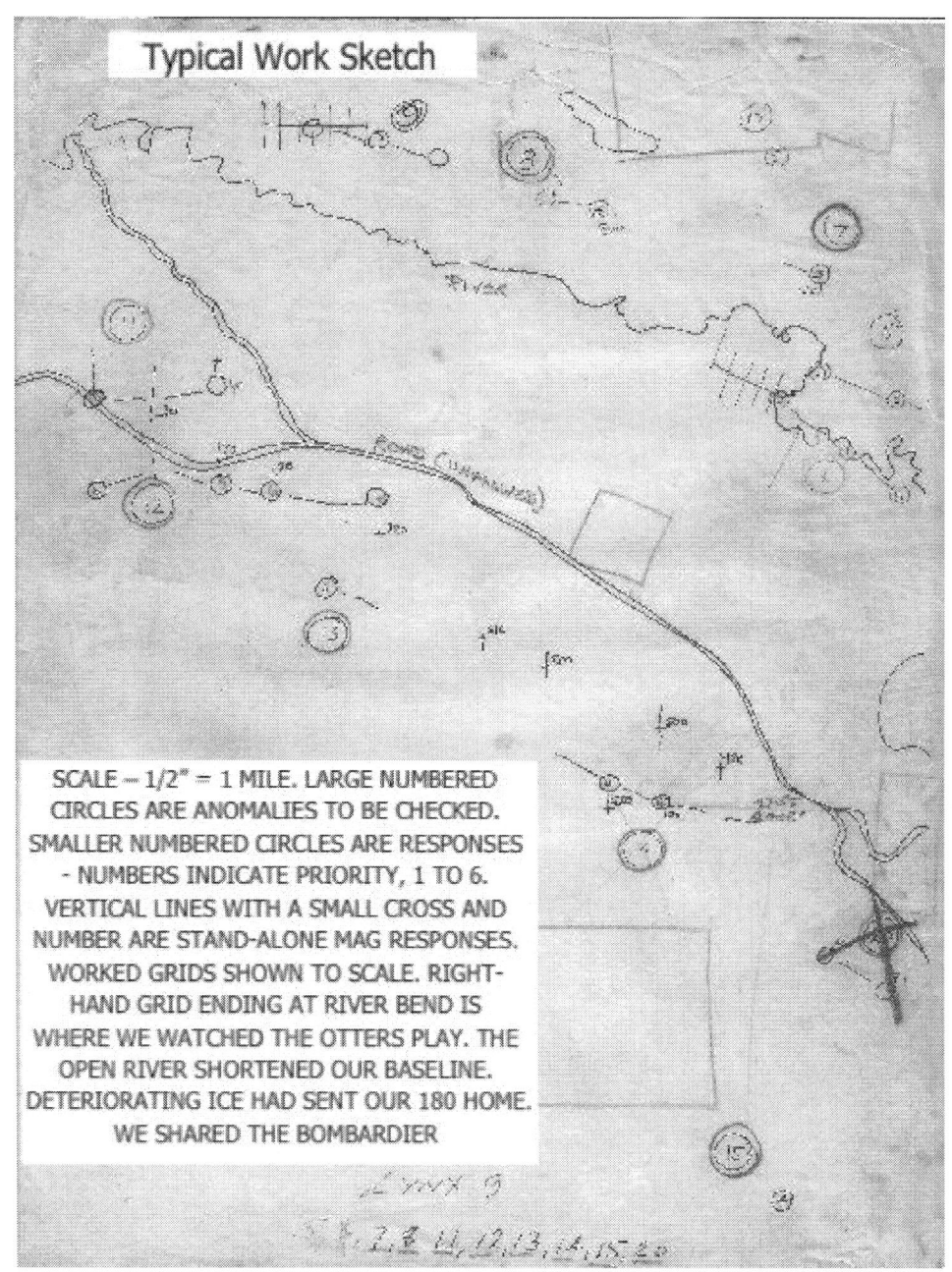

Typical Work Sketch

SCALE – 1/2" = 1 MILE. LARGE NUMBERED CIRCLES ARE ANOMALIES TO BE CHECKED. SMALLER NUMBERED CIRCLES ARE RESPONSES – NUMBERS INDICATE PRIORITY, 1 TO 6. VERTICAL LINES WITH A SMALL CROSS AND NUMBER ARE STAND-ALONE MAG RESPONSES. WORKED GRIDS SHOWN TO SCALE. RIGHT-HAND GRID ENDING AT RIVER BEND IS WHERE WE WATCHED THE OTTERS PLAY. THE OPEN RIVER SHORTENED OUR BASELINE. DETERIORATING ICE HAD SENT OUR 180 HOME. WE SHARED THE BOMBARDIER

This is a typical work sketch, carried by us as we accesses responses. It was self-generated on "tracing paper." We had no handy printer/copiers back then.

Chapter VIII Summer '63 – Northwest Territories

I arrive back in Copper Cliff May on 1st to learn that my summer will be spent in the North-West Territories, 230 miles north of Yellowknife where Inco has a high-grade gold prospect on the south-west shore of Contwoyto Lake (Con-TWOY-toe.) An ongoing aggressive exploration project will be continued this summer, and I am pumped!

This is sort of uncharacteristic, I'm thinking. Inco is nickel first, copper second – and remember, in '61 we had walked away from Hemlo. I guess it was like the Texan who kept hitting oil when all he really wanted was water. You just go with the flow.

This is going to be magnetometer work only. The gold is associated with amphibolite/iron formations, and EM surveys would add little to the picture.

We are changing over to the new fluxgate magnetometer. The old Sharpe's A-2 mag is being retired. We will have one in camp, but it will serve as a base mag only. Eventually the A-2s will be packed away in a corner of the field office to collect dust. I am to be fast-tracked on the new mag.

The instrument chosen to replace the A-2 is the Sharpe's A-3. There is no more tripod – the A-3 hangs from a strap around my neck. A battery pack on my belt behind my back has a cable leading to the instrument. There will be no more conversions from scale to gammas – the A-3 gives a direct gamma readout. There is no quartz bearing and fulcrum to worry about – just push a button, take your reading and move on. And there is no more little compass to orient the mag. You still have to face north to take your reading, but every old prospector knows where north is.

There is one problem, though. The damned thing is so easy to use that production will be easily tripled, especially in the treeless terrain we will be encountering at Contwoyto. Two hours of field work will require one hour at the map table. Eight hours in the field plus four in camp does not compute – however, I will come up with a solution eventually.

I spend a week in the field office honing my contouring skills – or more accurately, learning how to contour. Bill Aronec, our head draughtsman/contouring expert, gives me a large sheet of uncontoured mag readings from the previous season's work at the main Contwoyto showing. I am to contour it and have my work critiqued by Bill. I am pretty cocky. After all, I have been-there-done-that, so I whip her off in less than an hour. "Feast your eyes on a work of art, Bill," says I.

Everyone in the office is smiling. They know what is coming. Well, by the time Bill is finished editing there are more red marks on that map than black pencil lines. "You didn't catch this high magnetic zone," and on, and on. My face is beet-red.

A fresh map comes out, and this time I spend four hours on it.

"That's better," says Bill

The next three days are spent on different maps. By Thursday, even I can see my improvement, and that day I receive my BBC (Batchelor of Bush Contouring). I also reaffirm my theory of proper workplace environment, a decision I made after my first full-time job as a bank teller six years before. I will never work inside again. I need fresh air!

With my contouring up to snuff, I am ready for the summer. I will leave in two days, but tomorrow I am heading to Capreol with Freeman Marshall. He is an old-time prospector who now spends spent most of his time in the Sudbury area, staking the odd claim and helping out with camp gear coordination. He is a quiet, calm, competent man and very likeable.

The next morning we load a 16' cedar strip canoe on the roof racks and head out to do some field sampling. Spring is here: the snow is gone, and it is a lovely day in early May.

We drive up the old Whistle Mine road north of Capreol to a point where we will cross the Vermilion River. I had camped at this spot the previous fall and we had done some geophysics and line cutting on the property where Freeman and I are headed. The Vermilion roughly parallels the road on the west side, beyond that is the CN line, and beyond the tracks is our day's target. It is not a long walk, perhaps a mile, and conditions are great. It is warm enough to be wearing only a light bush jacket, and there are no flies yet.

Last fall we had crossed the Vermilion using the twelve-foot skiff/cable arrangement. We hope it may be back in the water, but have brought the canoe with us in case we might need it.

Sure enough, some jerk of a hunter has shot through the cable and the boat is gone. Well, we have the canoe, but we have neglected to bring along a coil of rope. We scrounge around in the company truck and piece together every scrap of rope we can find – it looks like we have enough.

The spring run-off has raised the river level but the banks are high enough to contain the Vermilion at this point and the water is rolling through at a pretty good clip. Our plan is to tie off the rope on our side, paddle over, tie off on the other side and pull ourselves back across on the return trip. We study the angry looking current for a bit and finally talk ourselves into it.

I get in the front, Freeman jumps in the back and we paddle like crazy. We are almost at the far side and I am reaching for an overhanging tree branch when we run out of rope! Into the water I go, and Oh Boy, is it cold! The canoe tips and throws Freeman out also, but he falls forward and I am able to grab his jacket. With a good hold on the tree branch I am able to keep him from being swept down-river. We climb out, drenched to the skin, and watch as our canoe, still upright and tied off on the other side, drifts across the river and nestles up to the far bank. We take off our boots, wring out our socks and sit in the sun for a bit while we dry out somewhat, counting our blessings and looking to the future.

The CN line disappears around a bend a mile south and we are not sure if it crosses the river any time before Capreol. There is zero chance of anyone else appearing where the truck is parked, so it looks like a ten mile walk down the track to Capreol, then beg a ride back to the truck.

So we head for the tracks. The terrain flattens out a few hundred yards down river, and the high water has flooded quite an expanse of what was probably a hay meadow in

191

days gone by. As we skirt this body of water to get to the railway, we see something hung up in the willows along the river's normal path. Could it be the skiff?

We are already wet, so we wade out to take a look. Sure enough it is the boat, mostly underwater, but otherwise undamaged. We are able to tip most of the water out and find a couple of broken boards in the flotsam collected by the willows, and we paddle over to the other side where it is a relatively short walk back to the truck. We rescue our canoe, and even get our axes and our packsacks back. We return to the field office sans samples, but with a good beer parlour story.

I am about to embark on a summer to die for: new experiences, new flora, new fauna – things to see and do that other people never have access to without a well-stuffed wallet – and I am getting paid to do it. Amazing!

So five of us leave for Edmonton/Yellowknife/Contwoyto. As usual we cash in our sleeping berths – we will share a drawing room. There are only four beds, so someone will have to sleep on the floor. This is no problem. One guy has his own bottle and by nightfall he is already asleep on the floor. The rest of us talk, read a book, and most important – we can eschew the bar car. We detrain at Edmonton, taxi to the airport and board a Pacific Western DC-6 to Yellowknife, where we overnight.

My first impression of Yellowknife is of a clean, modern, bustling town. There are two gold mines still operating, one of them the Giant – a scene of terrible tragedy a few years down the road.

It is May 15th, five weeks to the summer solstice and already there is more daylight than darkness (actually duskness,) leading to my second impression: "For God's sakes, doesn't anyone sleep around here?"

We've all been eagerly waiting to sample Hudson's Bay Overproof Rum which is sold only in the Northwest Territories and the Yukon. All hard liquor elsewhere in Canada is 40 proof bit Hudson's Bay Rum is 80 proof. Well, I sample it – and it tastes like crap.

I go to bed at 11 pm – pretty early, since the bar is still open. I drop off to sleep, but not for long. Some joker in a Volkswagen Bug with no muffler starts cruising the town. He must be making up for a long winter, because he is sure enjoying himself. Every ten minutes he pulls up to the stop sign under my window and gives it a couple of revs, then bangity-bangs on down the street, pulls a U-turn and bangity-bangs on back. A closed window doesn't help. A pillow over my head doesn't help. Once in a while he goes away for a half hour, but that cat always comes back. He finally runs out of gas, I guess, and I get a full three hours of shuteye.

After breakfast we go to the airport and board a Wardair Bristol Freighter for the flight into Contwoyto. The Bristol Freighter is a high-winged, big-bodied two-engine jobby. It is a unique aircraft, but not the only strange duck I will fly in this summer. It is of British manufacture, and the only one flying in Canada, I am told. Two big doors open at the front and heavy machinery can be driven up a ramp and right into the plane. Today the five of us will share the ride with a muskeg tractor and some drill rods. We settle into some fold-out canvas seats along the sides and watch the pilots climb a ladder to the cockpit, way, way up there. The cargo area is huge, sort of like being in the belly of a whale, but the flight is comfortable. It is also quiet in the whale's belly. The two motors are so high above us that we hardly hear them.

BRISTOL FREIGHTER
LOADING A BOMBARDIER

We land on the ice strip at the Contwoyto base camp, where we will spend two days gearing up for our summer. I am going on to Fuz Lake, 30 miles north-west of the northern tip of Contwoyto. One other man will have a fly camp, and two will stay to work the main showing. One of our group, the legendary Ed Gatien, will be geophysical supervisor, with his own office tent at Contwoyto. He will check our work as it comes in and will do some co-coordinating. The area supervisor is John Mullock, and the area geologist is Phil, a nice guy, but scatter-brained. He carries a book in his pocket all summer entitled "How to Improve Your Memory." I'm not kidding! He has clip-on sunglasses on the bill of his baseball cap, and he can never find them. We are a strange mixture, no doubt about it.

Table Scrap

The strangest of all was a junior geologist at Bobby Zadow's fly camp.

The guy had no concept of camp living – no "all for one – one for all" esprit de corps.

He worked on his own and worked his own hours, often late in the midnight sun. I did the same thing and I was always careful to respect my tent-mates' sleep time. Not him – he would return late and crank up the volume on his personal radio.

He rarely ate with the group – cooking his own meals at odd hours. That would have been OK but for one or two things.

Say he wanted bacon and eggs; he would fry a dozen eggs and a whole slab of bacon, eat some and throw the rest out. The next morning Bobby and his men were bacon-and-eggless. Meat was the same deal, likewise potatoes. Bobby adjusted the weekly grub order – the jerk upped his wastefulness. And, to add insult to injury, he refused to do his own dishes.

The guy would <u>not</u> be swayed by Bob's intervention – he worked for Phil. Bob finally told Phil, who conferred with Mullock.

The jerk was not fired – he was ostracized. He got his own tent which was set up 200 feet away – always 200 feet. (The chain came in handy.)

He got his own wanigan and his own weekly grub order. It was sufficient for one man – eat sensibly or starve for three or four days.

He got a restraining order, enforced by three young totally pissed off bushwhackers.

And that's how he spent the rest of his summer.

(This is a true story. Any embellishment or exaggeration on my part would only be closer to the actual truth.)

Contwoyto is a good-sized lake, more than 55 miles long, up to 20 miles wide and runs north-west/south-east. The northern end will appear in the news a few years later. (Google Martin Hartwell).

The base camp is a collection of wood framed, comfortable buildings: cookhouse, dining room, office space, dorms, core shack and mechanical shops. Along the shore just north-west of these buildings are a row of five or six standard issue framed tents. One of these will be Ed's office/sleeping quarters. Ed likes to be alone.

The next day a PWA DC-6 comes in (direct charter from Edmonton) and a whole slew of people get off. A DC-6 on an ice runway? No problem. A few weeks before a three-man crew had come in to fire up some stoves, take delivery of the summer's fuel and flood the runway. There is now six feet of ice to land on!

(One of these men is a guy I knew from Moak. He says their first job was to shovel snow out of the buildings. It doesn't snow much in the barrens, but what little there is moves for miles on the wind. It is very fine and sifts in through even the smallest of cracks.)

Coming off the DC-6 are drillers, support staff and summer students – lots of summer students. Most of them are third year geology boys, but other disciplines are represented.

But the really weird thing is that the Eskimos show up just as the passengers get off the plane!

Maybe it is pure coincidence, but I have seen strange things in the north. Were the spirits speaking? Were there sonic vibrations when flying arrangements were made? For whatever reason, a large group had left with dog teams from Coppermine, 200 miles away on the Coronation Gulf of the Arctic Ocean and meet these kids before they walk half the quarter-mile distance from the airstrip to camp. Out come paintings, soapstone carvings and students' wallets. Less than an hour later the Eskimos go on south to meet the caribou herds migrating north. The students come into camp showing off their goodies. Crap – nobody but us, yesterday, but we had spent our last dollar in Yellowknife, and I'm sure the Eskimos knew that.

I am told the Eskimos trail the moving herds for weeks, harvesting the males and freezing the meat. Then they transfer the meat to Coppermine in two trips, sometimes three. The weather is warming rapidly and the meat cannot be left for long, At home the meat will be cut into strips and hung in the sun to cure.

I wonder if they still do this. It would be easier with Skidoos, but I doubt it - nothing stays the same forever.

The next day we are off to Fuz Lake. There are six of us in the Otter. On board, along with our personal stuff and groceries, are hundreds of laths to be used on this job. The camp is already up. It had been pre-built for us and I like that. Diamond drillers had used it for two weeks in early May, then moved on and left the kitchen and two bunk tents intact.

We off load, carry everything the short distance up to camp, light up the oil stoves and survey the situation. There are three well-built 14 x 16 wood framed tents, lined up on a nice, flat, two hundred foot wide plateau about ten or twelve feet above the shoreline. The wind would have kept the snow from accumulating there last winter and the warm spring sun has already melted what little snow cover remained.

Ed has told me I am to be in charge of this camp, but I have some doubts about that. The crew is made up of me, four students and one geologist. Gary and Wayne had spent the previous summer at Contwoyto, while Ian and Mark are rookies. The geologist will be doing his own thing. I am to do the mag survey and am the only one in camp with any geophysical experience. The other four will be putting in the grids. I'm not sure how things will play out, but I forge onward.

There will be three men in each bunk tent. Steel cots with mattresses are already set up and waiting, so we chose our spots and move in. Gary, Wayne and Ian take the end tent, Mark and I share the centre tent with the geologist. The third tent is the kitchen/eating room.

First I set up the radio. No more WWII stuff – a nice compact transistorized jobby - no battery charging. One end of the aerial is attached to a 2 x 4 upright nailed to the back of the tent and it slopes down to the ground 75 feet away, held in place by a few rocks. I build myself a good, large map table and decide to check out some surrounding countryside.

During my construction activities I noticed Gary and Wayne exchanging raised eyebrow glances. When I return, the radio is gone, as are most of the maps! The radio is now beside Gary's cot, and so are the maps showing the grid layout. No one says a word. My nose is definitely out of joint, so I sit on my bunk to figure things out. I decide to let things ride – play it by ear more or less.

As it turns out, Gary is right on. He has had a year's experience and knows exactly how to put the grid in. He is a gung-ho worker, always cheerful and pleasant, but I wish he had talked it over with me. He does his work and I do mine, but there is always a wall between us. Later on I found out that John, the area supervisor, had told Gary he would run the camp and that I would work independently. John never told me the plan, and Ed didn't have the authority to put me in charge. Once again, the brass screws up and it's the men who have to work things out.

But it's no big hairy deal – when the going gets tough, the tough get going.

Gary and Ian make up one picketing crew, Wayne and Mark the other. First they cut the laths in half. The pickets here will be two feet in length. With little or no underbrush there is be no need for four-footers. Then they establish an azimuth and chain the baseline. The first grid is three miles long and one and a half miles wide. Unlike a bush grid, one boundary will be the base line, with the opposite boundary being a tie line. The side boundaries will also be surveyed in by transit and chain. With the boundaries established, Gary and crew fill in the centre lines. There are ways to maintain control as you do this, similar to putting in an ice grid so I won't bore you with the details. I'll just say that with proper, patient establishing of the rectangle the rest goes lickety-split.

When putting in cross lines the boys spend a couple of hours each evening pre-marking pickets and tieing them in bundles. With lines spaced at 200 feet and stations every 50 feet, 160 pickets per line are needed. This grid will use 13,500 pickets, meaning a minimum of 13,500 mag readings. A smaller grid to the north-west will add another 8000 pickets/readings.

I remember the good old A-2 days, when a hundred readings was a good day's work. I will have to find a way to produce an average of 700 plus readings a day, just to keep my head above water. Also, eight hours of field work will require at least four hours at the map table. Read on – we will overcome.

The sheer volume of work is not the only problem facing me. The first day in camp I had taken a few mag readings in our front yard, and after establishing that there are no mag highs in the proximity, I set an old A-2 up to act as a base mag. During the second day, while we are settling in, I take a number of readings throughout the day. The diurnal drift is very erratic, increasing as the sun rises higher in the sky. I assume, correctly, that the sunspot activity will have more effect on my work as the day wears on, but I try the old arboreal forest routine at first – old habits are hard to break.

The next day I read the A-2 every fifteen minutes starting at 7 am. It is fairly flat at first, but magnetic drift starts to pick up, and at 9 am I decide to head for the field anyway. (I can't head for the bush this summer – there is no bush.)

Gary and crew had gone out earlier to shoot the grid perimeter, and since they have a head start I figure on getting at least a mile of base stations done. The routine? Read a half-mile, sit for fifteen minutes, and read back to the starting point. Compare readings, and make corrections in the field then read the half mile again, this time extending another half mile, then back that half mile, and so on. By lunch the picket crew has completed three miles. I work on that all day, and in the evening work on all my notes and decide that although it is not perfect, I have pretty good initial control.

By the end of the first day the picket line boys have almost completed the perimeter and tomorrow I will be able to work down the west side to extend my base stations to the tie line. It takes me half the day, and by the second day I still have a mile-and-a-half to read. Back and forth, back and forth is a slow process, but necessary. Good control means that I can do more work, and more importantly, more accurate work once I start on the cross lines which the boys are putting in, getting farther ahead of me every day.

I start reading grid lines – problems again. I can read an 8,000-foot line in just over an hour – the Fluxgate is fast –thirty seconds at each station, but the diurnal drift changes too much in that 8,000 feet. First it climbs 300 gammas or more, followed by a drop of 500 or more. My readings are impossible to control. I have managed to do so only by checking across to the perimeter line, but I will soon be too far away to make that walk and keep up production. I can check back to the previous line, but correction errors will be multiplied. I decide to read another control line in the middle of the grid, giving me three check points on each line; base, tie and centre. Much better, but I am still not moving fast enough. I have to keep my map work up to date as sometimes the map will tell you things you don't easily see in the field.

I had set the mag at an arbitrary background value of 1030 gammas. Mag highs will be above that figure, and under 1000 gammas will be considered a low, which is also important for interpretation. Since the cross lines are at a theoretical 90 degrees to strike, the contours should be generally north/south. Sometimes during the day the drift is pretty wild and I end up with and east/west magnetic trend. This is not an impossibility if the rock has a lot of folding, but it is highly unlikely. I make note of these areas and re-read them the next day and 99% of the time the readings come back into line with the others.

This is also time consuming. Some sort of fine tuning is needed, so I start to experiment.

Diurnal is daytime – right? Nocturnal is night-time. Well, we never get night-time this close to the Arctic Circle. (It is 30 miles north of us. I never did cross it, but I like to say I could see the dotted line from a rock on the hill behind camp.) Summer solstice is four or five weeks away and the sun now sets for about an hour with dusk, not darkness resulting.

So I get up at 3 am, have a bowl of oatmeal and hit the tundra before 4. I work until 8:30, return to camp, have another breakfast, sleep until noon, have lunch and hit the maps for three or four hours. Then I grab a sandwich, get back to the grid by 5 pm and work 'til ten o'clock. Things are going well now, but I am leading a strange life. But enough about work. There are a whole lot of other things going on in this strange solitude.

My first perception of the barrenlands was the lack of perception. Below the tree line there is always something to relate to in the distance – a grove of trees or a clump of willows. Look across a lake sometime. There is always something to compare for judging distances. Up here? Nada.

The day we came into camp I looked across the lake at a pile of boulders. I knew Fuz Lake was five miles wide, "But how big are those rocks?" I wondered.

On June 20th when we left Fuz, we flew over that boulder pile. Some of those rocks were as big as our tents!

The terrain around Fuz is a mixed bag. Areas like our campsite are fairly flat with a short, wiry sort of ground cover, I guess you would call it sedge. Everything is sort of undulating – low rolling hills, with lots of boulders, large and small, scattered around. Just south of camp is a boulder bed beside the lake. The rocks sort of fill a shear zone that runs down into Fuz. You can jump from boulder to boulder – some are as large as a car. At one spot I look down between gaps in the rock, and twenty feet below is a caribou skeleton. Perhaps the unfortunate creature had been crossing the boulder field on snow and had broken through, or he may have been cornered there by wolves. Certainly the wolves could not have stripped the carcass, but there are many hungry rodents to do the job.

South of camp the land rises gradually and along the shore of the lake is a cliff about 100 feet high, the only cliff I saw that summer, although low outcrop ridges are common in some areas. There are also flat semi-bog stretches which are not really wet, but sort of spongy underfoot. The summer sun has melted the surface of the permafrost, and these areas are akin to a wet cow pasture. There is short grass and the occasional small bush on the hummocky surface. It is not hard to walk through, but you have to pay attention to every step to keep from stumbling. It is also a good cardio workout because the darn ptarmigan in their summer plumage blend in so well they are invisible. They sit in a depression between the hummocks and won't move until you almost put a foot on them. Then they explode into the air, fly 50 feet or so and disappear again. I never get used to it – it puts my heart rate sky-high every time.

They say the Eskimos hunt them by throwing rocks, so one night I lay my mag aside and try it. Sure enough – I don't have a good arm, but there are so many of them I can't miss. I bag four and stuff them in my lunch pack to have for supper the next day. I never do it again, though. Inco supplies us with enough grub so I let the ptarmigan be. They are more important to their own food chain. I don't need them in mine.

197

One day I come across a melt-water pond about 200 feet long and 200 feet wide. The long days of warm sunlight can cause rapid change in the barrenlands. Two or three days ago this little depression had been full of snow and the picket line crew had crossed it. Today it is two feet of water, and I can see the pickets floating. I don't want to leave a small gap in the middle of the survey, and besides, a little wade in a pond never hurt anybody, right? Wrong! By the time I cross that 200 feet of cold, cold water, my legs are dead from the knees down. I return to camp with needles of pain shooting through my shins as the blood starts to re-circulate. I change my socks and jeans, put on my spare boots, and never do that again.

Fuz Lake is slowly melting. The lakes open up in rotation, from the smallest and shallowest to the largest and deepest, and Fuz now has a twenty-foot border of open water around the lakeshore. Base camp has flown in a v-stern canoe to make access to the south end of the grid easier. The canoe is for work – recreation is absolutely forbidden. The word is, if you get dumped in the water you have a maximum of three minutes to get out and my pond experience backs that up.

One day I catch a ride with the crew to the south end of the grid. I have never seen such crystal-clear water. I can look straight down to the lake bed, fifteen feet or more below. The odd fish swims by, lake trout or arctic char, perhaps. How can fish live at that temperature? A rock appears ahead, just under the water surface. I wave to the guy running the 3-horse kicker. We glide over the rock, way down at the bottom of the lake. The refraction in that unbelievably clear water had made it appear to be barely submerged.

We chug past the cliff. There is an osprey nest tucked into a ledge about thirty feet below the cliff top, and Mom and Dad Osprey are keeping an eye on us. Later that day as I work my way up the eastern grid boundary I go over to the cliff edge to check the nest from above. I carefully look over the edge, but quickly back off! Mom and Dad are letting me know I have breached their perimeter. Okay folks, I won't bother you again.

Our groceries are still coming in by ski plane. The ice is still four feet thick away from the shoreline. The drill crew had left their muskeg tractor behind and we back the trailer into the water just north of camp where the slope is gradual, and carry our groceries across the bridge, so to speak.

Later on a driller comes for the tractor and trailer and some gear. He drives it around the bottom end of Fuz and on to base camp. We use the canoe for one more grocery trip on the ice and then wait ten days for the ice to go out. In the 70 degree farenheit days the ice soon candles and a brisk warm wind opens the lake in one day: morning – ice, evening – open water. Wardair has the routine down pat. The Otter goes back to Yellowknife on skis, returns on floats and we have our groceries the next day.

Sidebar: "Into Yellowknife on skis and out on floats?" You (and I) say – how is this possible?

I am told that it is easy-breezy, but surely more easily said than done. The pilots landed on grass beside the runway. The otter was dollied in and the changeover made.

I was never crystal clear on the rest of the concept. They may have then dollied the aircraft down to the seaplane base, but I heard that they, (and other barren-land air services) used a dolly take-off on-the-tarmac deal. When the plane achieved lift-off the dolly coasted to a stop to be retrieved by the ground crew. (Wheel floats at that time were the opposite of little children – heard of, but never seen.)

Not that the weather doesn't bite us once in a while. One evening on my late split-shift, a south-east wind blows up some snow and freezing rain and I hustle home. As I reach the tent I see a sleeping bag on the ground. It is Geoff, the geologist. "What's the deal?" I ask.

A muffled voice answers, "I'm OK, see you in the morning." Weird, huh?

The next morning I have to kick the tent door open. The wind has driven the snow and rain horizontally and the tent fronts have over an inch of ice build-up. I give Geoff a nudge to see if he is still alive. He is, and again I ask him later, "What is the deal?

He tells me he is planning a back-packing tour of Nepal and wants to become used to sleeping outside. Well, I have never had a yen to see Outer Mongolia, but I have slept outside in similar conditions, and believe me, it doesn't require practice.

Geoff is a little strange, but he is alright. However, my experience has taught me that over-educated people, especially those of the English "upper class" can be a Royal pain in the ass.

Sidebar: Some years later I will be in charge of 15 or 20 men spread out in five camps north of Red Lake, Ontario. One camp will be semi-permanent and we are building two frame tents. A graduate geologist, a new arrival in our country from you-know-where, is standing around watching us peons work. Now, he does not answer to me. My jurisdiction is geophysics and line-cutting, but he will be living in this camp. I hand him a shovel and ask him to dig the biffy hole.

"I don't dig shit-holes," he says, "I'm a geologist."

"Well," says I, "I've noticed that geologists crap just like everybody else. Grab the shovel, grab a hammer, or grab the next plane south!"

He took the shovel.

Table Scrap

A story is told of an Englishman who was stranded on a northern shore for two days. When the search party found him he said how hungry and thirsty he was.

"Hungry for sure," said one of the rescuers, "But at least you had lots of water."

"But I didn't have a cup," said the Englishman.

Just before they take away our muskeg tractor we move a tent down to the south end of Fuz. Wayne and Geoff are going to do some mapping down there, and take the tent and all they will need with them. We tear down the tent and load the whole shooting match on the tractor and trailer and head down on the lake ice.

It is at least 75 degrees and most of the guys are in shirtsleeves and shorts. The sun reflecting off the ice warms us from above and below. We are in a festive mood as we trot on down. It is seven miles or so, but we are all in good walking shape by now, and we will be riding home.

199

Before we get to the end of the lake we are met by a herd of migrating caribou. This is the first encounter with the beasts for all of us newbies. What a joy it is to watch them. They are not fearful of us, just wary. They circle us, trotting counter-clockwise and taking a good look. Some females have calves at heel, which I guess they drop on the trip. The little guys must hit the ground running. It seems like a large herd to us, I'm guessing 800 to 1000. They go by for 2 or three hours – not single file, but not en masse, either – and they all have to check us out. We are like an eddy in a caribou river.

(Later at the Summer Solstice party the other fly camp crew tells us that a herd of caribou passed their tent one morning at breakfast time. When they returned to camp at 5 pm the last stragglers were still going by.)

We reach the new campsite and help set it up. The Otter comes by with groceries and cooking gear, and when we leave the camp is liveable.

We share our camp area with a family of Richardson's Ground Squirrels, commonly known as chick-chicks because of their cheerful call. It's their universal language. They can convey alarm, contentment, scolding, and (I'm sure) laughter by the tone of their call. They are fun to have around. They live in burrows on the tundra in places where the gravel content makes burrowing possible. I've often wondered – do they dig down below the frost line, or do they insulate their houses for winter hibernation?

We are always up to something in camp. First, we decide to play volleyball. For some reason base camp had sent over a volleyball, but no net. No problem. The same unknown reason has provided a cone of strong string for the kitchen. So we spend an evening tying our own net, kind of like a quilting bee – do a good job, too. Then we make two stands out of spare 2 x 4s and put the net up on a nice flat spot south of our tents. Now we need court lines. We consider using small stones, but there is a danger of twisted ankles, I suggest flour. The diamond drillers had a cook, and when they pulled out they left us their groceries, including two ten-pound bags of white flour. So we punch a small hole in a bag and make our lines. All right! Just like uptown!

For two evenings we play volleyball (like we need the exercise.) On the second night we get a light shower and the next day the sun bakes the flour. Our third night on the court, we disturb the lines, and the flour becomes chips, more or less. The chick-chicks now join in the game. They are hauling the flour-chip cookies to their food stash. Every time you go for the ball, there is a chick-chick, going for his next load, giving you hell even with his mouth full. We finally give up and sit down to watch our volleyball court disappear. Score: chick-chicks one, hairless apes, zero.

Our next project is a steam bath. The drillers had left their coil heater behind along with some plastic water line, valves, water pump, etc. Once again I am the instigator. I know how a coil heater works and I figure that even though we don't have the materials for a bathhouse or a way to heat rocks, at least we can have a hot shower.

The coil heater is a steel tube about five feet high, about 14 inches in diameter with copper coils wound around the inside perimeter, top to bottom. On top is a chimney with a controlled draft. An oil burner is in the bottom. The heat flows up the large tube and out the chimney, heating the water in the coils. Cold water goes in the bottom and hot comes out the top.

200

It isn't hard to rig a ten-gallon drum for fuel supply, but the plumbing takes some time to get it right. We need a way to control the water temperature. With lots of valves and fittings on hand, we design a system that bleeds off some cold water and feeds it into the top. With careful adjustment, we can now blend the temperature of the outgoing flow to individual tastes. For a shower head we punch holes in the bottom of a 32 ounce tomato juice can. It takes an hour every night to get the temperature up, but everyone enjoys the luxury of a hot shower. Two or three guys, are doing the semi-steam bath thing, standing under almost scalding water and then running down to jump in the stretch of open water between the shore and the ice. I want no part of that. I am satisfied with a nice shower.

Not long after, it all goes back to base camp, but we had fun, fun, fun 'til Daddy took the coil stove away.

One day a couple of arctic hares visit the camp. They are at least as big as prairie jackrabbits, and their fur is a mixture of white and brown. It is amazing how their coats change with the seasons. There is still snow in places where it is shaded by rocks or where there had been large drifts. If the hares are standing still on the tundra, they are hard to see. I think we are camped on one of their grazing spots.

So we decide to herd them on to a little point south of the tents. Do we think we can catch one? I guess we just want to see what they will do. We form a semi-circle and close in on them, forcing the hares towards the narrow point and closing ranks until we (the apes) are five feet apart. The two animals start running back and forth, and then break for freedom. We don't have a chance. They are through our perimeter and gone in a flash – no dumb bunnies these boys.

With Wayne absent, Gary's crew is down to three men (including Gary) and they are quite a bit ahead of me by now. I talk to Gary. Why not give me Mark as a mag man? So Mark becomes my right hand man.

It only takes a day to train him, he already has the grid system down pat. I show him how to keep the notebook, how to do base station checks, how to detail highs, how to record significant topography changes, and so on. Any reading 400 gammas above background is considered anomalous. It might be a trend, or perhaps just a magnetic boulder. To detail the area, the operator fills in a square around the station by pacing an intermediate one-hundred-foot line, bracketing the anomalous reading.

Mark is a sharp kid but he is Gary's buddy, and being aware of the undercurrent of animosity between Gary and me, Mark is a little cool. Well, tough titty, Miss Kitty. I now have confidence in my own abilities and need no Oscar to whine to. Mark will be my mag man for the rest of the summer and we will eventually get along.

We now have 24 hours of daylight. I start to read base stations at night, leaving camp at 9 pm and returning at 3 am. Mark does the regular survey in the daytime and spends an hour or so in the evening correcting his notes. At night I have no diurnal drift at all to worry about, and our base station checks are good, solid readings. A soon as I detect drifting, I shut it down in the field. I work sometimes to 3:30 or 4 am, other times no later than 2:30. I read cross lines every thousand feet, giving Mark a base check every

12 to 15 minutes. The maps are going to be bang on, and back readings for unreliable results are no longer necessary.

Also, as I grow more confident in Mark's ability, I start showing him a few tricks: how to keep four or five readings in his head, and how to zero the mag before starting each line. This eliminates some notebook correction time, giving Mark less work in the evening. I sleep until 10 am and then spend the rest of the day at the map table.

We are really rolling now. By summer's end Mark and I will generate over 40,000 mag readings, not counting base stations. Not too shabby!

(We will use over 10,000 board feet of lumber for laths this summer. Assuming equal production elsewhere in the Contwoyto project, that number can easily be doubled. We are more than a hundred miles north of the tree line, but we plant our own forest.)

I really enjoy the night work. It is so quiet out there, never windy and never very cold. It is just cool enough to wear a bush jacket, which is a good thing because the flies and mosquitoes are thick as soup. In the daytime the hot sun and winds keep the mosquitoes away and even slow the black flies somewhat, but at night they both have a hey-day. I carry a can of insect repellent in my lunch pack and stop every half hour to spray my face, neck and the backs of my hands. I'm sure I will smell of "eau de off" until Christmas.

One night in the wee hours I top a little rise and stop dead in my tracks. A big old arctic wolf is loping toward me! Well, maybe not so old – he may look old because he shedding his winter fur but he sure is big, and getting bigger! I stand motionless, and I tell you, every hair on my body stands straight up, even my nose hair! The old boy hasn't seen me yet, and I wait until he is 200 feet away before I chicken out and holler.

The lobo stops short, (thank goodness) stares at me for the longest time, testing the air a little. I guess I don't look tasty, because he does a 90 degree turn and chugs off over a little knoll.

I consider my options. He seems to be travelling alone. Perhaps he is too old for the pack and will soon find his final resting place in the midnight sun. I decide to keep on working, but my hair won't lay flat and I am looking over my shoulder more often than ahead. I pack it in – tomorrow is another day. I will never see a wolf again that summer, but once is enough.

I am now finished reading base stations. Mark has two days' work left, the grid crew has one day, and I have lots of map work ahead of me, but that night I settle down to read a book. It is James Michener's "Hawaii," and that sucker is thick. I have already read a hundred of the 900 or so pages, and I nestle against my rolled up sleeping bag for an evening's enjoyment.

Well, I love that story, and with 24 hours of daylight and no gas lamp necessary, I just plain lose track of time. I read the whole thing, finishing at 3 am!

Thanks, Mr. Michener, wherever you are, thanks for screwing up my next day. I have to work until midnight to finish up the maps. We will be moving on in the morning.

Table Scrap

I am concentrating at the map table about 2 pm one afternoon when someone knocks on the tent door, startling the hell out of me. I hadn't heard anyone coming and besides, who knocks on doors in the arctic?

I open the door, and find I have a visitor! A middle-aged man stands outside, nodding and smiling. "Hello," says I. He nods and smiles.

"Where did you come from?" I ask. He nods and smiles.

I don't think he is an Eskimo, more likely Dene. He has a medium sized dog with him and the dog carries saddle bags. I say "Hello" to the dog and the dog nods and smiles (wags its tail.)

The gent points to the kitchen, so we go there. He sits at the table and I pick up the coffee pot. He shakes his head. I pick up a frying pan, again he shakes his head. He points to the groceries on the shelf. "Aha," I say to myself.

I start pointing at things. Klik? Head shake. Tomato juice? Same thing. I point at the 10 lb bag of flour. A smile and a nod. Now we are getting somewhere.

I put myself in his moccasins. What else would I need?

Five lbs of sugar? Nod. Two cans of pork and beans? Nod. I line stuff up on the table. Two cans of peaches, two cans of pears – not a whole lot of stuff. He holds his hands up, nods and smiles.

We take the food outside. He puts the flour and some cans in his packsack. Then he undoes the dog's harness and removes the saddle bags. Doggy runs around a bit, wagging his tail. The gent puts the sugar in one bag and some cans in the other. Then he puts the harness and saddlebags back on the dog. What a performance! The dog whines and cries like a baby. "How cruel," I think, but with the pack secured and with a nod and smile, they head north-west along the lakeshore, the dog, the phony cur, trotting along wagging his tail. I guess he is a natural-born moan and groaner.

I wonder if I have set a precedent and there will be more visitors tomorrow, but I never see him again. At the Summer Solstice party I tell John about it and he says I did the right thing. People would show up from time to time, take a little grub and disappear. Nobody ever took advantage of the Free Food Convenience Store, just hit it in times of need, I guess.

On June 19th we are up bright and early to tear down. The Otter comes in at 9 am and we start our move to the next camp. Gary and Ian go with the first load, Mark and I stay behind to finish the teardown and load the next two trips. We are taking the kitchen and one bunk tent with us, as there will only be four of us on the next grid. Contwoyto will send out a couple of guys to retrieve the tent left behind. We make sure no stuff is left lying around outside, and we nail the door shut before we leave. (This is not to discourage human visitors, we just don't want Mother Nature blowing the door open and inviting our furry friends inside.) Our last load includes the canoe, motor, life jackets and mixed gas. These items will stay on the Otter and go back to base. We will no longer need a canoe because we are moving to a small pothole 20 miles southeast of base camp. It is perhaps five miles inland from the south-east edge of Contwoyto.

Two things about that trip stand out in my memory. As we fly along we can see the marks made by the muskeg tractor when it was driven back to Contwoyto. We could see that the driver tried to stick to rock as much as possible, but half the time tundra is unavoidable. The tracks stand out like a sore thumb, and they will last for years. I'm sure you can still see traces of them now.

Also, as we fly over large and small bays along the Contwoyto shoreline we see rings extending inland as if high water had left little ridges as it advanced and receded. I will later realize they are caribou paths. Like cattle, they follow each other in their migrations. How many years have the caribou followed those same paths? Makes you think.

Our new lake is a perfect two-mile circle, completely surrounded by the same sedges that made our Fuz Lake camp so comfortable to live in. There is also a wide beach of fine sand circling the lake between the sedge and the water. The only things missing are palm trees and naked ladies, so we christen it Lake Bora Bora.

Behind camp is a gravel esker, a.k.a. moraine. This is a new experience for me. Eskers were formed beneath the huge ice cap as it melted and moved, scouring the uneven rock surface beneath it. Rivers beneath the ice deposited the ground up gravel along their path, leaving the eskers behind – think speed bump. Now picture it forty or fifty feet high, and a hundred wide at the base, and perfectly symmetrical, covered with the same short sedges that cover our campsite.

The esker gives us shelter from the west wind. It also is between us and the job, and gives us a good workout morning and evening. That sedge is tough. Four of us will cross it twice a day for over a month, and never leave a footprint.

We throw up the tents and will fine-tune the camp tomorrow. We now have five foot walls. Each time the tent frames are dismantled you lose some height to splintering at the ends of the uprights, requiring a trim job.

When we offloaded the Otter the previous day we had been greeted by a family of four chick-chicks who help us build the tent – they are always underfoot. These guys are super friendly. In the evenings we leave the door open for fresh air, and they are always in the tent, just like curious kittens.

The oil stove is still sitting outside, it will be the last item installed in the tent. The pipes are lying beside it, with one longer 3-pipe section on the ground. The chick-chicks are having a ball running in one end and out the other, probably thinking what a great burrow it would be. I am working near the pipe when two chick-chicks dash into it. I give it a good roll, maybe six or seven feet. The two little guys stagger out the end like drunken sailors, sit on their haunches and quote me a few verses from the chick-chick bible, then go off somewhere to sulk.

We never do install the oil stove at this camp. It is summer in the high arctic, and supplemental heat is unnecessary. We will need the stove again in our last camp of the summer.

On June 21st we go out to check the job, and Mark and I help Gary and Ian start the grid. On the evening sked the night before, we were told not to cook supper the next day, the Otter would pick us up about six pm. Gary knows what is happening, but won't tell us.

204

So, on the evening of the 21st of June we return to camp, put on our cleanest dirty shirts and board the Otter. We are going to attend that pagan ritual known as the Annual Summer Solstice Bash.

It is only a 20 minute hop to base, and we land along the shore of Contwoyto, beside the base camp. Contwoyto isn't ice-free yet, but there is lots of open water near shore. The ice is no longer resting on the lake bed, and has candled. It is waiting for a north-west wind.

There must be 75 or 80 people milling around. All the fly camps have come in, all the base people, off shift diamond drillers, and even a few suits from Yellowknife are here. We renew friendships. Wes Marsaw is here, Bobby Zadow, everyone!

At 7 o'clock we have a roast beef dinner. We are the second sitting. The dining room was built for 40 men so we have lots of room. After supper we sit around and drink coffee and burp and fart. At 9 pm we watch a John Wayne movie, and after that is the volleyball tournament. I don't remember who won. There are five teams, but we are pretty non-competitive, we are just having fun. It is a dry party of course, but no one cares about that. At 1:30 we all go in for pizza and by 3 am we are on our way home. We sleep 'til noon and take the rest of the day off.

Sidebar: Contwoyto was John Mullock's baby, and knowing the man, I would have to think the solstice party was his idea.

When you think about it, the cost of the party was insignificant – we ate food in camp anyway. The fly camps were not all that far away, so the Otter taxi fare per man was negligible and Yellowknife V.I.P. contingent would have shared an Otter.

John threw the party, and he was repaid in full. We hit the tundra rejuvenated, and showed our appreciation by seldom taking a day off until season's end.

Sidebar: During the period from one week before to one week after the summer solstice the sun never touches the horizon, it just makes a big circle overhead. My base station reading time is limited to four hours – 10 pm to 2 am. This job is the same size as Fuz Lake, but on one grid. It's a long walk to the far end, and the area is rife with amphibolites, the highly magnetic iron formation associated with the gold discovery at base camp. I still read cross lines at 1000 foot intervals for Mark's check stations, but in the amphibole areas base stations are useless. In 25 feet the reading can change drastically. The upside is that Mark can check in at the next 1000-foot control. (A 200 gamma drift is negligible in an active magnetic area.) The downside is that 25-foot detail is necessary to track the structure accurately. Total mileage is down, readings are up, and the map work is endless.

Table Scrap

Inco did some work in the Australian Outback in the mid-sixties. Barry Krause was there. He said magnetic float would screw up the readings because small magnetic rocks just under the surface turned the mag into a coin finder. To put some distance between the mag and the float, the operator carried an eight-foot wooden stepladder and climbed it at each station to take the reading. Must have looked weird, eh?

On July 1st the ice goes out on Contwoyto. A strong north-west wind comes up about midnight and when I return to camp at 3 am the lake is grumbling and groaning. By breakfast time I join the crew to climb the esker and we share my binoculars to watch the phenomenon. The lake is five miles away and the noise has been building all night long. It is as if the lake is in agony – groaning, shrieking, crashing and smashing. No movie special effects could come close to imitating the sound, and we can see the ice piling up higher and higher – 50 feet high – but at that distance it could be a hundred. What a spectacle that would be at close quarters. The ruckus continues all day, although in the field it is muffled somewhat by the esker. The wind dies down that evening and Contwoyto goes back to sleep. In two days the ice pile disappears, but I'm sure the shoreline must be altered a bit on a yearly basis. Mother Nature can show her muscle sometimes.

Other than the chick-chicks, we see little wildlife. There are some birds around. We see ptarmigan on the tundra, along with the odd arctic hare and our little lake is visited occasionally by arctic terns and smaller, long-legged shore birds. They don't stay long. Our lake freezes to the bottom every winter, so there is little food for them here, and we are probably a bit too urban for four-footed creatures.

Falconbridge is working a concession between us and base camp, and there may be other prospectors in the area. (You can stake claims in the Territories just like anywhere else, but companies with deep pockets merely lease large concessions from the N.W.T. Department of Mines, so claim staking is unnecessary.)

Base Camp moves a drill into our grid. They haul the rig down by muskeg tractor and trailer and the first load doesn't show up on schedule. No aircraft are available, so Contwoyto asks Bobby Zadow, whose fly camp is between us and base, to take a canoe down the shoreline of Contwoyto to look for the guy.

Bobby pulls into a bay where a Falconbridge tent camp is located. Two men are sitting in front of the tent and as Bobby beaches his canoe they get up, go inside the tent and close the door. "That's strange," says Bobby, so he knocks on the door.

One of the guys opens it and looks at him. Bobby asks if they have seen a muskeg tractor. The guy shakes his head and speaks not a word. Bobby gives his head a shake, too. Time for some R and R, boys! (The missing tractor shows up at the drill site at midnight. He wasn't lost – he just didn't know where he was.)

Sidebar: Drillers never carry a compass: ergo, the muskeg tractor driver got lost. Even if he had carried a compass, he would have to know how to use it. The magnetic north pole and true north are never in the same place. The earth wobbles on its axis and the magnetic poles change a little every year, moving in an oblong path. A compass always points to magnetic north, and has to be adjusted to read true north (declination.) In the early '60s Thunder Bay was at zero degrees declination and Contwoyto was at 31 degrees east declination. If you didn't know that, your compass was not of much use to you.

The Last Camp

We finish the Bora Bora job in mid-August, send the maps to base camp and move on to our last camp of the season, a short hop to the south end of Contwoyto. We set up on the north-east shore less than a mile from the end of the lake. Our plane loads get smaller with every move. The tent frames are now down to four-foot side walls with no floors. The flooring materials have been taken back to base camp, with some smaller pieces left at Bora Bora. In our defence, I must point out that kitchen waste is flown back to base with every backhaul to be disposed of at their landfill, but we do leave some detritus, including human waste. This would be unacceptable today, but at the time we were considered very environmentally conscious.

We set up the new camp on a small gravelly area near a sand beach. An Arctic Blue Fox shows up immediately, and he has absolutely no fear of us. He is like a chick-chick – every time you turn around you almost trip over him. It is a little cool today and we are wearing work gloves. To hold a nail, of course, you lay your left glove aside – bye-bye glove. There is a boulder bed just north of us, a good place for a biffy. The poop can drop between the boulders – out of sight, out of mind. Of course you remove a glove, and make another donation to the foxes' junk drawer. By 3 pm we are all down to a single glove, and that goes into a pocket when not in use.

We lose our hammer. Someone had gone to the beach for more materials and had laid it down. When we need it, it is gone. Did that rascally fox take it?

Well, I mentioned that the mag was like a coin finder, and the sand is very dry and fine. I unpack the mag and run it over the surface of the beach. On the third pass I find the hammer where it had been buried as we hauled our lumber. Mr Fox is off the hook.

The fourth night in our new camp, we are all snug in our sleeping bags at 11 pm when something hits our tent roof and slides off. It startles us somewhat, but I think it may be a ptarmigan, who, flushed out by a predator such as the fox, had taken flight in the darkness and has blundered into the tent. But then it happens again, and yet again. What the heck? I grab a flashlight, pull on my boots and go outside to check it out. I am a little nervous, to tell the truth.

There is the fox, having the time of his life. He takes a short run at the tent, jumps as high as he can and skis off the tent fly to the ground. We knew he was full of the devil, but I am amazed that he can make such a human-like choice of late-night recreation. Oh well, he's having fun, and soon he'll get tired of the game.

But the little bugger won't quit, and we need our sleep. Finally, I get a short piece of 2 x 4 and the next time, I give a little shot to the tent roof to discourage him. It takes four jabs before he finally gets the message, but he finally quits. Mamma Fox didn't raise no dumb kits.

We have been feeding scraps to the fox, but he becomes a pain in the ass around camp. We hate to be mean to him, but we start carrying the scraps 200 feet from camp and chase him away if he comes closer. Once again, he gets the message. Smart boy. (Girl?)

The days are rapidly shrinking. We lose 15 minutes of light every day, if not more. Night-time base station work is out of the question, but the diurnal is not all that bad and Mark and I split the days. Map-work is now done on a small piece of plywood

207

balanced on our knees. No problem, we just do a rough map, readings only – no contouring. We just need to know if a significant trend deserves more detailing, and it soon becomes apparent that this grid has nothing going on. By August 23rd we are almost done.

Yesterday a bull caribou appeared on a slope a half mile north of us. We put the binoculars on him – he seemed to be in distress. He looked old and shaggy and carried a huge set of antlers. He was feeding on lichens on a flat outcrop area, and he would toss his head, feed a little bit, run a few steps and toss his head some more. It looked like he was going goofy.

We finally figured it out - he <u>was</u> going goofy. The bulldog flies were driving him nuts. He must have been ahead of the herd now migrating southward. I hope the flies left him with some breeding energy when his harem returned.

On August 31 we tear down and go back to base camp. The students all go home, and I will spend the next three weeks at base.

Sidebar: There is now a diamond mine on the north shore of Contwoyto. The Diavik Mine is a huge open pit/underground operation, and I believe it is the richest of several in the Canadian Shield. It was found in '91 and I know there was the gleam of diamonds in some prospectors' eyes in 1963, but if you had mentioned diamonds around the base, you'd have been laughed out of camp.

Table Scrap

In Cochrane during the fall of '62 I dated a very nice girl a couple of times. I met her dad, and of course, he knew I worked with the Inco crew. He pulled out a letter to show me. A junior exploration company was looking for investment dollars (there's gold in them thar wallets) and extolling the exciting possibilities of diamond bearing kimberlite pipes in the Cochrane area. I told him I would ask my boss's opinion.

When I told Barry about it he laughed. He pulled out some air photos and showed me what the venture boys were flogging. The air photo showed the supposed kimberlite occurrences – areas where tall trees encircled a boggy Labrador tea depression. (think coral reef around a Pacific Ocean atoll.) A diamond mine was quite a stretch of the imagination in Barry's judgement. Well, guess what? Today, diamonds are being mined not too far north of Cochrane. Those circles really were kimberlite pipes, and diamonds are still being found in the area.

Mindsets are hard to change. In June of '69 I was in a Cessna 180 flying from Thunder Bay to Red Lake with my boss. Highway 599 runs from Ignace, Ontario 200 miles north to Pickle Lake. We crossed 599 forty miles north of Ignace, and I had been looking down at miles and miles of what seemed to be untravelled boreal forest. I tapped the boss's shoulder, pointed down and said, "Maybe someday we'll be chasing a mine down there."

He scoffed. "Nothing but moose pasture in this area."

Can you imagine? Even as we passed over, a drill crew may well have been emptying a core barrel of massive sulphides into a core box. Three months later

208

Mattagami Lake Mines, a division of our own company, would make an announcement'

Another rush was on!

Another phenomenon! As I pointed out, we no longer had a tent floor and had set up on what we thought was dry gravel. Well it may have been dry for an inch or so, but within a week we were walking in a soupy mixture of gravel and clay on our way in and out of the tent door. We had to lay down a couple of pieces of spare plywood to walk on. It seems that constant travel eventually melted the permafrost lying just beneath the surface. The mystery of the caribou ring paths is solved.

I mentioned my observation to Ed at base camp a couple of weeks later. He told me that back in July a drill crew had left a caterpillar tractor idling on the permafrost while they were setting up the drill. Two or three hours later they came back to find the cat mired up to its radiator in soup. The vibrations had melted the permafrost and the cat was headed for China.

Seasons End

It only takes me two days to redo the last maps. It is sort of a treat to work in a brightly lit office with other people around and I can stand it for a day or two. One of the draftsmen is my friend Wes. I don't know how he does it, but he always has a joke for suppertime – one joke a day and never the same joke twice. Let's see – 4 months x 30 days = 120 jokes. I must have missed all the good ones, because it looks like Wes is getting to the bottom of the barrel.

Example: I heard on the news that a woman was shot in Edmonton and the bullet is in her yet. What's a "yet?" …Groan.

There are always some real characters in these camps. At the Summer Solstice bash during our 2 am pizza feast, a little diamond driller jumped up on a table and it was obvious he'd been into a private stash of overproof. He gave us a complete, unabridged, concert-hall-worthy performance of "The Cremation of Sam McGee" by Robert W Service. I was told that if you added up all the other words he'd spoken that summer the total would not have exceeded the poem.

Sidebar: I have known many drillers over the years. A lot of them swear that this year they are going to save their paychecks and go to Mexico. They never make it past Thunder Bay or Winnipeg. That fall the little guy never made it out of Yellowknife. He sobered up in the slammer and went to work in one of the mines.

I work in the core shack for my last ten days. My job is to split core. The geologist leaves me a row of core boxes every morning with the sections to be assayed marked. I use a hammer and a broad chisel to attempt to split the core in half. Sometimes it splits easily, but often the rock fractures diagonally. Then I have to remove each small section, split it and put it back in the exact order it came out. Later on, the geologist bags one half to be sent out for assay. The other half stays in the box for backup. Sometimes, (but not likely) the assay office will screw up, or investment regulators might want proof that an assay actually came from that intersection. (Also it is highly unlikely that a big company like Inco would be running a stock market scam.)

209

Nevertheless, half the core always stays in the box. Sometimes the assays can be surprising and a re-logging is in order.

These days, the core is cut with a saw. But no matter how you cut it, it's a boring job. I wish I was out with the black flies.

I do get a break occasionally. From time to time the drill holes are checked for angle and direction. Say, a hole is started at 60 degrees due west. At depth every hole will drift, left or right, or at a flatter or steeper angle. A simple angle test can be done with a mild acid/water mixture in a glass test tube, but this only gives the angle of the hole. A Tropari test is more accurate.

The Tropari unit is a gyro compass housed in a six-inch-long brass tube the same diameter as a core barrel. To isolate it from the steel rods it is mounted at the end of a fifteen foot brass barrel. This is screwed onto the end of the drill string, the wind-up timer built into the instrument is set, and it is lowered down the hole to the test depth. Wait a bit for the trigger to trip to give the gyro a spin. Wait ten minutes more for the mechanism to lock the gyro. Pull it out and take the whole thing back to camp for the geologist to read. He now has an accurate measurement of dip angle and direction at that point in the hole. The only limitation of the Tropari is that down-hole magnetism can skew the results – it has to be used in non-magnetic sections of the hole.

There are three drills working the main showing, so I lug the thing out almost every day, sometimes twice. My shoulders are getting sore, and the core splitting is gaining lustre in my mind. Crickey, it's impossible to please some people, right mate?

Table Scrap

Why did I lug that unit back and forth, rather than leave the fifteen-foot barrel on site? Well, let me tell you a story. (re: Bayfield Ventures and Bob Marvin – 2010.)

One day I arrive at the Bayfield core shack to find Bob sitting at his desk, scratching his head. "Got a problem?" I ask.

It seems there is a hiccup, although not really Bob's problem. Their contract driller owns their own Tropari – not a Tropari as such, but similar – a brass alloy tube with a few thousand dollars of electronics on the end in place of the gyro-compass. The drillers had moved to a new hole and had laid the instrument on the floor of the drill shack. The drill was moved as whole unit, pulled by a cat. The brass tube is usually carried by hand to the new hole, but the foreman, who usually totes the Tropari, was not on site. As the drill bounced over the rocks, the tube slowly slipped through a hole in the floor, breaking off a few feet at a time. The drill arrived at the new site and the foreman showed up carrying $15,000 worth of junk and swearing like crazy. Maybe that's why I had to lug that thing back to camp every day at Contwoyto.

I notice an odd thing at the drills. Each drill has an open topped tank beside the shack that looks like a big horse trough. At the top of the casing is a tee fitting, the return water is routed to the tank, and the drill water is taken from the top of the tank. I know there is a supply pump and coil heater back at Contwoyto, since I follow a water line from base camp every day that tees off to each drill site. Normally there would have been another coil heater at each drill, but there isn't. What's the deal?

The problem is the permafrost. Drilling with heated water is the norm in cold weather in the boreal forest. If the supply pump is too far away, the water cools off, so you must set up another heater at the drill. But drill holes can also be water wells, and this is the arctic. When you pull the drill string, the static level of the water may be above the bottom of the permafrost. Before you can empty the core barrel and send the string back down, the water in the hole will freeze. So you need anti-freeze.

Barrels of methanol sit by every drill, and methanol is added to the water in the horse trough. The methanol/water mixture doesn't freeze, but to add methanol to every gallon of supply water would be prohibitively expensive, so the return water is routed to the horse trough and recycled. The sediments sink to the bottom of the tank, and pressure pump water is taken from the top of the tank. Water and methanol is added as the hole progresses. The strength of the solution is controlled by a simple specific gravity test, just like an antifreeze tester. Methanol evaporates easily, so hourly checks are made. Methanol is used instead of the more stable ethylene glycol for environmental reasons.

Table Scrap

Before we leave Contwoyto I've got to tell you about Ed's chick-chick. When Ed arrived in Contwoyto in May he was greeted at his office tent by a chick-chick that had a burrow not far from his front door. Ed started to bring a snack back from every meal and Chick-chick would wait for his treats, giving him hell if he forgot. Over the summer the little guy packed on the pounds, and when I arrived back at base he was no longer a little guy. He looked more like a Buddha, sitting there beside his burrow, which had been under constant renovation. I can't remember what name Ed gave him, but a few years later he would have definitely been Jabba the Hut.

Ed also told me that a cook had planted a little garden along the south side of the cook shack, protected from the westerly winds. The peas did not do well, but the radishes and carrots flourished, shooting up in the 24 hour sunlight. However, when they pulled the veggies there was nothing underneath, just tops.

I've often wondered – we've all driven past fields of sunflowers. It's amazing how they greet the morning sun and follow it until sunset. Well, what if you planted a sunflower in the arctic? Would it follow the midnight sun day after day, eventually screwing itself into the ground?

I used to ask people's opinions of my sunflower theory, but I don't anymore. Their eyes glaze over, and they say something like, "How 'bout those Maple Leafs?"

Done and Done

We are heading home! Eighteen of us are waiting at the beach with our packsacks and sleeping bags and some of the base camp sucks have suitcases. (They have been sleeping on clean sheets all summer, too.)

Our plane arrives, settling gingerly down on the water amidst an impressive spray show. Now, there are strange things done 'neath the midnight sun, but the strangest of all is the Stranraer. It is a biplane flying boat with no landing gear, just a hull with pontoon floats at the ends of the lower wing. Two big radials sit on the upper wing and fifteen miles of piano wire criss-cross between the wings. This is to make sure they all stay on the same page while in the air, I guess.

211

The Stranraer is an endangered species. This may be the only flying example left in the world. I think one or two other relics are parked in the corner of an airfield somewhere, salvaged for spare parts and scrap aluminium.

The plane can only come within a hundred feet of shore due to the slope of the sandy bottom. It turns into the north-east crosswind and anchors are deployed – no kidding! Then we watch as twelve to fifteen fuel drums are unloaded. It is pretty neat. The air crew opens the cargo doors and attach a little rail to a bracket on the underside of the lower wing. The rail meets another one attached to the ceiling of the cargo hold running front to rear (or I should say fore and aft – his thing was like a submarine with wings.)

Now the barrels appear, hanging on a barrel clamp suspended from a little barn door roller thingy. When the barrel clears the fuselage, someone pulls a rope, trips the clamp and the barrel falls into the water. The wind blows the floating barrel slowly to shore where a muskeg tractor awaits. As the barrels hit the sand bottom, a man wearing hip waders attaches another clamp and the muskeg tractor pulls the barrel up onto dry ground. It sounds kind of complicated, but in a half hour the plane is empty and now we can get on board.

(You know, we don't get many north-east winds. If we had missed this weather window I might have been splitting core for a few more days.)

SUPERMARINE STRANRAER

We shuttle out to the plane by canoe and step through the door, down into the belly of the beast. I feel a little more secure about the Stranraer's structural integrity once I am in my seat. The 18 seats are split into three sections, six to a section. Strong aluminium bulkheads separate the sections, with a passageway big enough for passengers or freight. It sure looks strong and the resemblance to a sub is even more apparent. Every seat is a window seat, and when I look out, (big mistake) the lake level is only a few inches below the window ledge – I swear on a stack of Robert Service poems!

We taxi out onto the lake, turn left, and the pilots pour on the power. "Now this is strange," I think, "A plane always takes off into the wind, and Contwoyto is more than four miles wide here."

I had seen the Stranraer coming in crosswind, and we are leaving crosswind! It doesn't take long for me to figure out why. The water below my window starts to pass by slowly as we pick up speed, and then the lake drops away from my window just as slowly. Faster and faster we go, and finally break free. How far we went to gain flying speed I do not know, but I think we took a left over Coppermine to head south to

Yellowknife, where we will overnight and then catch the DC-6 to Edmonton the next day.

I don't think the Stranraer made many more trips. A few short years later I read that Pacific Western donated it to an Aviation Museum in England. Rest in Peace, Stranraer. You got me to Yellowknife in one piece, and I thank you for that.

Two days later I get off the train in Winnipeg at the end of a strange, wonderful summer. Like the guy said, I've seen and done enough to write a book!

Table Scrap

Inco spun off the Contwoyto find, reputed to be at an ore grade of six ounces per ton, and it later became the Lupin Mine.

In 1980, with gold at $850/oz, Echo Bay (Lupin's parent company,) picked up the property and flew all the construction material to Contwoyto – an unimaginable feat at the time.

Using a Convair 640 and a Hercules C130, and flying 24/7 into a gravel airstrip for a year, 64 million pounds arrived on site (and we thought the Bristol Freighter and the Stranraer were heavy duty!)

The Lupin Mine went into production in 1982 and produced 3.3 million ounces of gold over the next 20 years.

In 1982 a 360-mile ice road was built from Yellowknife to Contwoyto to service the mine. This became the basis for the present winter road to the current diamond mines, one of which is on the north shore of Contwoyto, not far from our last camp in 1963.

And here I was thinking that the twinkling along that shore as the sun went down was Mr. Arctic Blue Fox, winking at me as he chewed on my glove!

Chapter IX   Fall of '63 - Season's End

After three weeks of R&R at the home place I bus to Sudbury. My Meteor has died an unusual death. An overheating problem has resulted in a snapped oil pump shaft and I will never drive the old girl again. I think of her once a month though, when I send my payment to the finance company.

I learn I will be taking a crew of two into Bull Lake, forty miles north of Sudbury. We will take an EM and a mag of course, but we will also have a packsack drill with us. I still hate that vibrating devil, but we will be drilling our own conductors, and I look forward to it. To my knowledge, Inco crews had never done this before. Maybe I will hit an ore body, get to sink my own shaft and put my own name on the security gate – Nah. Once again I am dreaming of champagne on a beer budget.

My partners are Kenny Wolfe and another guy whose name I have forgotten, let's call him Ron. They are both good guys, but only Kenny has any experience. No problem.

Kenny is a wannabe hood – wears his hair slicked back in a duck tail and talks tough. Trouble is, he has a great sense of humour, and is too good-hearted to be a bad guy, but he loves big cars with big motors and 4-speed transmissions. We get along fine.

So we get our gear together, pick up some groceries and head to Ramsay Lake. We fly in on an Austin Airways Beaver, and Kenny and I are on the first load. As we circle Bull Lake I think I spot a road, and sure enough, our campsite is going to be on an old landing, and as we tie up the plane I see a road heading south. It has obviously not been used recently, but a road has to go somewhere, right?

Now let me tell you – a fly camp on the ground in October/November doesn't warm the cockles of anybody's heart, but sometimes you have to bite the bullet. We start unloading, but my mind is working overtime.

"Hold it right there," I say, and pull out the maps. Our four targets are off to the southeast, and the road seems to go in that direction. The first target is four miles in and there are two more targets beyond that – too far to walk to and still do a decent day's work.

"We're going to do a little sight-seeing," I tell the pilot, and we reload the plane and take off again. We follow the road less than five miles and find men and horses working. Another five miles and we fly over a good-sized lumber camp, and beyond that is a better road continuing on down the northeast side of a river that runs southeast.

I am getting pissed off. "Back to Ramsay," I tell the pilot. He smiles and banks the Beaver toward Sudbury.

I bit the bullet to go into Bull lake, but I am about to spit it out. A year ago I would not have had the cojones, but c'mon, man! I have come within 30 miles of earning my sourdough badge, and you are going to put me in a fly tent camp ten miles away from a lumber camp?

We unload the plane at Ramsay and I call a surprised Herb. Someone picks us up and I stride into Herb's Copper Cliff office, lay the maps on his desk, and we have a short one-sided confab. Herb grudgingly pulls out a few air photos and there, in black

214

and grey, is an all-weather road leading down to the town of Field, where we can pick up a secondary route to Highway 11, just 50 mile east of Sudbury. A few phone calls later we find out that the lumber company cuts mine timbers for Inco, and, in fact, is owned by them! We are welcome to stay there for the cost of our meals!

The next day we load our stuff into an old Land Rover station wagon and head in. A little foresight on Herb's part could have saved us a whole lot of hassle, but I have a feeling in the back of my neck that this is going to bite me sooner or later. Less than three months down the road, Herb will get the last bite.

We head up the road to the lumber camp. The road sure looked better from the airplane. It is a real turkey track, to my surprise – a wood haul road should have been much better than this. It turns out that they don't haul the wood by road. The logs are piled beside the Sturgeon River until spring, then floated down the river to the sawmill in Sturgeon Falls – a good old-fashioned river drive.

After catching some air on the first two bumps I stop and get down on my belly to see if we have broken a shock hanger. Nope – no busted hanger – we have no shock absorbers! To keep the axle from heading south, this Land Rover has two snubbers of strong webbing fastened between the frame and the axle. This might work on an African safari, but such a system has not been used in North America since the Reo Speedwagon.

Sidebar: Lots of people are in love with the old Land Rover, but I am not among them. I have read articles extolling their merits – tough, dependable and they last forever, so they say. Well, they are tough on the people inside, but certainly not dependable. They are rough riding, noisy, and hard to start in the winter.

And as for lasting forever? Later on I worked for Noranda Explorations for three years, and they had a Land Rover. It sat in the corner of the shop waiting for brake parts from South Africa that never came. As far as I know it is still sitting there. That one might last forever, for sure.

The Old Lumber Camp

So we slow for every bump, and there are lots of them. High gear is out of the question, and we arrive at the lumber camp well after dark. I pick out the best looking building and knock on the door. The boss opens the door and we introduce ourselves.

"Sorry, you missed supper," he says, "But breakfast is right around the corner." Turns out he isn't kidding, breakfast is at 5 am!

He shows us our beds and we hit the sack – it's lights out at ten pm and we had just made it. At 4:30 the boss gets up. At 5 am there is a good loud clang, clang, clang. Someone is beating the hell out of a triangle with a big steel crowbar.

We jump into our boots and join a line of men heading for the cookhouse. It is a long building wide enough for two tables with attached benches running the full length, as if they are going for a Guinness record for the world's longest picnic table. We grab three places near the door and are lucky to get a seat. There are few empty spaces.

There is food on the table – lots of food, and good stuff: pancakes, porridge, fried eggs, bacon, ham, toast – even bowls of canned fruit. As fast as a bowl is emptied, a cookee replaces it with a full one.

215

I am full as a tick. I pour myself a cup of coffee, light up a smoke and lean back – and the cookee taps me on the shoulder.

"Put out your cigarette, put your dishes in the sink and get out," he says sternly.

Oops! I am a little embarrassed, but I am not resentful. It definitely is a working camp here – no dudes allowed.

We go back to our quarters and have a chat with Hank, the camp boss. He shows us a storage shed to store our drill in and gives us a few rules – not many, but not bendable. It is a dry camp – no problem, we seldom take booze into the bush anyway. Breakfast is at 5, coffee is on at 9, lunch is available at noon if we want it, supper is at 6 pm, no appointments necessary – and it's lights out at 10 pm. This is not a request, he turns the lights out himself. If you are halfway through a sentence in your reading material, tough titty - pick it up tomorrow.

Turns out his bark is worse than his bite. He always gives a two-minute warning at lights out. Hank is a throwback to the good old days, and so is the camp.

Hank must be 70 years old, and he is a man-mountain. He is all angles and planes – not a rounded corner on his body. His face looks like an unfinished work of chainsaw art. You could split wood with his nose. It looks like the two rocky outcrops of his eyebrows need some finishing work, the same with his chin and cheekbones. His hands are out of proportion. Big as he is, they are still too big for his body. It looks like he is getting a little stiff in the joints, but he stands and walks ramrod straight. This is a man who, although he commands respect, does not demand it. We are in awe. Naturally, he is soft-spoken – everyone knows you don't give no lip to Big Hank.

The camp, like Hank, is rough-hewn. The buildings are of rough-sawn lumber, inner and outer walls. The ceilings, doors and even the sinks in the cookhouse are made with unplaned boards. The only smooth surfaces are floors, benches and table and counter tops. In addition to our large cabin, which we share with Hank and the government scaler, there is the cookhouse, cooks' quarters, large bunkhouse, large horse barn, the mechanical shop and the office/commissary. None of the buildings are painted, but paint would look out of place, anyway. They have all weathered to a soft grey, and the camp looks exactly as it should. It is and old-time lumber camp.

There are maybe 60 men in the bunkhouse, probably 45 of them cutters. The rest are truck drivers, equipment operators, and other support personnel. This camp is certainly not very mechanized. There are three or four old International 5-ton single-axle trucks and there may be a couple of cats and loaders back on the slash. The men use power saws, of course, but the logs are skidded out with horses. Each three-man crew has a team. One guy cuts and limbs, one man hauls the trees to the landing, and the third man cuts them into sixteen foot lengths at the landing and piles them. The trucks will then haul the logs to the riverbank to wait for the spring drive to the sawmill.

As rough as lumber camps are to work in, this one is rougher than most. The terrain is hilly, rocky and dangerous. The men are paid piece-work – in other words they are paid for what they cut, and the truck drivers are paid per load moved. Few are on an hourly wage. The scaler tells me that only three or four of the top producing crews get decent strips to cut in. The others have a tough time to make a go of it. One guy worked a month and when his commissary bill was deducted he had ten dollars left. Also, if you injure a horse, say, let a trace chain rub an open sore on his leg, the barn boss takes the horse out of service and the crew has to pay for its feed until the injury heals.

The camp has an old school bus divided into passenger and cargo area. It runs twice a week fifty miles to Sturgeon Falls for supplies, and the front half is usually full of disgruntles men going out and hopeful men coming in.

The camp clerk runs his own commissary selling tobacco (cigarette, pipe and chew,) rolling papers, shaving cream, razor blades, etc. and chocolate bars and pop. The first day in camp Hank tells me to put my cigarettes away. No one is allowed to smoke tailor-mades in camp as they have a chemical in them that makes them smoulder, while a roll-your own will go out if you drop it. So I go to the commissary and buy tobacco and papers.

The next day we go to work. We locate the first conductor right away and blaze a small grid. Since we are doing our own follow up drilling, an elaborate grid is not necessary. We merely trace the conductor far enough to establish the strongest part, check for magnetic association, and then spot the bore hole. We spend two days on the grid and bring in the drill on the third day.

On the first day of bush work we hit the cook shack at 5 am, but it doesn't really get light out until 7:30 and you can't do instrument work in half-light. We nap until 8 am and it really screws up our day. So we go to plan B - we decide to sleep in until 8, get ready for work and hit the morning coffee break at 9. Our job is not far off the road, and by working nonstop until five we can still get in a good day's work.

on the second morning we go into the cook shack at 9 am. There are five or six guys at the table, sitting up near the huge oil fired cooking ranges; Hank, the scaler, the barn boss, the commissary clerk and a couple more, probably mechanics. Besides coffee and tea, the table is loaded with goodies – muffins, tarts, cakes and pies. This camp eats good, let me tell you!

So we sit down to a continental breakfast and I sit next to the barn boss – big mistake. This guy must wallow in horse piss! My eyes are watering and I can't breathe. It doesn't seem to bother the other camp guys, but we can hardly stand it. After that, we make sure to sit on the other side of the table as far away as possible without being obvious about it. We really don't want to offend the guy.

In the barn boss's defence, I must point out that this camp has no shower, bathing or laundry facilities. He works with the horses all day and cleans the barn. He probably also has sleeping quarters in the barn. By the end of the week we are pretty much used to it. We don't smell so good either and he is only a little more stinky than us.

We set up the drill and pressure pump and start drilling. The rock is not too hard and we make good progress. At the end of the day we have the casing collared in bedrock at fifteen feet, ready to start coring. Day two, we drill 30 feet, not bad for a packsack crew. On day three we make another 25 feet – drilling goes slower at depth.

Back in Copper Cliff, Herb had shown me a formula to use to spot the hole using the EM info and I wish I could remember it now. By measuring the angles on the EM readings on each side of the conductor you can determine the dip of the rock structure, and also calculate the set-back of the hole. You can also determine the angle of the hole and can predict the depth at which you should intersect the conductor.

217

Using Herb's formula we are supposed to intersect the conductor at 75 feet. I am sceptical, but the next day we drill five feet and pull two inches of solid graphite – bang on. I am amazed! Old Herb does know a thing or two after all. We run another five feet, determine that we are on the other side of the zone and shut the hole down. We are very pleased with the work so far. It is October 21st and the first hole is done.

The next hole looks easy, but we have some rain and it takes us three days to locate the conductor and spot the hole. We set up on an outcrop – that is a bonus because we can put in a rock bolt for the Johnson bar anchor and no casing is needed. We get pretty wet starting the hole, as water sprays out until the bit is down six inches. The first ten or fifteen feet goes well, but then we hit quartzite and the packsack drill doesn't handle quartzite very well. It is very hard rock, and we can't put enough pressure on the bit to keep the diamonds from being polished. We manage to drill through the two-inch hard section, but three inches farther on we hit quartzite again. The rock is also fractured and very blocky. We seldom pull more than six inches of core at a time. Pull and change the bit – pull and clear the block – it is all very frustrating.

By the third day we have used up all the bits. We had brought ten in with us and had only used two on the first hole. Now we are only thirty feet down the second hole, and have polished eight bits. We must go back to Copper Cliff for more. We take the core boxes for the first hole back at the same time. The trip takes us two days, but we do have a chance for a shower and a beer. We come back with twelve bits with coarser diamonds – maybe these will cut better – nope. Herb suggested throttling back a bit, perhaps a slower speed will prevent polishing. I know this is common practise on the big drills, but they can handle it. Our packsack drill has a centrifugal clutch and needs high rpm to keep the clutch properly engaged. I try the throttling back thing against my better judgement – result – a burned clutch and housing. We have spare parts, so we borrow some tools and bench space in the shop and replace the clutch. I make a decision to abandon the hole and go on to the next one.

That evening Kenny goes over to the commissary to buy something and comes back white as a sheet. "What's wrong?"

"There's a dead man lying on the floor over there," he says. "I've never seen a stiff before."

It turns out some poor guy had died in his sleep – ticker problem, I guess. They laid him out in the commissary to await the coroner's arrival. Tough guy Kenny is freaked, and doesn't rest easy that night.

The next hole is even worse. We find the conductor on top of a high hill above the Sturgeon River and we spot the bore hole fifty feet down the side of the hill. Our water pump is 125 feet below us, and by the time we have water lines in place and the drill anchored, we feel more like mountain goats than men. It is ridiculous. I am standing below the drill, reaching over my head, and Kenny is up above, trying to put pressure on the Johnson bar. We do our best, but three days later I pack it in. We had put down five feet of casing, collared the hole in bedrock and once more are into quartzite almost immediately. We take the packsack back to camp and finish off the other two anomalies without trying to drill them.

We battle bad weather, but finally finish around Nov 25th. We return to copper Cliff, unload the Land Rover and turn in the keys. Herb isn't too happy, but we had given it our best shot.

Seasons End

I will be working with the land surveyors until Dec 15th when my Christmas break starts, so I check into the Frood Hotel.

My old friend Carl is at the Frood. He has spent the summer in the Nemiscau River country of northern Quebec, chasing anomalies. The area is now part of the reservoir for Long Grand One and Long Grand Two, Quebec's huge hydro-electric dams. Carl has then spent the fall season in the Sudbury basin observing and reporting on a new portable drill that Inco is testing. Carl's report will have some far reaching consequences, both for Inco and for Carl and I, but more about that later. In the meantime it is back to "do a little work and drink a few suds."

The land survey crew are a great group to work with. There are four of them and they have all put in their share of bush time. Now they have scored a basin job – home every night and weekends off. They have earned it.

Table Scrap

One night Toisto, one of the survey crew, invites Carl and I over for supper and drinks. Toisto has a nice homey place in West Sudbury, and he also has a nice family. We have a pleasant evening.

Toisto's street is on a bit of a hill and Carl parks his old '57 Meteor on the street, leaving the standard transmission in second gear. (The parking brake had rusted solid many years ago.)

After supper we are sitting in the living room sipping a cool one and telling lies. I look out the window and there is the Meteor, heading downhill. The engine has cooled off and lost compression, and Meteor has decided to go home without us.

"There goes your car," I say.

"Yep, that's my car." (Carl is in the middle of a story.)

"I'm not kidding," I say, "Your car is taking off!"

Carl jumps up and looks out the window. The Meteor is slowly picking up speed. Carl runs out in his stocking feet and catches it halfway down the block. Luckily it hasn't hit anything.

This time Carl parks it in reverse. We all laugh like crazy.

Inco is testing the new Winkie drill and Carl is doing the evaluation, thanks to a recommendation by our old friend, J-J-Jim. The Winkie was invented by Fred Wink and Fred had formed a partnership with J.K. Smit and Sons, a Toronto based drill manufacturer and supplier. Fred has come to Sudbury and he and Carl really get along well. He joins us in the Blue Room Lounge at the Frood and we have a long evening telling lies.

Fred has a proposal for us. He wants us to take a Winkie to Chattanooga, Tennessee later this winter. A company there by the name of Hensley-Schmidt is interested in the drill, and Fred needs competent men to demonstrate it. He has already had preliminary

219

discussions with Carl who told Fred he knew someone who should go with him, and Fred wanted to meet me.

Fred came up through the ranks from stope miner, to shift boss, to underground foreman, to management, to entrepreneur and inventor. He is a world traveller and has hob-nobbed with royalty, but he has never lost his connection with the working man. His enthusiasm is impossible to resist. By the end of the evening we all shake hands on a deal. The upshot of Carl's evaluation is two-fold: Inco will buy several drills and Carl and I will spend the winter drilling in North Carolina (but not for Inco.)

1963 has come to an end. Carl goes to his home south of North Bay and I spend Christmas with Aunt Thelma. It has not been a bad year at all.

Chapter X   Winter of '64 –North Carolina.

Jan 2, 1964: I go to the field office in Copper Cliff, where a whole slew of guys are milling around in the lower level chatting and comparing camp assignments. One by one we go up to Herb's office to receive our orders. This guy is going to Red lake, that guy to Smooth Rock Falls, etc. Carl goes in before me and learns he is going to Smooth Rock Falls with Barry.

Before Carl and I meet with Herb, we have a discussion. We are scheduled to be at J.K. Smit in Toronto on February 1st and I think it is only fair to tell Herb our plans. To let them put me in a camp for one month as party leader and then find out later that I'm leaving, seems kind of irresponsible on my part. There will be extra travel expenses for Inco, and some manpower shuffling will be necessary. Carl tells me to keep my mouth shut. As usual, Carl is right.

So I go into Herb's office and find out I will be going to Red Lake, the farthest away of the winter camps. I feel bad about it, and tell Herb I will be leaving at the end of January. I suggest that I work closer to Sudbury and save Inco some money and manpower hassle. Herb's mouth puckers up. "If you're going to quit, you might as well go right now," he says.

I am astounded! I am speechless! I look at this jerk who has been a thorn in my side for two and a half long years, then turn and walk out.

To return to the basement I have to walk through the building to the back stairs, passing the desks of Bill, Wes, and others who I had worked with and socialized with. They have all heard Herb – and are all very busy concentrating on whatever is on their desks – not a word – no one will even look at me. I feel very alone.

As I reach the top of the stairs, Fred – Fearless Fred of the Chibougamau saga stops me. "I overheard the conversation, Bob," he says, "That is very unfair. Come on back. I'll go in with you and we can fix this."

Now, this is a man I had locked horns with in the past, yet he is willing to go to the mat for me – the only one offering to do so. I feel like crying, yet I am strangely comforted.

"Thanks, Fred," I answer, "But I just can't kiss that man's ass." We shake hands and I continue down the stairs. (I will never see Fred again. You were a good man, Fred. May your peas and potatoes never touch each other on your heavenly plate.)

I walk through the basement to Nick's office to tell him I am done. The group is already very quiet - the moccasin telegraph has preceded me. I collect my packsack and sit on the stairs leading outside. Someone will drop me at my Aunt's house, I hope. Conversations start to pick up again, but I am not part of them. I have been cast out of the inner circle.

Sidebar: I didn't rat on Carl, and he didn't say, "I told you so." Carl waited until mid-January to tell Barry he was quitting. No problem, Barry wished Carl well and told him he would do the same thing – that opportunities to expand one's horizons should never be ignored. Different management strokes for different management folks for sure!

Table Scrap

Inco had switched to the Winkie drill and Mr. Packsack was retired. For two years they ran the Winkies haphazardly, then finally realized they needed to streamline the operation. A formal Winkie drill division was set up and they needed a supervisor. Carl was approached and he accepted the job on one condition. He knew he would be reporting to Herb, and he wanted a five year contract with the provision that if he and Inco parted company at any time in the first four years for any reason, he would be paid for the full five – and he got it. In other words, Carl could quit, but Inco couldn't. Pretty sweet, huh?

So it came to pass that Inco downsized the exploration department. Carl was called back to Sudbury, walked into Herb's office and sat across from him to learn his fate. Carl knew what was coming – the moccasin telegraph – right? Herb, with his usual smirk, handed Carl a letter. "Here's your notice of severance," Herb said, "You're all done with Inco."

"Well, not really, Herb," said Carl, "Do you have a copy of my contract?" After a short search, Herb pulled it out of a drawer.

"Maybe you should have read it," said Carl, "There's still a year left on the early layoff penalty."

"Give me back that letter," said Herb.

Carl replied, "Not on your life!" and walked out the door. He would spend a year prospecting and partying on International Nickel's nickel.

Herb didn't know it yet, but his days were also numbered. When there was no one left to fire, he was gone. Maybe he had to fire himself.

Back in the day when Copper Cliff was young, Inco had built two residences for single professional male employees. The Engineers' Club has 25 rooms, the Algoma Club 15. I am told to choose the Algoma and since I don't want to hang around my Aunt's for a month, I move into the Algoma Club.

What a great place; a big old high-ceilinged living room complete with TV, card tables and over-stuffed chairs and sofas, a spacious dining room and private rooms upstairs but no private baths – this is an old building.

I don't tell them I will be working for the government now (unemployment insurance,) although it wouldn't have mattered as tenants are getting scarce. I am anything but professional, but I am single and male – two out of four ain't bad.

Inco supplies the furnishings, bed linens, towels and utilities. The tenants pay the groceries and wages for the cook and a full-time maid. A tenant volunteer keeps the books and at month's end the costs are split 15 ways. They need a full house to keep per capita costs down. There is even an honor system beer fridge in the basement. If you take a beer you hang a tag on a peg with your name on it. I love the place!

On Jan 3rd I go straight down to the unemployment insurance office to file a claim. I tell my tale of woe, and whether it is my sad-sack look or Herb's reputation, I am put on full benefits, starting right now! Normally there is a two-week waiting period, but I receive two checks before heading off to Chattanooga. I have an enjoyable lazy month while Carl is slugging the snowshoe trails at Smooth Rock Falls. Who's the smart guy now?

222

Finally the anticipated month's end arrives. Carl picks me up in his brand new Chevrolet Bel Air two-door sedan and we head for Toronto. A couple from Carl's home town goes with us. They are going to Toronto to shop for a new car, and Duffy says, "I'll buy you a beer before we say goodbye."

Table Scrap

There were three ways to drink in public premises in Ontario in those days; men only (beer,) ladies and escorts (beer,) and, in an uncharacteristic lighter shade of the blue-balled Liquor Board of Ontario rules, cocktail lounges, with mixed drinks and beer in mixed company. The only difference between a beer parlor and a lounge is that the lights are lower and the prices higher. There is no dancing, no laughing, and in fact no fun allowed. Just shut up and drink! So we stop at the Jolly Miller on the Don Valley Parkway and go in the "Ladies and Escorts" door.

The Jolly Miller is semi-famous. We have all heard of the Jolly Miller and expect something fancy. Nope – same old beer parlor you can find anywhere. It is 3 pm and the place is not too busy. We pick a table and sit down, and are totally ignored. The waiter passes us by without even looking at us. We check each other's faces for drool, and Carl, Duffy, and I make sure our pants are zipped. Finally I get up and go to the bar. "Any chance for a beer?" I ask.

"Nope."

"Why not?"

"It's the law," says the barkeep, "No more than two men to one woman."

Can you believe it? We can't either, and we stomp on out. Jolly Miller my ass!"

Carl and I check into the Conroy Hotel and the next morning drop in to the nearby J.K. Smits office/shop/showroom to meet Fred and some other suits. We check out the gleaming new Winkie drills, pumps, etc. and receive some indication of what is expected of us and how we are to be paid. I guess the statute of limitations now covers my butt, so I'll spill the beans. We are to get $550 per month from Hensley-Schmidt Engineering of Chattanooga Tennessee (H-S). We don't have green cards, so H-S pays J.K, who sends us a check, and we send the checks back to our bank in Canada. It is a slow process, but with one small advantage: the Canadian dollar is a bit below par, so we receive a $12 premium each month. (Big plus – we pay no taxes to anyone.)

H-S expects us in two days and has a room booked in the Ross Hotel in downtown Chattanooga. The drill and various goodies will be shipped out the next day. We are already checked out of the Conroy and leave immediately.

We drive straight through, switching drivers when we are tired. We get our story straight before we reach U S customs – we are going to Winnipeg, Manitoba and want to see some northern U S scenery on the way. The customs officer looks at Carl's golf bag in the trunk and says, "OK, boys, but don't work while you're down here."

We made it! We think we are pretty cool. Try that nowadays, after 9-1-1.

(I was bragging on this the other day and was told that Neil Young did the same thing on his way to L A, the only difference being that Neil was driving an old hearse, and became a world famous singer/songwriter/gazillionaire. Well, if you must be out-cooled, it might's well be by the coolest. Besides, I wouldn't want Crazy Horse living in my gatehouse.)

We decide to take the scenic route, down through Pittsburgh and into West Virginia, catch a little bit of Virginia and continue down through eastern Tennessee to Chattanooga. It is slow, but definitely scenic. We are half way through West Virginia when the sun comes up, and it sure is beautiful. I assume this is coal-mining country, but we don't pass any mining activity, just pretty winding roads, soft hills, and barns with "Clabber Girl Baking Soda" ads painted on the sides of the tin roofs. We pass dozens of Burma Shave poem ads, on fence posts, no two alike. Virginia is more of the same. We parallel the west side of the Smoky Mountains, on down through Tennessee. It seems more prosperous here, and we see a few Kentucky-style horse farms with neat white-painted lumber fences.

Two things stand out in my memory. The highways, although twisty and turny, are smooth – no frost heaves like on our Canadian roads and we also discover the world of Local Option. We drive many miles without a bar or saloon in sight, then bang – six or seven honky tonks, all in a row, on both sides of the highway. We have cut through the corner of a wet county – counties on either side are dry.

We come into Chattanooga at 3 am, immediately get lost and find ourselves in a black ghetto. We are not at all worried. To us the desegregation turmoils are news items in a faraway land. We pull up beside a group of young fellas and ask directions to the Ross Hotel. They check out our license plates and politely give us the needed information.

The next day we tell Leland Grant, our new H-S boss the story. He shakes his head and says we are lucky to be alive. We don't feel lucky, we think it was a normal reaction under the circumstances. We will have a few more culture shocks over the next three months.

Table Scrap: (As told to Maj. Rod Durnin by a US Special Forces Agent))
There were a number of Middle Eastern men and women being cleared by Canadian Customs at a Canadian airport. I.D. Cards were being checked, and the women refused to lift their veils for photo comparison. After much argument, the Customs officers finally agreed to let them board the plane. The Middle Easterners were talking in their own language and laughing.
"Stupid Canadians," they said, "What if we were terrorists?"
The American Special Services officer, fluent in Arabic, was behind them. "Canadians are not stupid," he told them, "They are just decent people."

We check into the Ross Hotel. It's an old-timer, all mellow-toned varnished wood, with an elevator from the 20's – folding criss-cross metal door, and just big enough for two people and luggage. We feel right at home.
The next day we call H-S and Leland comes down to meet us. He is the head engineer on the job we will be doing and a partner at Hensley-Schmidt. H-S has a wide range of operations; architectural, construction oversight, etc. Leland's department deals with soil and rock analysis pertaining to airports, highways, bridge abutments, and

foundations. We will be sampling the centerline of a proposed interstate access road in the Smoky Mountains.

The drill is having problems clearing customs, so Leland tells us to take a couple of days to check out Chattanooga, and we do so.

(Carl and I think this is hilarious. We had jumped the border, but a hunk of innocent steel is held hostage.)

We start with our immediate surroundings. First off, we meet the owner of the Ross Hotel. It turns out that he is a transplanted Canadian, hailing from a little town 30 miles east of Carl's hometown. Small world, eh?

We eat in the hotel restaurant/bar, another anachronism, all the same mellow surroundings as the rest of the Ross, with comfortable booths and a small rectangular bar with perhaps ten stools – not really a big-time drinking place, just good food and quiet surroundings. Old-timey, sepia-toned, enlarged and framed photographs line the walls, depicting street scenes from the early days – street cars, square automobiles, and horse-drawn delivery wagons. Next to the restaurant/bar is a soda fountain with a long marble counter, swivel stools and soda jerks. We feel like time travelers.

We sit at the bar and meet Iggy, who leases the restaurant complex from the Ross and is very much a hands-on type. We never visit his establishment without seeing him, either running the bar or acting as maître de'. He is a rotund, jolly, late-middle aged gent of German lineage and of course, the place is called "The Rathskellar."

"What'll it be?" says Iggy.

It will be beer, naturally. "Regular or premium?" Michelob is the premium beer on tap, and we are flush at the moment so Michelob it is. When we need to tighten our belts we'll switch to Falstaff.

"Bottle or schooner?" Jeez, we don't know, let's try a schooner.

Iggy takes two huge goblets out of a small freezer – they look expensive, sort of like cut crystal, and are very heavy. He sticks them under the tap and draws two Michelobs. The keg sits under the bar at room temperature, but when the beer hits the schooner a film of ice immediately forms on the inside and the outside frosts up. As he sits the two schooners on the bar, the film of ice floats to the top and disappears. The beer is now ice cold, droplets of water form on the outside of the glass and run down to be absorbed by the coaster. How delightful!

So we relax to enjoy a few Michelobs -and I save Carl's life.

ME AND CARL AT THE RATHSKELLAR

Carl and I are joined at the bar by an older gent from Alabama and since we are the only ones at the bar we introduce ourselves. He is an iron worker, built along the lines of Norm, my Moak Lake Beaver pilot – all shoulders and muscle. The beer flows and the conversation comes around to race relations, a dangerous subject for us outsiders to be involved in, especially when Carl is in his cups.

Carl is an ardent socialist, always has been, always will be. He is also a capitalist, retired now, after a pretty darn good career as a hardware/plumbing supply store owner, but he always has been aware of his responsibility to the less fortunate. The iron-worker is a white guy from Alabama. Need I say more? I am a fence-sitter. I can always see the other guy's point of view – maybe wishy-washy is more accurate.

So I am in the middle, both figuratively and literally. On my left is a bleeding-heart liberal and on my right is an ultra-conservative.

Three things save our butts that night: Number one – Carl is a reformed stutterer.

(Perhaps that is why he had formed a close relationship with J-J-Jim in Bancroft. Carl told me that when he was young he could not answer the phone. He couldn't say "Hello." He'd grown out of it by his teens, but one of his compensations was that he'd talk very fast after a beer or two.)

Number two – the Alabaman has a deep southern drawl. I can understand him, but he might as well be speaking Greek as far as Carl is concerned.

Number three – because I am sitting between them, I get to translate, modifying the conversation as it passes through me, left to right, right to left.

I stick to three Michelobs – I have to have my wits about me. We finally call it a night and go to bed, and I am completely drained. The next day I tell Carl he has to be careful, we are strangers in a strange land. He agrees, and I never have to save his sorry ass again.

"You must take the Incline Railway to the top of Lookout Mountain," everybody says, so we check it out.

Lookout Mountain is on the edge of the city and it looks pretty impressive. It is more than 2800 feet above sea level, and I'm guessing about 1500 feet above the surrounding countryside. We drive down a street which is actually a route out of town leading down towards Atlanta. The lower station of the railway is located right where the street turns south and we stop to take a look. It starts off flat, but it soon climbs sharply and it looks pretty darn steep to me. Half way up the track divides for a short distance to allow the car going up to pass the one coming down. "Well, at least they've got that covered," I say to myself.

We give it a good study, and decide to drive to the top to check things out further. (We may be international travelers now but we are not exactly intrepid.) We drive up a long, winding road and by the time we get to the top, Carl's six-cylinder slush-pump Chevy is gasping for air. Maybe it is afraid of heights, too. There is a community of nice homes at the top, definitely up-scale in more ways than one, and given the long climb to the top, we assume the Incline Railway is the preferred method of travel.

We walk to the observation point on a rock knob. You are supposed to be able to see seven states from here, but they are not colored like on a map, so you can't prove it by me. We go over to the upper station just as a car is arriving, popping over the top of the mountain onto a flat surface in the station enclosure. A few people get out and I notice that no one is kissing the platform as they disembark.

We walk around and study the wall art. Cutaways of the cable show the different materials used to build a thick, rupture-proof, absolutely safe system. A section of rail shows how an automatic block dealie senses a sudden increase in speed, making a runaway definitely impossible. I start to feel I am in a hard-sell TV commercial. "Call within the next 20 minutes, and we'll double your order absolutely free!" I point out to Carl that of course it is all good news from up here. They are not about to show us pictures of wrecked cars and mangled bodies in coils of absolutely indestructible cable.

The car is about to leave on the return trip and a few people get on board, strangely calm and unconcerned. Maybe they are androids – they certainly have not thought this through. We walk to the edge of the platform to watch the car leave. Holy cow! The first 200 yards is virtually vertical! The railway lines up with the street at the bottom, and I can just see what will happen if we get on. That damned thing will scoot all the way to downtown Chattanooga, and we can get off at the front door of the Ross! We get back into the Chevy and carefully drive down the mountain.

"So what do you think of the Incline Railway?"

The worldwide internationally famous travelers shrug condescendingly. "Been there, seen that."

The next day we walk a couple of blocks past the town square to check out some local shopping. We browse through a department store and step out onto the sidewalk right into a mini-stampede. We had heard of the Jimmy Hoffa trial, but had not paid much attention. It seems the trial is on right up the street and we are almost run over by three or four guys stiff-arming their way up the sidewalk clearing the way for backwards-trotting TV and newsreel cameramen. Behind them comes a phalanx of U S marshals escorting short, unsmiling, pudgy Jimmy Hoffa – a square of marshals with Hoffa in the centre. Behind them, drawn along by the current, is a school of sharks carrying briefcases. We head for the livery stable and duck behind a water trough. We know all about cross-fires – we saw the movie.

On the way back to the Ross we pass through an uncontrolled intersection near the town square. It is probably not usually a busy corner, but with the trial on, traffic is heavy. A black cop stands in the middle of the intersection directing traffic and he is pretty to watch in his immaculate uniform with white gloves, performing a ballet right there in front of us. We stop to admire, while locals pass by uninterested – they see this every day, I guess. The cop holds up one hand, blows his whistle, points at a line of cars and waves them on. He now pivots 90 degrees and repeats; spinning, turning, blowing his whistle and gesturing, all with a fluid dance-like motion. I wish I had a movie camera.

We have supper at the Rathskellar. It is full of young, pretty girls, and other young, pretty people of indeterminate sex – the Ice Capades is in town. We introduce ourselves to the bevy of beauties at the next table. They are all Canadians, and we tell them we are Canadians also. Their eyes light up – we are suddenly very interesting dudes. "You must be hockey players. We love hockey players. Come to our party tonight, we are staying right here in the Ross."

Once again we prove we are not the sharpest diamonds on the bit. "Sorry, we're drillers," we say, and decline the invitation with regrets – oh, the regrets! We tell them are leaving for North Carolina in the morning and need our sleep.

227

Dumb, dumber, dumbest! Why weren't we hockey players? We could have been hockey players, as far as they knew. We might have even scored!

And, we might as well have gone to the party. The damn thing lasts 'til 4 am and we don't get much sleep after all.

Leland picks us up right after breakfast and we head for North Carolina. The Winkie is still tied up at the border. I guess it can't lie as well as us. We are going to check out the job anyway.

Interstate 40 is being pushed east through the Smoky Mountains toward Asheville and an access road will lead south along an interior valley to provide a link to the Blue Ridge Parkway. Our job will be to sample the soil and rock on the centerline of the proposed access. The Department of the Interior (DI) will use this information to determine cut and fill/rock removal ratio on the right-of-way. It is important for both the government and potential bidders to have a good idea of what lies beneath the hillside. Going in blind can be expensive for either party.

Sidebar: As a case in point, consider this: In the late '60s a seven-mile access road was being built to the Kenora, Ontario airport. The Ontario Department of Highways did not test the centreline. They estimated 75% rock (expensive), 25% cut and fill (dirt movement only – dirt cheap.) A local contractor was the successful bidder and he later bragged that he put a million free bucks in his pocket on the contract. It had actually been the reverse – 25% rock, 75% cut and fill.

The four-hour drive to North Carolina is pleasant. Even in winter the scenery is beautiful. The only downer is a few miles in the southwest corner of the state where we pass some pretty muddled-up countryside – all gashes and ridges with little vegetation. Leland says this was a strip mine, and back in the day, reclamation was not mandated. Current environmental laws no longer allow a company to leave such devastating scars behind.

We drive along the Tennessee River, an old wooden pipeline visible on the far bank. It is a penstock feeding a hydro-electric generating station somewhere downstream. We pass a Cheyenne Indian reserve with small, neat houses dotted along the riverbank.

Pat, our official local liaison, will later tell us that a young Cheyenne was charged with the murder of a young lady on the reserve. At the trial, none of the band members would testify against him. They didn't know anything, or had forgotten this or that. The charges were dropped and the young fellow returned to the reserve, never to be seen again. They have their own justice system.

We drive through the Maggie Valley. The highway is lined with rock shops, antique shops, all the touristy stuff, and all shuttered – it is wintertime. A big sign says, "Visit the Ghost Town in the Sky." We will later do so. Leland tells us that Maggie Valley real estate is the most valuable land in North Carolina.

We see glimpses of the Blue Ridge Parkway. It is also closed for the season, unfortunately. There is no snow, the trees are all gray except for the conifers, and it is now cloudy. But I'm sure things will look better in bright sunlight. We turn onto a gravel road and travel through farmland to the foot of a mountain where the road continues upward, with three long switchbacks before we reach the gap. The mountainside had been logged off years before, but there is no erosion – low bushes and brambles cover the steep hillside. Here and there are old fences, but not a cow in sight. Maybe this is mountain goat pasture.

We reach the top of the gap, which goes on twenty more feet to dip down into our valley. Snow covers the road at the very top. It had been drizzling a little, and the elevation here is high enough that it has turned to snow, just enough to cover the road at the top of the gap and twenty feet down on either side.

Leland says "Oh, Good," hits the brakes, puts the car in park, and jumps out.

"What's wrong?" we wonder.

"I've got to write my name in the snow!" says Leland, and he does.

Carl and I think it is pretty funny, but we get out and autograph the gap also.

Now, every Canadian boy has autographed a snowbank, and when we were younger, we even crossed the "T" and dotted the "I" We realize this is the first snow we've seen since leaving Pennsylvania. Leland doesn't often get the opportunity.

Over the gap we go and down through heavy forest. Leland tells us that we are now in Smoky Mountain National Park. This side if the valley had been hi-graded back in the days of horse logging – only the huge old growth was taken. We will be working on the opposite slope in mostly virgin timber. The way down to the valley floor is steep, but not long, with only one switchback. We come out into a meadow, and at its end is a neatly painted old farmhouse and a large barn, weathered, but in good shape.

This was once a working farm and is now part of the park. Park rangers use the house in season, and this accounts for the overall neatness of the grounds and structures. The old barnyard will be our parking spot/marshalling area.

A crystal-clear creek runs between the road and the steep slope on the west side of the valley. The buildings are on the right, the creek is on the left, and the gravel road curves on westward through a notch, following the creek. The creek is spanned by a little log bridge which we will cross every day for the next two months.

The access road centre line has been surveyed, cut, and marked with pickets, starting at the trees just across the creek and continuing along up the side of the mountain for six miles. We will be sampling at two hundred foot intervals for two miles at this end, and for one mile at the other.

H-S has already done the three miles in the middle with a Boyle Brothers Series 2 drill (BBS-2,) a large, six-cylinder, screw-head machine – anything but portable. In Canada they are used mainly as development drilling rigs in areas with road access, and if moved by chopper, require extensive tear-down and set-up time. They are very expensive to operate. With a 2000 to 4000 foot capability, they are hardly the machine to be using for holes averaging less than fifty feet. To add to H-S's problems, the contract stipulates that the forest along the right-of-way be left undamaged – hard to do when you are bulldozing a road along a mountainside. Some collateral damage has already been done and the Department of the Interior is not too happy.

Also, and no doubt most important, is the fact that while the BBS-2 has done the easiest part of the job, H-S has already gone ten grand in the hole on the contract. Carl

229

and I are to don our superman costumes, and with the Winkie, pull their asses out of the fire. We will do that very thing.

We return to the pavement and continue on ten miles to the little town of Waynesville, North Carolina, where we will spend the next two months. We meet Pat, our helper/translator/liaison guy, who has already checked in. The drill has finally cleared the Customs paperwork hurdle back in Toronto and will be air freighted to Asheville, a city 35 miles north of Waynesville. (Knoxville, Asheville, Waynesville – there sure are a lot of "villes" down here.) The drill is expected in three days, so we head back to Chattanooga to spend two more days before heading back north.

Leland is enthused about the Winkie, and on the way back to Chattanooga he discusses the possibilities with us.

For instance, in 1964 the railway system in the southeastern US is still mostly intact, with few abandoned lines. Along these lines, many streets and rural side roads pass beneath the rail lines and problems are becoming apparent. Back in the day, these underpasses had been built using reinforced cement for the bulkheads, and stone work abutments formed the roadbed behind the bulkheads. Now heavier and faster rail traffic is starting to destabilize the stonework – what to do?

Highways and high traffic city routes can justify major construction, but to redo every little underpass would be prohibitively expensive. You also can't divide rural or city neighborhoods without expecting severe citizen backlash – folks just won't stand for it.

So, Leland has the answer – our Winkie drill. It is light, portable, and can easily drive a two-inch steel casing through 25 feet of stonework and/or boulders. After reaching a pre-determined depth, cement grout will be pumped down the casing, the casing will be pulled back up a few feet and pressure gauge readings will indicate that the gaps in the stone have been filled with cement. Pump, back off and pump, until you are back up to the surface of the rail bed. Now move a few feet and repeat the process until you have created a nice, solid, cement and rock roadbed, able to handle the weight and vibration of modern rail traffic.

Leland is full of ideas for us. He is a visionary and always upbeat. It is obvious this is a man to watch and learn from. We will have our problems in the Smokies, but Leland will be behind us all the way. He already has future plans for us. After the Waynesville job is completed, Leland wants us to set up a drill division at their Marietta, Georgia office. We decide to check it out.

We drive to Marietta and meet the man in charge. He is an ex-football player and we are told that he had played a couple of seasons in the Canadian Football League (CFL), possibly for the Edmonton Eskimos. This guy is no Leland. He is unfriendly and curt – a bad sign. He shows us the building which supposedly will be our drill shop/office. It is full of junk to the ceiling!

"You can clean it yourselves," he says, another bad sign. Maybe the future is not so rosy after all.

It is still early when we leave Marietta, so we decide to have a look at Atlanta. For some reason I don't recall, we knew there was an immigration office there. We decide to drop in, say "Hello, eh?" and get some information on immigration procedures. We look up the address. It is on Peachtree Street. No problem – we go to Peachtree Street to locate the consulate and we can't find it. The number doesn't exist. We drive past buildings with lower numbers and higher numbers – no consulate. We retrace our steps and try farther out with no luck. We give up and headed back to Marietta where we can pick up our bread crumb trail back to Chattanooga. Holy cow! We cross Peachtree again. We finally stop at a drug store, buy a street map, and enlightenment ensues. Atlanta is built like a wheel with a hub at the centre. Five or six main thoroughfares radiate out from the hub, and they are all Peachtree Streets - Peach Tree NW, South, etc. etc. It is too late to find the consulate, so we headed back to the Ross. We don't tell anyone our story. We feel dumb enough already.

(A few weeks later Carl and I are having a Bud and laughing at ourselves. What if we had found the office? A couple of Jimmy Hoffa's US Marshals would have escorted us to the border, that's what. Our thick skulls had saved us from a massive brain cramp.)

We have another day to kill so Leland sends us down into the northeast corner of Alabama to do some soil sampling.

We pick up the H-S auger with Carl's car. It is not a huge unit and fits in the trunk. We take four three-foot auger flights with us – a green sheet of plastic protects the trunk floor and we take no extra gas. This will not be a big job, and will turn out to be very short and sweet.

The unit is simple and straightforward. A McCullough chain saw motor has a direct drive to the auger flights. Just below the engine base is a plate with spokes to a three-foot diameter tubular outer ring. This is the sole method of control. With no centrifugal clutch, you just fire up the motor, hang on to the ring and let the two-inch auger spin. Only twelve-foot holes are required. The community airport wants to extend the runway and H-S only needs to know the soil type they are dealing with. This will be a piece of cake.

On the first hole we auger three feet and take a soil sample. There is a bit of reverse torque, but with a good grip on the ring, one man can handle it. We put on a second flight, make another foot-and-a-half, and the auger does not want to continue – it wants to take a break and the torque is too much to handle.

We try it with both of us on the ring – no go.

We try another hole – maybe the clay will be less compacted – no go – at 4.5 feet the auger goes on strike.

One more shot 25 feet away and the same thing happens. This time we are determined to hold the merry-go-round stationary. We dig in our heels and they plow a circular furrow. We finally collapse on our backs on the warm Alabama grass and laugh like crazy – and crafty Leland drives up.

We think he may be upset, but he is also laughing as he exits the car. He knew darn well the thing wouldn't work – he was just making a point Leland-style!

The rascal tells us the McCullough never did produce a dollar's worth of work (that's why it was so clean and shiny.) He must feel a bit bad that he has hornswoggled us, so he takes us to an upscale establishment in Montgomery and buys us a prime rib dinner.

231

Over coffee and pecan pie he tells us that a backhoe will sample the airport and he just wanted to demonstrate how valuable the Winkie will be to H-Sand he shares another plan he has for the Winkie.

The US economy is expanding, and so are towns and cities. Business districts are growing. Towns that had never had metered parking want to control the all-day parkers and increase revenue. To do so they must install parking meters, and to jackhammer a perfectly good sidewalk and refill around the meter base is expensive and will leave a ugly looking patch job.

H-S will mount a Winkie on the rear of a pickup, back up to the sidewalk, drill a hole the same size as the outside diameter of the meter standpipe, drop in the meter, and wait for the nickels to flow in.

It is impossible not to get caught up in Leland's enthusiasm. Maybe Marietta will work out despite the ex-jock.

The next day we head back to Waynesville. The drill had arrived in Asheville yesterday, Pat had picked it up and we will head into the valley tomorrow. In the evening we settle in at the hotel and socialize with Pat. We also meet our Department of the Interior (D.I) inspector. His last name is Putnam, so of course, we shorten it to Putt. (We should have cut a record – Pat, Putt, and the Two Canucks.)

Putt will go out to the job with us every day to make sure we stay on the centreline and he will take soil samples from our auger holes, one sample every four feet, plus one at bedrock if we hit it before the planned depth. If we hit rock we will core five feet to verify it is bedrock, and not a boulder. Putt will take the samples and cores back to the D.I office every day. He is a long-time resident of Waynesville and will prove to be a great help to us. We like him a lot.

Pat is from Tennessee, not far from the famous Cumberland Gap. He normally works for Gib, the BBS-2 foreman, but is on loan to us as an assistant, and as I have already mentioned, he acts as local liaison. Most drillers on the larger rigs hate packing equipment on their backs, and Pat is no exception, but if prodded does his share. We think he had also been told to keep us out of trouble, which he does. He is an easy-going kind of guy and we get along OK.

Our motel rooms are actually a small annex on the side of an old, fairly large hotel, the name of which escapes me. The two-room annex is built shotgun style – that is, the first room leads on to the next with only one exterior door. Each room has two beds. Carl and I take the first room and Pat the second, which also holds the three-piece bath. We don't mind, it is sort of like a bunkhouse. It is off season and the hotel has few customers. We are told that in tourist season, the annex rents for $50 per occupant. We pay $10 per week each, with H-S picking up the tab. The hotel dining room is closed in the off season, but we have a $10 a day food allowance, and find a good, reasonably priced restaurant to eat in. Ten dollars is sufficient; e.g. a bacon and egg breakfast costs $1.99 including coffee and grits.

Everybody eat grits: grits with milk and sugar, grits with salt, grits al dente (thank you, Cousin Vinny.) I don't eat grits. When we were young, we had three basic breakfasts on the farm – oatmeal porridge, Red River cereal porridge, or Cream of

Wheat porridge. Grits is like Cream of Wheat; I didn't like Cream of Wheat then, and I don't eat grits now.

So Pat hauls out a half-gallon mason jar of white lightning, his welcoming treat for us, and we partake.

"What do we use for mix?" we ask.

"You don't mix," says Pat, "You sip."

So we each take a sip, cough, snort and dry our tears.

I go out for some Coke. "Don't mix it," says Pat, "Use the Coke for a chaser."

That goes a little better – at least we don't cry. We start to mix our drinks – a little moonshine – a lot of coke. Things smooth out.

We pour a little into a saucer and light it up. The flame is almost invisible, and when the fire dies there is nothing left in the saucer. Pat shows us how to test the quality of your white lightning purchase. You give the jar a shake and check "the bead," the little air bubble that forms on the inside wall of the jar on the surface of the liquid. The higher the bubble floats, the more powerful the batch. We try it out – the bubble sits almost at the surface. Hudson's Bay Overproof is baby formula next to this stuff. We continue with several more of our heavily diluted drinks, bragging on how little it affects us. Pat don't say nuthin' – he knows!

The next morning we head into our valley to uncrate the equipment. We let Pat drive. It is a cloudy day, but we wish we had sunglasses. We reach the old barn, unload the crates and take a break. The sun comes out and it is a beautiful day (to some folks, somewhere). Carl and I sit very quietly. Sudden movement is dangerous. Our eyes hurt; in fact, we hurt all over. Stubborn leaves still cling to the trees nearby and a small breeze carries some of them earthward. When they hit the ground, the crash intensifies our headaches -we never touch white lightning again.

We tried to be good old boys – we really did! Pat suggests we are maybe more Jack Daniels types, so a few days later we buy some Jack Daniels Black Label (the premium grade). We try a couple of snorts, but don't like it. We switch to Budweiser.

Our brand-new Winkie, upacked and assembled in the old farm yard, in front of its packing crate.

So the drill is on site, and we can finally get to work. Leland had told us JKS would be sending down the complete package, and did they ever! In addition to the basic stuff they had sent all the necessary tools – pipe wrenches, flat wrenches, etc. – everything except shovel, axe, grub hoe and water line. We will seldom need water, and garden hose will fill our needs. We have 18 flights of three-inch augers in three-foot lengths. (Our deepest hole will not exceed 50 feet.) We have five feet of core barrel and 50 feet of drill rod, plus some spare

233

rods and augers. Pat had already gathered together the required hand tools before we arrived. He also has a gas can, a tool box and a credit card. We are all set.

Putt gives us an overview of the job. Our first hole is right at the edge of the valley and only four feet deep. The holes will deepen as we climb the side of the mountain.

(We are all on page one, here. Pat and I are rookies as far as the Winkie goes, Carl has seen it work and has done some hands on, but the Winkie is not exactly rocket science.)

The Winkie is head and shoulders above the packsack drill. It is much easier on the operator, who does not have to hold the drill in place. He also has more control of down pressure with less exertion, and without the vibration feedback of the packsack drill.

We will be auguring dirt most of the time, so drill string down-pressure will not be much of a factor. The chain and sprocket system makes it easy to pull the augur up to clear dirt samples from the auger flights.

Pressure pump and spare parts, up close and personal. Note as yet un-sweat-drenched packblard to left of the water hose.

The compact drive unit is a West Bend two-cycle, one-cylinder motor driving through a centrifugal clutch to a two-speed transmission. The West Bend is a screamer. It is designed for go-karts and the red line is stratospheric. In stock configuration it has a compression ratio of 20 – 1 or higher. During initial field tests it was found that a kickback when pulling the starting rope would destroy the rewind assembly, and, in fact, one over-zealous operator who wouldn't let go of the handle broke two fingers. A thicker head gasket detuned the motor a little, but we are still very careful when starting it.

The noise this thing produces is harmful to the ears. A 20-foot flex hose on the exhaust carries the noise far enough away from the crew, but anyone within a mile of us can tell if we are working.

There is a 4 to 1 reduction attachment to slow the auger down to 300rpm. It is not too big, only about a foot long, and it will turn out that it is not too robust, either. It will soon prove to be problematic.

The first two holes we finish in no time flat. The second hole is 200 feet up a gradual slope, and is only eight feet deep. From here on the holes will get deeper.

Right from the start we are unimpressed with the reduction gear. The housing is the size of a football, and to prevent its natural tendency to spin with the output shaft, a steel strap is attached to the bottom of the housing, resting against one of the side rods of the unipress. The problem is, the torque tends to force things off-centre that were not built to withstand side stress. When we finish the second hole the reduction gear housing is almost too hot to touch.

HI-TECH SWEDE SAW BUILDING A SET-UP.

We go on to the next hole, a twelve-footer. At eight feet the reduction gear is smoking at its input shaft bearing. It is after 3 pm so we decide to shut down for the day to let things cool off a bit. Maybe if we take it slower tomorrow things will work better. We cover the set-up with a small tarp and go home.

The next day we run the auger down the existing eight-foot hole and the reduction-gear housing is already hot. Before we advance one foot, the reduction gear packs it in. We remove it, pull the auger and take the reduction assembly back to our little operations base and tear it down. The bearings are obviously toast and we take the top cover off to check the gears – there are no gears! There are shafts for the gears to turn on and there are little pieces of steel in the gear oil, but nothing we can positively identify as a gear.

We go back to Waynesville to the D.I. Office with Putt and call Chattanooga. They call Toronto who will send out a new reduction unit by air freight.

The next day we go to the airport in Asheville - no package. We wait for the next flight - still nothing. We head back to Waynesville to the D.I office for some more telephone tag. Paperwork is the culprit again. The new unit will be in Asheville in three days. Very frustrating!

So we check out Waynesville, and it doesn't take long. Thirty-five hundred people live here, with no visible means of support that we can see. We are told that Waynesville is a bedroom community for various mills and plants in the area; tobacco, paper and textile. There is a vigorous looking three-block main street, and of course, the summer tourist season keeps the town buzzing.

235

We check out the local bar. There is only one, and it is a real dive. Local laws mandate a 9 pm closing time, but it opens at 9 am! As a result, you can't stop in for a quiet beer after work – by then everyone in the joint is hammered. It is definitely not a spot for socializing.

We go to the local drive-in restaurant for a burger and notice that root beer is as popular as it is back home. Wait a minute, it's not root beer, it's the real stuff! In a town where the bar closes at 9 pm, you can have a beer on your window tray! Weird, huh? The culture shocks are multiplying.

We also notice a large number of young men hanging around at the drive-in, apparently unemployed. We ask Pat if jobs are scarce.

"Heck, no," says Pat. "These boys have received their draft notices and will be called up sooner or (they hope) later. A company won't hire a young man of draft age, as they will soon have to be replaced."

I suddenly realize that the Viet Nam conflict puts young men's lives on hold in more ways than one, and too often permanently.

When we tell Pat how surprised we are about curb service beer, he scoffs, "That's just 3.2 sissy stuff."

I don't have the heart to tell him I like 3.2 and I don't think I am a sissy. We do have a couple, but it just doesn't seem right to be drinking beer in a car at a drive-in restaurant.

Carl and I go down to the Maggie Valley one day. We want to check out the "Ghost Town in the Sky." Pat says it is closed for the winter, but we think there might be something to see.

There is a winding gravel road leading up a steep hill, curving off behind a stand of pines. There is a big parking lot at the bottom with an old-timey railroad station and a sign saying, "Incline Railroad to the Ghost Town in the Sky." A board with a big "Closed for the Season" is nailed across the sign. It doesn't hurt our feelings any, we know all about inclined railways. Another sign proclaims, "Gunfights Every Day!"

We drive up the switch-back road and a town appears behind a locked gate at the top of the hill. Since we have come this far, we climb over the gate and go on foot for a quarter mile more to Main Street.

What a neat place! We know it is all bogus, of course, but everything for a wild west movie set is here, and they are real buildings; saloons, bank, post office, barber shop – everything!

An old two-story hotel with a full-length second-story balcony sits beside a livery stable with hay visible in the loft. The dusty street is bordered by planked sidewalks, and being deserted, the town does give an eerie feeling.

Oops, someone is coming!

Around a corner strides a bearded gent in cowboy boots, a cowboy hat, cowboy shirt and a bandana, with not a sequin in sight. This guy looks authentic. Are we in trouble?

"W-w-w-welcome, f-f-folks," he says, "What c-c-can I do for you?" This guy has a real bad stutter, but he is also very friendly, and when we tell him who we are and why we have come up to bother him, he graciously shows us around. He lives here in the winter as a security guard and seems glad of the company. He is one of the bad guys in the summertime.

He points to the hotel balcony. "I g-get sh-sh-shot off of there f-f-four times a d-d-day, every d-d-d-day." and he laughs like crazy.

He gives us a postcard showing his death dive off the balcony. He says it's not really dangerous – he lands on a huge air bag hidden behind the water trough. This sure qualifies as an odd job to us. We thank him for his hospitality and leave. As we drive down the mountain we opine that if we got shot off a balcony four times a day, every day, all summer, we might st-st-stutter, too.

This is typical – everybody standing around while the guy dies with his boots on.

The replacement reduction gear comes in and lasts one hole. We are getting pissed off. Like Willie and Paul, we came to play and not just for the ride. More phone calls, and we ask for two units, this time.

Leland pays us a visit, and we are a little nervous, but he reassures us. "Things will work out," he says, but – another two holes result in another two trashed gear drives.

This cannot go on! We think maybe H-S is not getting the point across.

So Carl calls Chattanooga, gets the JKS number and calls Toronto, billing the call to Chattanooga. (Bet you take cell phone technology for granted, eh?" That's how you had to do it in '64.) Carl tells Toronto they have a lot of nerve, sending us to demonstrate a drill with an unworkable attachment. They admit their error and tell us a new 10 – 1 assembly is being tested, promising delivery in ten days.

We don't really want ten days off. We have been here almost a month and have only completed twelve of the 75 holes planned. Also – remember the old saying, "All work and no play makes Jack a dull boy?" Well, all play and no work can also make Jack a broke boy!

We talk it over with Putt. If we keep water handy we can put water down the hole from time to time to help the auger turn easier. He says we can't do it. If we hit water in a hole it is duly noted in Putt's report, so adding water is out. Can we hop from hole to

237

hole, drilling eight to ten feet and come back when the new gear arrives? No, for the same reason. A heavy rain could put water in the hole. We will just have to wait.

Don't get me wrong; it is a nice place to be with nothing to do. People are very friendly. The project engineer treats us to a good dinner at a fancy restaurant in Asheville. We are invited to parties – lots of stuff.

About now Gib, the BBS-2 foreman also visits us and we spend an evening sipping beer and chatting. Pat and Gib sip from the mason jar, passing it back and forth. (Nothing unsanitary about that – germs cannot survive within three feet of that stuff.) It seems that Pat and Gib are pretty close, and Carl and I and the Winkie may not be not entirely welcome. Are these two guys worried about the Winkie replacing the BBS-2?

Table Scrap

We buy all our gas from a Sinclair station next to the annex. The owner is an older man, very nice, and an ex-dirt track stock car racer. One day he takes me to his backyard shop and shows me a red '57 T-Bird, 312 C.I. two four barrels and an all-black interior. We give it a test drive with the top down. Man, this is livin'! When those back barrels cut in, this baby squats and scoots!

There are 25,000 miles on the clock and the owner needs money. I can have it for $800 – no kidding! These T-Birds go for $80,000 plus nowadays, and although I don't have a crystal ball, I know 800 bucks is a bargain.

But, there were two problems. Canadian protective tariffs at the time will add at least $1000 to the price, and I just plain don't fit in the car. I am 6'5 and need lots of leg-room. I regretfully decline – Oh so regretfully now.

To keep from getting bored we spend a few days pre-building set-up sites. Every day on our way in, we drive through a pleasant looking area of small farms with neat houses and cultivated fields of different sizes. I mention that everyone seems to do a lot of gardening, and Pat tells me that these plots are called tobacco base. It's like our milk and egg quotas. Everyone is allotted a one-eighth acre tobacco base for their own use. Those who sell tobacco have larger plots in I/8th acre increments. A farmer might have a 1/8 to 3/8 acre plot, but seldom larger. Each tobacco growing farm has a high barn-like building with gaps between the boards and with good tin roofs. The tobacco leaves are hung on poles to dry – all very organic and old-timey.

We pass a couple of little white churches with cement pools behind them. I think they are sewage retention ponds, and ask Pat if septic fields are not allowed. He laughs and tells me they are baptismal pools. Oh, my! Faux pas extraordinaire!

We stop every morning at a little country store to buy lunch-meat and bread. (This is where I bought Prince Albert in a can).It is a time warp. There is a long counter on one side with shelves of canned goods on the wall behind. Shovels, pitchforks, axes and the like lean against the other wall. Various milk pails, buckets, ropes and what-nots hang from the rafters, and bolts of bright cloth are piled on a table. In the middle is a pot-bellied stove, and five or six guys are sitting around it on stools and chairs. There is even a pickle barrel!

When we enter, all conversation stops. When we leave, we can hear them talking again. Pat tells us they think we might be "revenoors." Moonshine is still a cottage industry here and we are accompanied every day by Putt with his Department of the Interior logo on the door of his truck. All "gummint" people are suspect.

One day we walk in and an old fella spits in a can and says, "How are y'all today?"

"Jes' fine," we say, "Hope it don't rain none." It took a month, but we are finally accepted. We are now honorary good ole boys.

Another day, Pat points out a big brick bungalow on a country lot south of Waynesville. "Those bricks set that ole boy back 60 grand."

Seems like bulk purchases of sugar by the bigger moonshine operations are a dead giveaway. The guy had gone two states over, and arranged for a truckload and had paid in advance. The seller knew darn well what the sugar was for, and that it was unlikely the Better Business Bureau would be contacted, so when the guy opened the bags, he found sand instead of sugar.

Well, when you have lemons you make lemonade: he mixed the sand with clay, fired his own bricks and built a house.

This smells like more of Pat's BS and I don't say anything, but I can only handle my chain being yanked so many times. Someday I'm going to call Pat and make him show his cards.

We always come back to our marshalling yard to eat lunch and pick up some gear to take back onto the hillside. One day we are relaxing in the noon-day sun, and I decide to scope out the old barn. It is built in the style of what we Canadians call an "Ontario Barn." The second floor hay loft has an earthen ramp sloping up to a large door, and back in the day, hay wagons could be driven right inside, and the hay unloaded into lofts on either side. I peer through a crack in the door, and way back in a corner sits a 39-40 Ford. I can just make out the grille and front fenders. I go back and tell Carl and Pat of my find.

"That'll be an old moonshine runner," says Pat. "It'll have extra springs at the back, a spinner knob on the steering wheel and the gearshift will be on the left side of the steering column. The springs are to compensate for 200 gallons of white lightning in the trunk. The gearshift is moved over so the driver can shift with his left hand, leaving his right hand on the spinner knob. If the Feds are on his tail, he wants to be able to handle those switchbacks."

I've had enough. I laugh and laugh. "Bullshit," says I.

Pat isn't even insulted, and tells me to see for myself. The loft door is padlocked, but I find a couple of loose boards and squeeze inside.

The car is a '40 Ford two-door sedan, dusty, but obviously in good shape. The rear end sits pretty high. I check underneath, and sure enough, it looks like extra leaves have been added to the transverse spring and there are coil-overs in place of the standard shock absorbers. The doors are not locked and I look inside. There is a spinner knob on the steering wheel, and the gearshift is on the left. It is exactly as Pat said, dad gummit! I've listened to 99 BS stories and I have called him on the 100th. Either I'm unlucky, or all his stories are true.

I wonder what the Ford is doing here. Was it here when the government bought the farm? Or maybe the park rangers have a sideline of their own in the summertime?

239

Sidebar: The car is certainly genuine and not a recent mock-up. It has surely run a gallon or two in days gone by. The access we are working on will pass near the old farmstead. It is conceivable that artifacts such as the car are being collected and the old farm may become a historical display.

Another day we explore a road over a higher gap to the north. We come out of the trees and the interstate (I-40) construction is going on far below us. Bulldozers are working on the steep terraced side hills and it sure doesn't look like a job for the faint of heart. Twin tunnels under construction are visible a mile or so east. Pat says both jobs are very dangerous and if a worker is killed the whole crew goes home for <u>one</u> day.

Just then a four-engine prop passenger plane flies over, perhaps a Martin 300 or a Convair or such. "There must be a forest fire somewhere," says Pat. "That's a water bomber." Pat knows he has me now, and keeps piling it higher.

We meet a cowboy, a friend of Pat's, at his horse ranch not far out of Waynesville. He has a string of horses and runs sort of a dude operation, with riding instruction in the winter and trail rides in the summer. He is a pretty good sort. (I also found out, 50 years later, that he was also the source of Pat's mason jar supply. Carl knew it but I didn't. No one tells me anything – I am a blabbermouth.)

The wrangler also told Carl that while horses were his main interest in life, moonshine put money in his sock. He figured in five years he would sell out, move to Kentucky and breed race horses. I hope it worked out for him.

I decide it would be cool to have a picture of me holding a jug of white lightning, so one day when Pat is out somewhere I borrow his shine and Carl, Putt and I pose for a couple of snapshots in the parking lot, holding the shine up and grinning like the idiots we are. I guess the gent at the Sinclair station is concerned, and rightly so. He rats us out and that is the only time I ever see Pat angry. He chews us out royally – we deserve it – and here's why: Less than two days after our viewing of the I-40 construction, a small article in the local paper tells us that a Cadillac carrying an unstated amount of illicit alcohol was apprehended at that very spot. The driver was charged with transporting shine and the car impounded – and we were carrying the stuff outside in broad daylight!

The new reduction gear comes in, and now they have it right. It fits into the unipress and slides up and down with the rest of the power unit, so the torque is now evenly distributed and there is no side thrust. The gearing is ten to one, (120 rpm) supplying lots of power for auguring. Thirty-foot holes are a snap. It is also invaluable when driving casing. (This set-up is still in use today.)

Now we are rolling. It is still hard work, but when you are producing results you don't mind working hard. Because the holes are only 200 feet apart we don't tear down completely. With a man on each end, the drill can be moved largely intact by doing a good imitation of stretcher-bearers.

The wrangler with his plucky horse

Putt lashing a load on the pack horse. As DI, he was supposed to be hands-off. But truth be told, he was more help than our H-S assistant.

But the drill is heavy and the mountainside is steep. We pull ourselves up by grabbing small trees with one hand, and in one particularly tough spot I reach up and grab a rock ledge. Pat is on the other end of the drill and tells me never to do that in the summertime, as a rock ledge is a favourite sunning spot for rattlesnakes.

I make a careful entry in my memory bank. Pat also tells me that if I hear a rattle, to stand very still until I identify the source and then move away very, very slowly. I enter that info, but it keeps coming up –"Error, cannot compute." I figure at the sound of a rattle I'll be ten feet in the air anyway, hoping I won't come down.

We have finished the first two miles; now we have to move the drill and associated

Me. Carl. Putt and White Lightnin' at the Annex.

gear three miles up to attack the last mile. It is a daunting prospect. We have Gib's bulldozed road to walk on, but it is steep, and a three-mile carry will take many trips and at least three days of hard slugging. We talk it over and have a brain wave. What about the wrangler/moonshiner? Carl and Pat visit him that night and he is all for it. He has a good pack horse he uses on trail rides and his price seems right. Chattanooga OK's the deal and the next day the pack horse hauls our drill up the road. It is steep even for him, but he is a windy little guy and gets the job done. (A fartin' horse will never tire, a fartin' man is the man to hire.)

In one week we are finished and pack out with no horse this time, it is all downhill.

241

L eland comes up from Chattanooga to say goodbye. We had told him we were heading straight home from Waynesville and would start immigration proceedings when we get home. Leland, as usual, is very upbeat about things and is looking forward to our return. He tells us that although H-S lost ten grand on the first half of the contract, we have recovered the loss and put them ten grand in the black on the whole job, and we have brought the job in on time, avoiding a late penalty! That means we would have made forty grand for H-S if we had done the whole thing. Not bad in 1964 dollars, and H-S has bought the Winkie! We have been successful on all fronts!

We say goodbye to Leland – he has been a good boss. The next morning we clean and pack up the Winkie and put everything into the company truck. Pat will take it to the Chattanooga warehouse tomorrow. It is our last night in town and we decide to have a party – no big deal, just drinks and good times. We invite everybody who has been so nice to us during our stay.

It is a pretty good party. We keep running out of beer and we make a couple of runs to the package store. First Carl goes then an hour later I make a trip. The liquor store closes at 10 pm, and just to be on the safe side, we decide to get another 24 Buds. Carl and I are getting a little the worse for wear, so we send Pat in Carl's car. He is stopped by the local sheriff's deputy.

"Tell those Canadian boys to stay sober or stay out of the car."

We do so. We can take a hint and we appreciate the heads up – they have been watching over us.

Table Scrap

We have all heard the stories and seen the movies about Big-gut Bubba with his badge pinned on a sweat-soaked uniform, preying on unsuspecting strangers, but the media stereotype does an injustice to the men behind the star.

A case in point: In the early '90s my wife and I owned and operated a country convenience store/restaurant/gas bar. A nearby town had an excellent dirt track and we were big stock car fans and sponsors. A couple nearby were avid race fans and took an annual trip to Daytona for the February Speed Weeks.

One year they took their two grandsons with them. The little guy's father drove a dirt track modified and the boys were old enough to be race fans. They checked into their usual down-scale economical motel and partook of the many events associated with Daytona.

The boys were getting a little tuckered out, so one day Grandpa went chasing dirt while Grandma stayed in the motel with the young ones. Grandma decided to take them out for breakfast and as they walked down the sidewalk they were stopped by a sheriff's deputy. He told them that this was not the safest place to be walking with young children, and suggested she wait for her husband to return and then drive to a better section of Daytona to eat.

She was surprised, and told the deputy that they had been there the year before and no one had seen fit to warn them off the street.

"We know," said the deputy, "But we figure adults can take care of themselves. This year you have children with you. Please be careful."

Thousands and thousands of race fans converge on Daytona every year, yet these cops can and do recognize people from year to year. Amazing!

Table Scrap: Good Ole Boys abound – another Speed Weeks story.

A local racer and his family are at Daytona. Their five-year-old boy (now a fine Modified driver) is with them and he wants Dale Earnhardt Sr.'s autograph. Dale's booth has a long, long line of autograph seekers – Dad and son are far back.

Ken Schrader's booth is beside Dale's and Ken's lineup is short. Ken sees the young fellow, leaves his booth, takes the boy by the hand and plunks him down in front of Earnhardt.

"Little boys don't have to stand in line for an hour." Says Ken.

The next morning we say our goodbyes and head home. As we hit the road Carl tells me he has no intention of returning. I think I will – it may be interesting.

We cross back into Ontario and feel at home already. We've been out of Canada too long. Carl drives me to North Bay and I take the bus home. It has been an eventful winter.

Chapter XI Back to the Bush - May – Dec '64

Back at the farm I start the Marietta ball rolling. My first step is to get a Passport, which takes a month. I could have fast-tracked it I guess, but this is before Fed-Ex and Priority Post – letter writing takes time. In the meantime I am helping Dad out and enjoying the quiet life. My next step will be to contact the American Consulate office in Winnipeg to apply for a US work visa.

Cattanooga has sent me two letters:

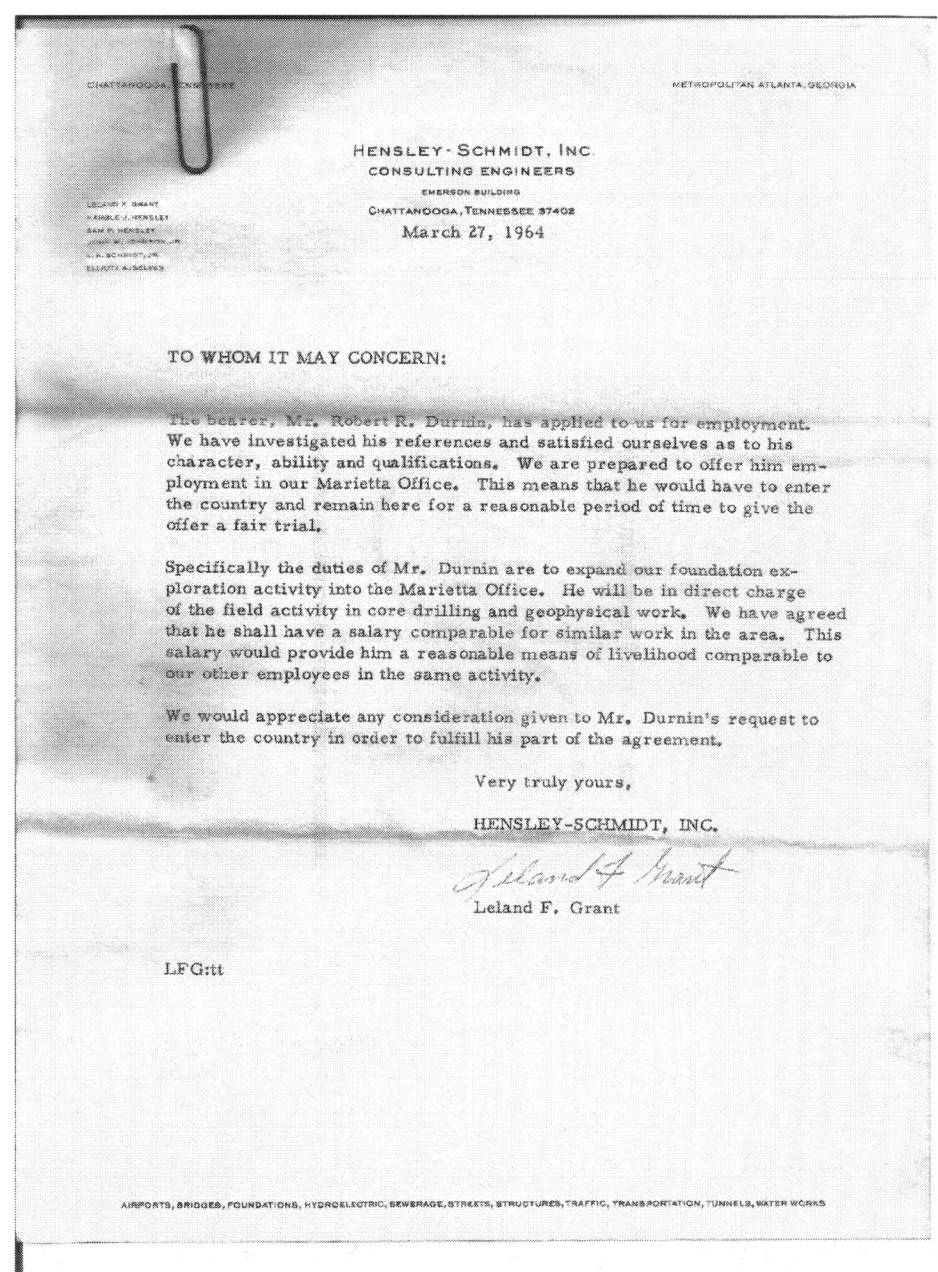

HENSLEY-SCHMIDT, INC.
CONSULTING ENGINEERS
EMERSON BUILDING
CHATTANOOGA, TENNESSEE 37402
March 27, 1964

TO WHOM IT MAY CONCERN:

The bearer, Mr. Robert R. Durnin, has applied to us for employment. We have investigated his references and satisfied ourselves as to his character, ability and qualifications. We are prepared to offer him employment in our Marietta Office. This means that he would have to enter the country and remain here for a reasonable period of time to give the offer a fair trial.

Specifically the duties of Mr. Durnin are to expand our foundation exploration activity into the Marietta Office. He will be in direct charge of the field activity in core drilling and geophysical work. We have agreed that he shall have a salary comparable for similar work in the area. This salary would provide him a reasonable means of livelihood comparable to our other employees in the same activity.

We would appreciate any consideration given to Mr. Durnin's request to enter the country in order to fulfill his part of the agreement.

Very truly yours,

HENSLEY-SCHMIDT, INC.

Leland F. Grant

LFG:tt

AIRPORTS, BRIDGES, FOUNDATIONS, HYDROELECTRIC, SEWERAGE, STREETS, STRUCTURES, TRAFFIC, TRANSPORTATION, TUNNELS, WATER WORKS

HENSLEY-SCHMIDT, INC.
CONSULTING ENGINEERS
EMERSON BUILDING
CHATTANOOGA, TENNESSEE 37402

March 27, 1964

LELAND F. GRANT
MARGLE J. HENSLEY
SAM F. HENSLEY
JOHN W. JOHNSON, JR.
L. A. SCHMIDT, JR.
ELLIOTT A. GRAVES

Mr. Robert R. Durnin
Box 362
Fort Frances, Ontario

Dear Bob:

We are forwarding the necessary letter for demonstrating that you
have a commitment for working with us at Marietta. We are pleased
to offer you this opportunity because we have faith in your ability. We
believe you can perform in every respect therefore we are willing to
commit ourselves to the limit of providing you employment for the six
months period we have agreed upon. We feel that with that commitment
you should show some initiative and responsibility of getting here to accept the job.

This office has provided travel expenses for you to get here and remain
long enough to size up our operations. Therefore we hope that you will
see fit to provide your own means of travel for returning to Marietta.

We are looking forward to you working for us as soon as you can complete
your arrangements in Canada.

Very truly yours,

HENSLEY-SCHMIDT, INC.

Leland F. Grant

LFG:tt
Attachment

AIRPORTS, BRIDGES, FOUNDATIONS, HYDROELECTRIC, SEWERAGE, STREETS, STRUCTURES, TRAFFIC, TRANSPORTATION, TUNNELS, WATER WORKS

One serves to indicate to the consulate that my Winkie expertise is unavailable in the US and the other deals with my employment contract. I am disappointed. They only offer a six months contract and expect me to cover my own expenses to Georgia. It appears they have cooled off and are not nearly so hot to trot as I thought, (or they are a bunch of cheap pricks.)

I talk it over with Dad. He is now 70 years old and still wants to farm, but has some minor financing issues. At his age borrowing money is impossible but he doesn't want to rent out the land and vegetate on his pension. He suggests that I buy the farm and work it with him. I agree to give it a shot and send a letter to Leland explaining the situation and respectfully refuse the offer.

245

So I spend the next three months putting in a field of barley, putting up hay, fencing, and struggling with ten pounds of paperwork to secure a $10.000 loan from the Department of Agriculture Farm Credit lending agency. By the time the loan clears it is also clear to me that the farm cannot support both me and my folks. Dad is very old school and I can't work with him. The farm is not really mine, anyway. I am merely holding it in trust for the family and Dad understands this.

While I am trying to figure out my options I get a call from Mining Corporation of Canada, the exploration arm of the General Engineering Company (GECO), a branch of Noranda Mines. Somehow they had found out that I am presently not involved in exploration, and they need a mag man. Dad and I are relieved – we are both off the hook. So in September, when the loan comes through, I deposit $10,000 in Dad's account, borrow $2,000 to buy a 1960 Oldsmobile and head to Manitouwadge. I am back in the bush!

In Manitouwadge I connect with my two new confederates. Gerry Daigle is the party leader, his helper is Raymond Paradis, and I round out the crew.

We muster at the Manitouwadge Motor Inn and fly out the next day with White River Air Service to a lake 30 miles north. (The Beaver picks us up at a small lake right in town.)

It is early October and the weather is nice. We are to do a small two-week job at this camp, move to a road camp, and then pack the tent away and spend the rest of the fall working from the Motor Inn. That sounds all right to me.

Another bonus is that Mining Corp works from cut picket lines instead of compass-blazed grids. We are following up on a contract airborne survey, and the company also contracts out the staking and line cutting. This makes for good working conditions and a more accurate survey.

Table Scrap

It took me four years to realize that with Inco I was bending the Mining Act while anomaly chasing. You are supposed to stake first and do the work after. Inco did the opposite, but they were soon to be forced to change their wicked, wicked ways.

One winter in the late '60s, Inco was doing the same old "work first, stake later" routine north of Red Lake, Ontario. They were so cocky that they even had a drill on a lake grid, and a drill on the ice stands out like a sore thumb.

Someone flew over, spotted the drill, and flew back to Red Lake to check out the claim maps at the mining recorder's office. The whole area was hot that winter, and he thought it might be worthwhile to tie up some adjacent real estate.

At the recording office, he was surprised to find that Inco was drilling on open ground. He hustled back north with a couple of men, staked the ground and recorded his claims. He then politely suggested to Inco that they should either deal with him, or get the effing drill off his property.

I'm pretty sure some folks were ducking buckshot back in Copper Cliff. A new company policy was drafted.

Table Scrap

Inco, as befits any huge company, often played their cards like Texas Tinhorns. How they ducked the bullet for so many years, I'll never know.

We all had staking licences – mine was H-9158. We just signed the application – Inco paid the ten bucks and kept the licences in someone's (Herb's?) drawer.

While at Waynesville I received an official looking letter from Inco. (How could they track me down when the US Immigration folks could not?)

The letter was straightforward and terse. If I would sign a form transferring ten claims (staked in Ontario in February) to Inco, they would appreciate it. Enclosed was a bribe – a cheque for $100 CDN.

Carl and I laughed like crazy. Someone had forgotten that Herb fired me and they had used my licence. I'll never know where the claims were located, but it must have been a hot spot.

I toyed with the pleasant prospect of demanding more ransom, but revenge is sweeter in the mixing bowl than out of the oven. I signed the transfer, endorsed the cheque and sent it to my folks.

Our 14 x 16 fly camp is pretty much same old – same old. The only difference is the tent manufacturer. Mining Corp's tents have a sleeve at each end of the peak that the ridge pole slips through, giving us an interior ridge pole that is handy to hang stuff on. Because the pole is fairly slender, it requires a support in the centre. This gives Raymond something to do. He is the type of guy who always has to be busy.

Every evening after supper, Raymond takes out his jackknife and carves. By the end of our two-week stay he has carved a cage in the support pole about four feet above the floor. Four slender columns of less than one inch diameter enclose a wooden ball which can move up and down in the six inch long cage, but will not fit between the upright columns – he has carved a ball inside a cage, if you can picture that. I'm told this is Raymond's trademark – he does this in every camp. The longer he stays at a camp, the farther the wooden ball can travel, although I'm sure there must be a limit to how long the cage can be. If it is a particularly good cage, Raymond cuts it out of the pole when the camp is torn down. He is from Amos, a small town north of Val D'Or, Quebec, and I'm sure that some of Raymond's "ball-in-cage" creations have a place of honour in many of his descendants' china cabinets.

Raymond is a very funny guy. A few years before, an accident had left him minus half of his right index finger. "No problem," says Raymond, "Now I can order a half-a-cup of coffee."

He also tells me, "When I start work for this company, I cannot even laugh in Hanglish." He is a hard worker and I like him a lot.

The party leader Gerry and I don't get along too well. That's OK though, I know where he's coming from. I've seen it before. Gerry has been recently married and I will go through the same thing myself a year later. They should pass a law to keep guys out of the bush a year after the wedding. It would save a lot of trouble, and maybe even a few marriages.

247

Mining Corp uses the Crone Geophysics J.E.M. Unit. It is much more portable than the Inco vertical loop system, as there is no teardown/set up. Two operators work on the same picket line at 200 foot spacing. There is no motor/generator to worry about. Each operator has a battery pack on his back and each man carries a transmitter/receiver coil about 2 ½ feet in diameter. The centre crossbar holds a simple angle/inclinometer and a transmit/receive switch.

On a north-south grid line the lead man stops at a station and hollers. He faces east and transmits. The assistant, meanwhile, who is 200 feet back, faces west and takes a reading with his inclinometer. The assistant then calls the reading to the lead operator and they both change modes – assistant to send, lead man to receive. Both readings are noted by the lead man. Together they result in a reading at the station halfway in between. In a non-conductive area, the readings will be zero. As a conductor is approached, the lead man will pick up the effect first. On the far side of the conductor, the readings will be negative. (Because it is battery operated, the J.E.M. had distance and depth limitations.)

The magnetometer used by Mining Corp is the McPhar MF-2. It is basically the same as the Sharpe's A3 that I had used at Contwoyto but with internal batteries. I prefer the McPhar.

The first grid is 25 line miles, normally a ten day job, but this is October, and weather happens in October. The air service is to come in on a weekly basis and does so. Our radio reception is spotty, either because of atmospheric conditions, or perhaps the air service-supplied radios are inferior, so contact with base is intermittent at best. This is not a problem grocery-wise as a basic food order is always left with the supplier to handle such situations. When the Beaver comes in with the second grocery run we tell him we should be finished soon. The pilot says he will await our call, and in the event of no call, will show up in seven days.

The next two days are nice and sunny, and we gave her heck and finish up. That evening we can't raise White River – nor the next morning. By noon we are a bit frustrated. If we can't contact the base we will have a long day wait with nothing to do, and five days off in a 14 x 16 tent is no fun. I volunteer to walk out. On the flight in we had seen a network of haul roads and know it is no more than seven miles to the Manitouwadge/Caramat road.

I skirt the lake and pick up a haul road at the far end. Well, shades of the Thompson Beer Run in '62! Once again, the view from the ground is much different from the aerial perspective. Roads lead off in all directions – the Old Prospector is lost again – not really lost, though, just confused. This time I am smart enough to carry a compass, but some roads go in a circle, and I can walk three miles only to end up a half mile from where I started. I feel like a rat in a big maze, scurrying around and hitting dead ends. Finally at 4:30 I hit the main road, having walked at least fourteen miles to get there.

The Caramat road is not much more than a turkey track. Marathon Pulp and Paper has shut down their bush camps for the season and don't haul wood on this road anyway. (Their haul roads lead to the rivers, where the logs are driven to Lake Superior in the spring and boomed to the mill.)

But it is moose season, and soon some hunters stop and I hitch a ride to Manitouwadge. I call White River, but it is too late to pick up the boys that day, they will have to wait until tomorrow. I get a motel room and crash. The next morning the Beaver brings Gerry, Raymond and the camp in and we head to the next job – still a tent camp, but this time we can drive in.

Mining Corp operates a little differently than what I am used to. They have two three-man crews in the bush, seldom in the same area. Each party chief has his own company vehicle (Pontiac station wagon) and keeps it even on time off. Pretty cool, I think, but the morning after set-up Gerry tells me he is driving home for a week or so, leaving Raymond and I to do the job.

"Like hell," I say, "I don't mind doing the job, but you are not going to leave me here with no transportation."

He reluctantly agrees to drop me at my car in Manitouwadge, provided I don't rat him out to Toronto. That's OK by me, I don't rat on people anyway, and what he does is his business. I bring the Oldsmobile in to camp. It seems to me to be advantageous to have some means of getting out in case of sickness or accident.

The job is not too large, only twelve line miles. By the time Gerry gets back, Raymond and I are finished. We store the camp gear at the Motor Inn and rent two rooms. We will finish two more road jobs and shut down for Christmas in mid-December.

It has been a relatively uneventful fall, but I have gotten my feet wet with a new company and am looking forward to the winter season.

Sidebar: I asked the Toronto supervisor why they used station wagons. I would have thought a four-wheel drive vehicle would be more appropriate for exploration. He said that they had gone that route, but the men tended to push the vehicles too far. They would go further into the bush, so when they sunk in a mud hole it was difficult to mount a rescue operation – better to park it and walk. They spec'd out the Pontiacs with northern suspension, and the Skidoos of the day would fit right inside. The wagons were good, reliable, versatile units.

Table Scrap

My old friend Sam told me this story in 1996. It seems a miner from Geraldton applied for an underground job at GECO in Manitouwadge. Now, Sam had started his career in Geraldton in 1938 and had worked in those mines, so he knew a thing or two about the earlier days underground. Anyway, the Geco personnel guy asked the applicant if he had any underground mining experience.

"Sure," the guy said, "I worked ten years in the (Name Withheld) at Geraldton."

"I mean in a real mine," the personnel guy said, "Not some filthy hole in the ground."

Chapter XII   Timmins - Winter '65

January 1st 1965 and naturally, I am on the road again. I am on my way to Timmins. which had been the scene of a huge staking rush last year. Texas Gulf Sulphur had hit the jackpot, and Mining Corp (GECO) had picked up two blocks of claims. I am to join a geophysical crew to do the work on those two claim groups.

I am expected in Timmins on Jan 2nd, so I decide to leave at noon on New Year's Day for the twelve-hour trip and drive half the night. Although the new causeway taking Highway 11 across Rainy Lake has been completed, I am unsure of the rest of the Fort Frances-Atikokan connection, so I take the U S route, heading south to Orr, Minnesota, then turning east through Ely and Tower, and catching the Thunder Bay-Duluth road at Grand Marais on the Lake Superior northwest shore. The road is good to Tower, but from there to Grand Marais it has many curves and frost heaves. I hit a piece of broken pavement and blow a tire. Rats!

No problem – I have a spare. Turns out there is a bit of a problem: I had forgotten to take an inventory of the Oldsmobile's trunk. I have a spare tire and a jack, but no wheel wrench – and no one else is dumb enough to be on this road on New Year's Day.

I search the trunk and under the seats, and even the glove compartment, but no wrench appears, so I dig into my trusty packsack. Right at the bottom I find a screwdriver and a ten-inch crescent wrench. Why I carry them in my packsack I do not know, but they do the trick. It isn't easy, but the screwdriver serves as a jack handle as long as I don't put too much pressure on it. I grab the rear bumper and lift while I put downward pressure on the makeshift screw driver/jack handle, and it works! I am lucky. If it had been a front tire I never would have been able to do it.

More luck – the lug nuts aren't rusty. Very carefully, so as to not round the corners, I break them free and put on the spare, which is not such a good tire, either.

I drive very slowly to Grand Marais, expecting to find a motel room for the night. Timmins will have to wait for a day.

I get to Grand Marais with four round tires, and holy cow! There is an open service station, bless his heart! He is about to close, but I run the Olds into the bay, up on the hoist, and he slaps a new tire on in no time. He even throws in a wheel wrench/jack handle on the deal. I am mobile again!

Now I am behind schedule. I pass through Fort William/Port Arthur and turn left onto the Trans-Canada Northern route (Hwy 17) at Nipigon. It is already 10 pm and I have a good six hours ahead of me, but my gas tank is full and the road is clear – very clear. There isn't any traffic at all – I am the only one on the road.

By midnight I am between Longlac and Hearst, having passed the road sign 50 miles back that reads, "No gas for the next 120 miles." (I loved that sign. I should have stolen it years ago and put it up at every camp.) It is cold and clear, with a full moon and I am getting very tired now. I play games with myself – shut the lights off and see how far I can drive without headlights. I roll the windows down from time to time to blow away the cobwebs.

Something is fishy! Every time I roll down the right hand window I can hear a clicking noise, and it is getting worse. Is a wheel bearing packing it in? I hope I can reach Hearst before the problem becomes a real problem.

Finally, I bite the bullet – I must get out in the bitter cold and investigate. I pull over and start removing wheel disks one at a time. I get to the tire that was changed in Grand Marais, pull the disk, and three lug nuts fall to the ground!

When the guy slapped on the tire, he had put the nuts back on with an air impact gun and hadn't checked them for tightness. The two remaining nuts are halfway off. The threads and wheel holes are damaged a bit, but they tighten up OK. I am very lucky I stopped when I did.

I reach Timmins at 5 am, check into the Empire Hotel and hit those beautiful, comfortable sheets.

At 10 am someone hammers on my door – it is Carl Branch, my new party chief. We go down to the Empire dining room and I have a bite to eat and get the lowdown. We will be setting up a plywood frame tent on Nighthawk Lake and will do a mag and vertical loop survey on a 96-claim block in Langmuir Township. (96 claims = 96 line miles) This is a good-sized job and will take us to the end of February. We will fly out of South Porcupine tomorrow, so we put together the lumber and food we will need and make sure it is at the Austin Airways base, ready for an early departure. We have a few drinks at the Empire that night and hit the hay early – tomorrow will be a busy day. I have already left the Olds at a service station in South Porcupine. They will make sure it doesn't walk away.

On Jan 3rd we move into Nighthawk just a little southeast of Timmins and not far from the old gold mining town of South Porcupine. Nighthawk runs north-south for 25 miles and is perhaps seven or eight miles across at its widest part. We will camp about five miles from the south end, so it is a short 20-minute hop by Beaver.

It takes a few trips to ferry our stuff, but by noon everything is piled on the lake near our proposed campsite and building begins. Within an hour I realize I am in a bit of a pickle. I had expected Carl to come in and help build the camp, but it seems that he has other plans. He will bring in the Skidoo and sled on the morrow. It is up to me and the two men with me to set up camp. They are young fellows from Val D'Or – no slouches, but neither one of them has been in the bush before, and I am the only guy who has ever set up a frame tent. Normally even this would not be a big hurdle, but we have a language problem. André has some English, but is not really too fluent. Jean-Louis has no English at all – or if he has, he refuses to use it. (I will learn later that he is an ardent Seperatiste.) My pitiful high school French is pretty much useless - but we soldier on.

The other problem we have, (problems seem to always come in bunches) is that one of those freak weather fronts had moved through the night before. The temperature skyrocketed and we had an overnight dump of fresh, wet snow. When we left South Porcupine in the morning it was plus 45 Fahrenheit, and now the temperature is plummeting. By tomorrow morning it will be minus 30, a 75-degree change in 12 hours. Unbelievable, eh?

So now we are wet to our knees from walking through the wet snow, our lumber has wet snow stuck to almost every piece, and everything is freezing. Every piece has to be cleaned before it can be used. Normally I would have called a plane to take us back to town before dark and return to finish the job tomorrow, but a radio had been deemed unnecessary, and rightly so. We are only 20 miles out and will have a skidoo in campso emergencies can be handled without radio contact.

251

By 4 pm we have the floor down and the frame up and throw the tent over. It is getting dark and we have no time to put the laths on to hold the tent tightly to the frame, and we have yet to build the door. We need to install the stove and we need firewood. I have to put up the stove as Andre and Jean-Louis have never done so. I ask them to cut some firewood. "Sec," I say, "Make sure it's sec." They look at me like I'm nuts. Maybe it's because "sec" only works with wine.

I touch a tree in front of the tent. "Comme ca," je dit, "But dry. Comprenez-vous?" Maybe that will work.

I put up the stove while they try to start the power saw – no go. It had been in storage since the previous winter, and these boys seem to know nothing at all about saws. I am really frustrated, but I don't get angry. If I am mad at anyone, it is myself and Carl. This whole deal could have been avoided with a little common sense and foresight.

It is etting darker now, and I still have to nail the tent down to keep the wind out. I hand them the trusty old swede saw and point to the bush, then continue to batten down the hatches.

I have just finished making the tent half-assed livable when my two helpers return with the firewood. Billy-be-damned, they have brought in green wood! It is pitch black out now, and another wood-hunting excursion is impossible. It is going to take a gallon of gas to get that fire going without kindling, and I'm sure if I'd had a can of gas handy I would have immolated myself just to get warm. I say to hell with it and crawl into my eiderdown fully dressed. The other two do the same. Wea are alltoo tired to cook supper.

It is a long night of fitful sleep. In the morning I poke my nose out and there is three inches of frost built up around my breathing hole. I don't really want to get up and while considering my options, a plane comes in. There are voices and sounds of stuff being unloaded. I assume a crew from another company is heading for another claim block. I hope they will think the camp is empty and leave us alone, but no luck. Someone pokes their head in the door and laughs. All I want is for them to go away, but I hear sounds of tinkering outside the tent, and then the power saw starts up. Within ten minutes a guy comes in with a load of dry firewood and the airtight heater is soon dancing and throwing off heat like crazy. I crawl out of my sleeping bag to go outside and thank them, but they are gone – leaving a pile of dry firewood in front of our tent.

"Who were those masked men, anyway?" It turns out they are our line cutting crew, coming in to finish up. They had gone back to Quebec for Christmas and have returned to complete the last 30 miles of our 96-mile grid.

It is impossible for me to put into words the result of that kind favour done for us on that cold, frosty January morning. I would have eventually made it out of bed and found firewood, but my attitude and confidence were both at rock bottom that morning. I was pissed off at Carl and myself, and wondering if I really wanted to do this. There were lots of jobs around the Timmins bush that winter, and maybe I would fit in better elsewhere. These were the options I was exploring in my eiderdown nest that morning, but it is amazing what a warm fire will do for an attitude adjustment.

We dig out our Coleman stove and our frozen bacon and eggs and have a hearty breakfast – the first food we've eaten in 24 hours. We are back, baby! By the time Carl arrives with the skidoo we have the tent pretty well completed.

We follow the line cutters' tracks to a firewood bonanza. Just a hundred feet down the shoreline, high water had killed a grove of black ash. They had stood tall and dry for a few years, and all the bark has fallen off. There is no wood that I am aware of that is as hard as standing, water-killed ash. It's tough on saw chains – sparks fly when you buck it into stovewood lengths, but Mother Nature has stored a heap of energy in those closely-knit wood fibres.

So, every night, just before crawling into the sack, I fill the airtight to the top with round, unsplit ash and close the draft. Overnight the wood quietly turns into anthracite coal. I keep a lath by my bed, and in the morning, I reach over and poke open the draft. Within a few minutes that good little girl is doing her dance act, and we climb into our clothes in a toasty-warm tent.

Sidebar: Airtight heaters are made to have a good time, but not for a long time. They are made of tin, and despite having an inner liner, a good fire can turn them cherry red in no time flat. When the fire is really rolling the tin lid dances on the stove top. It is like having a cheerful friend in the tent. Unfortunately, your friend does not have a long lifespan – repeated red-hot fires will eventually burn out the walls. They never survive more than one winter's use.

Sidebar: Northern Quebec has always turned out men with a strong work ethic. They stake claims, cut lines and do contract surveys in all sorts of bush, living in basic throw-up and tear-down fly camps. They have never received proper recognition for their contribution to the Canadian mining exploration effort. I wonder how many mines are a direct result of their untiring efforts.

Sidebar: Bob's Handy Hints for Housewives Housed in Horribly Hostile Habitats: Keep your eggs and potatoes stored in a snowbank outside – they will still freeze a bit, but at least they are not as hard as rocks. We seldom peel our potatoes in the winter anyway, and if you don't like your eggs burnt on the outside and runny on the inside, you can always boil them.

The topography along the southeast shore of Nighthawk is very flat and our claim block is rectangular, 16 claims wide by six claims high (four miles by one and a half miles.) The east boundary barely touches the west side of Nighthawk, so there will be no lake work.

We are camped near the south boundary, and will have a short walk to get to the eastern part of the grid. There is a good skidoo trail running northwest from camp. It angles up through the west half of the claim group, and we will use it as we work our way westward. The baseline runs east-west, with north-south grid lines spaced every 400 feet. Add in tie lines on the north and south boundaries, and my mag survey will exceed 100 miles. The south half of the block is barely above lake level, but there is a small ridge near the centre, and the terrain rises in the northern portion. The whole area has obviously been cut over for mining timbers years ago and only a few stands of larger trees remain. The rest is second growth spruce and cedar, and some lower spots are thick with alders. This would be tough bush to work in without the benefit of picket lines. There is sure to be heavy overburden, and for that reason, the vertical loop has been chosen for its ability to penetrate the soil covering the bedrock. This is also why Carl is here. He is Mining Corp's vertical loop guy and is the senior party leader. Gerry, Raymond et al are somewhere in Northern Quebec with their J.E.M.

Carl starts on the west side, I start on the east and the work moves right along. Carl and the two young fellows are doing their EM thing, and I am doing my mag survey.

In the middle of January, Carl has a talk with me. He has two receivers, but no extra operator. He can train André on one receiver, but with two men reading, that leaves Jean-Louis on the transmitter, and it's a bit more complicated than the Inco method I learned (more about this later.) Carl's French is not much better than mine and he has difficulty getting the two-receiver method across to Jean-Louis. André can translate, but he is trying to get his own mind wrapped around the concept. Carl thinks that if I take Jean-Louis with me, he can take notes and speed up my job. When the mag survey is done, I can join him on the vertical loop. In the meantime he and André will continue on at a slower pace.

I point out that Jean-Louis and I have a language barrier. "It'll work out," he says, and it does.

So, the next day I take "Ti-Loup," as Jean-Louis is called, out with me. (It's pronounced tsee-loo and is a contraction of "petit loup," l'il wolf in English.) No one ever calls him Jean-Louis, so Ti-Loup it is.

Ti-Loup and I quickly become a team. He understands the grid system, and I soon have the "un, deux, trois thang" down pretty much pat. In camp at night I practise on André. On the picket lines I write the reading in my book and Ti-Loup gives me the French version. After four or five days things are clicking pretty good, and Ti-Loup takes over the notebook. My basic French seems adequate. "Reste tranquille" suffices for "break time," and anyone acquainted with The Three Musketeers knows that "Allons-y!" means "Let's give 'er shit. Pal!" (I've lost most of the French I gained that winter, but I can still count from one to infinity. You want to try me? Go ahead, give me a number – any number.)

The whole area is staked solid around Timmins – not a sliver of open ground exists. The total area covered by the rush is probably 70 square miles or more, meaning there are at least 78,000 claims, and they all need work. If you can shoot a compass and swing an axe you are welcome.

(I will spend three days in Timmins in early March and run into Lynn Richardson, an old high school friend who is working for Ontario Hydro. He and five other guys had staked six claims that day 35 miles north of town, and they are celebrating. Boy, are they celebrating! He told me later that they optioned the claims for $1,000, which came to $150 profit per man after they paid their staking costs and the bar bill. It is not bad for one day's work by guys who had never swung an axe in battle.)

With that many men working in the bush we can't help but meet some of them. One day I run into Jack Hodge on our north boundary. He and his partner

Garth are doing a contract geophysical survey on the claim block just north of us. We have a smoke break and a chat. I like the cut of his jib, and vice versa.

(I will run into Jack more than once in the years to come, and in 1996 I will be sitting in his living room with my memories as he takes his last breath in his bedroom just down the hall. Jack was a honest, hard-working man, a good father and a good friend. He was quiet, dependable, and never self-important. He died surrounded by family and friends. What more can you ask from life? And how many life-long friendships have started with a smoke break on a snowshoe trail? More than a few, I'll wager.)

A couple of nights later I borrow the skidoo and buzz over to have a cup of tea with Jack and Garth.

"We are camped on an island," Jack had said, "You can't miss us."

Well, the headlight picks up the island, but no tent, just a big old boat perched high and dry on a rock, surrounded by a few scrawny spruce. I see a light inside and a Sno-Cruiser parked out front, so I investigate. I climb a short ladder to the deck, knock on the door to the cabin and walk in.

What a cool camp! This is a fairly large boat, probably used to move people and freight on Nighthawk back in the day, and someone has converted it into a cottage of sorts. Jack and Garth have rented it for a couple of months, and is it ever comfortable looking. A short stairway leads down into the cabin. Although not too spacious, there is room for Jack, Garth, and three line cutters.

The line cutters are from Dacre, a small town west of Renfrew in the Ottawa Valley. They are farm boys, familiar with hard work, and when a family friend mentioned that line cutters were needed at Timmins, they jumped at the chance. Not a one had any bush experience other than helping Dad cut firewood. They arrived in Timmins with sleeping bags, shiny new snowshoes and axes, had a two-hour tutorial over a cup of coffee, then were turned loose on Nighthawk. They were given what I'd call a tourist tent, a 9 x 12 footer with a floor – not something I'd like to live in in the winter, but with a staking rush on, lots of things were in short supply. They were also given an airtight stove and a pail of sand.

Now, when you set up an airtight, you can start out with small fires until you build up a bed of ashes, or you can cover the bottom of the stove with an inch of sand or dirt. They were not really sure what to do with the sand, so they spread it on the floor under the stove, lit a roaring fire and hit the sack. Two hours later the bottom of the stove melted and they awoke to find the nylon floor of the tent on fire. They managed to get their supplies and personal stuff out, but the tent went up like so much tissue paper. The next morning Jack and Garth came by and found them huddled beside an open campfire, and invited them to share the boat.

I run into them from time-to time on our shared boundary line, and I get to know and like them. They were neighbourhood buddies and ran free in their puppy-hood days. Now at eighteen, they are an amalgam of derring-do, come-what-may, with a good dose of work ethic to get them over the rough spots. I get a kick out of their attitude – had I been thrown into the shield with no Charlie McLeod Jr. life preserver I doubt that I would have survived as well as these boys!

255

One day one of the boys tells me he had lost control of his picket line and had cut a tangent – crossing two previously cut lines. He was just too stubborn to say "Whoah!" The next time I run into Tom I make a mild suggestion. Perhaps they should pay more attention to their back-sites – sometimes slower is faster.

Table Scrap

The leader of the wolf pack is Tom O'Reilly and the other two defer to him. They had been given a new Mercury Sno-Cruiser, and the first-edition Sno-Cruisers had a drive belt issue.

One day Tom and one partner go to South Porcupine on a grocery run, using the old tote road through our property. With the grub on the sleigh they stop at the pub for a jar or two and close the joint. On the way back they miss the turnoff to our camp – easy enough to do in the dark, and the tote road complex is well travelled this winter. They hit Nighthawk at the south end and the Cruiser blows the drive belt.

They have no axe and no way to make a fire. They must walk home – five miles up the lake to our camp – six miles to their island. There is a skidoo trail on the lake. It has drifted in somewhat but is walkable, although with no snowshoes it is a tough go.

Tom's partner is whipped and wants to rest. "Just for a little while," he pleads. Tom knows that if they sleep they may never wake up. All night Tom keeps the kid going, using threats of a licking when necessary.

At 7 0'clock I am gassing up our sled and I see two guys out on the lake. I toodle out to see what's up and find two half-frozen young pups on their last legs. In ten minutes I have them at their warm houseboat.

If their moms were there, those boys would have gotten a slap and a hug – I figure Tom deserves an "Attaboy!"

Table Scrap

You don't have to be a rookie to torch a tent. In 1970 during the Umex/Pickle Lake rush I hired a staker who told me this story.

He was on a small staking job in Quebec with two other guys. They expected to be there only three or four days, so they threw up a quick camp with no fly. The tent was sort of an oddball – canvas walls and ends, with a sailsilk roof. That night, a spark from the airtight set the sailsilk alight. Ray jumped into his boots, yelled at the other two, ran outside, grabbed a shovel and started to throw snow on the roof. The other two guys each grabbed a snowshoe and did likewise.

Well, you can't do much with burning sailsilk. The roof completely disintegrated, but the canvas walls were still standing. They went back into the tent to find their efforts had filled their open sleeping bags with snow. They laughed like crazy, shook out the snow, stoked up the airtight and slept under the stars that night – no problem. They used the remaining parts of the tent to build a wigwam and finished the job – plan B.

Our work goes along with one minor hiccup. Our skidoo blows a tire! One day the track separates and there we are sitting on the bogey wheels with the track laid out neatly behind us. No problem. We lay the machine on its side, put the track back on and lace the broken ends together with lampwick. (I use lampwick as snowshoe harness and always carry a spare roll in my pocket in case I blow a shoe.)

256

We make it back to camp, and since the next day is grub day we ask the pilot to find a new track and bring it in forthwith. It turns out it has to be ordered and we don't get it until three days later. We are working at the far end of the property by now, and don't relish a three-mile walk to work. We replace the lampwick lacing with strong nylon rope, and with careful use of the throttle, make our trusty sled last three more days.

The new track arrives and now the real fun begins. We are in the middle of a cold snap and have to install the track in -30F weather. The sled won't really fit in the tent, but if we lay it in its side, we can slide the back half in the door. At least we keep our bums warm part of the time. To get the one-piece track on we have to remove all the bogeys and the front and rear carrier axles. We put the front drive sprockets and bogeys back on without too much trouble, then struggle to get the rear sprocket assembly's mounting holes to line up. In the cold, the track is approximately a half-inch shorter than it would be if we were in a nice warm shop, and these early Skidoos have no adjustment for track tension at all. The bolt holes just won't line up and it takes us until 4 pm to finish. We had started the job with a full five-gallon pail of cuss words, but ran out and had to order extras. Once the job is done, we sit back and say. "No problem."

Ti-Loup and I finish the mag survey and we join Carl and André on the vertical loop. We are really picking up speed now, and anticipating a short break in Timmins before we move our camp to the last grid.

Mining Corp uses the McPhar One/Five dual frequency vertical loop (1000 cycles per second/5000 cps) It is similar to the unit I was accustomed to in my Inco days, but considerably more sophisticated. The two frequencies give you extra information for interpretation of results.

The McPhar set-up is also similar to the Inco unit with a unipod tied off with ropes and holding the transmitter coil. It is quite powerful, and since we are reading up to four lines on each side, voice orientation is impossible. We work on a time system, and it can be quite tricky.

An orientation board is clamped to the vertical pole of the unipod. This is an aluminium plate 18 inches square, with an arrow on a swivel in the centre of the plate. On two sides of the plate are marks indicating where the receiver guy will be as he moves along his line, so you have four lines with marks indicating 100-foot stations. (For example, if the transmitter is set up at 10 north and the reading on the next line is also at 10N, then the arrow will be at right angles to the setup line.)

First you plumb the board. A plumb line hangs from the apex of the transmitter coil, and you centre the board under the plumb bob and carefully line the board up with the pickets on the setup line. This ensures that everything is level and copacetic. If the arrow is pointing at the station where the receiver should be, you can be confident that the readings taken are faithful and true.

Now the real fun begins. With two receiver guys working, one east and one west, we must start at opposite ends of our lines. To simplify things, let's say we are set up on the base line: the receiving guys will read 2000 feet on each line, so I go to 10 north on my line and Carl goes to 10 south on his line. With the arrow lined up on the appropriate indication on the board, Carl and I can take our readings simultaneously, and he will then move 100 feet north and I move 100 feet south. Now, here is where a time factor comes in. André is on the motor/generator, Ti-Loup points the coil and makes sure it is lined up with the arrow and holds it steady. Carl and I head out to our first stations. Andre has a watch and gives us ten minutes to get there, then he starts the motor and flips the toggle switch on and off three times. This lets us know that we can expect the first reading. Now André runs the motor for one minute – no more, no less. We have 30 seconds to read each frequency, then we boot it to the next station. André gives us one minute to get there, and we repeat the process.

At the midway point on the line André gives the toggle switch signal again. This lets Carl and I know that we are still on track. When the lines are completed, André gives us five minutes to cross to our respective adjoining lines, and the whole deal starts again until we have each read four lines. An eight-line set-up takes four hours, start to finish. A quick lunch, another set-up and we have put in a good day with six miles of grid covered. (Talk about stress city!)

Of course, there is always a back line (set-up line) to read, so by mix-and-match, we can make some days less strenuous. Carl and André have done half of the EM survey while Ti-Loup and I were doing the mag, so in five days we complete the initial survey.

The time system works better than I expected, but stuff happens. You can blow a snowshoe, for instance. Lampwick tends to last about four days of steady snowshoeing, so to reduce blowouts, I put on fresh lampwick every morning. Another danger is that sometimes a shoe will snag on a cut off willow stump on the line, and you will crash and burn. No problem – you flounder around in the deep snow, take a quick inventory, and double-time it! Sometimes you may miss a reading before you catch up, but it's not the end of the world. If there is a conductor in the area an occasional missed reading is inconsequential. You will be detailing it later anyhow, and when detailing, you use sound orientation. (Detailing will be necessary because we have picked up an interesting conductive zone in the west/central portion of the claim group.)

The basic EM coverage is finished, and now only three days of EM detailing are left to do. While Carl, André and Ti-Loup are taking care of that, I get to baby-sit the gravity meter guy who has come in from Toronto. His survey will add info on the conductive zone.

The gravity meter guy is Dieter, who works for a geophysical contracting outfit. Dieter is very much a technocrat - highly intelligent, and more than a little weird. The gravity survey is very technical, requires a highly intelligent operator, and is also very weird - the job fits the man. Dieter spends the first night in camp explaining the theory of the gravity meter survey and I spend the evening with visions of sugar plums dancing in my head. We will make a good team.

All I can get out of the tutorial is that you have to imagine an ore body beneath your feet with a mass/weight ratio exceeding that of the lighter less dense grano-diorite country rock surrounding it. (See I caught you – you are thinking of dancing sugar plum fairies too, aren't you?)

Anyway, the gravity meter picks up info and Dieter logs it in his book of trolls.

I am made very aware of the delicate nature of this unit. I am told that the main sensor is a quartz coil spring sort of thingy, cut from solid quartz, hanging in a magical tube within. Infinitely small variations in gravity are measured by this dohicky, and great care must be taken when handling it. Evil spirits can be released if it is bumped, jostled, dropped or held upside down. It travels in an aluminium case shaped like a five gallon milk can with form-fitting styrofoam wrapped around its precious body, and it's worth two gazillion smackers! I am afraid to touch it.

We head out to the bush, Dieter carrying his baby in his arms, carefully supporting it's head. I carry my axe, a shovel, our lunches, the tea pail, and the collapsible tripod deal the gravity meter hangs from. I feel like I am the old prospector's burro. The only things missing are the pickaxe and gold pan.

The setup is carefully prepared. It must be level and firm. All branches and other detritus are carefully cleared away and a bed of snow is packed down and the gravity meter set up. I am told that if we are in a treed area that it must be a dead calm day. If a breeze causes a tree to sway, the root movement can throw off the reading, and I am cautioned to hold still and not shift my weight from one foot to the other. Holy cow! I back off about 15 feet, hold my breath and tighten up all orifices. One healthy fart could set that quartz spring bouncing for days!

We don't have very much real work to do, we will read 500 feet of grid line over the conductor on five lines, giving us 2000 feet of strike length to cover. It is the most boring three days I have ever spent in the bush, but the survey is successfully completed.

We are pulling out of Nighthawk! While Dieter and I have been gravity metering, Carl and the boys have been hauling stuff to South Porcupine on the Ski-doo trail. On the last day, we load Dieter and his gear on the sled and Carl takes him to South Porcupine with Dieter holding his baby in his arms. André, Ti-Loup and I pull the tent off the frame and leave its skeleton on the shores of Nighthawk. We are picked up by the Beaver and return to South Porcupine. André and Ti-Loup head for Val D`Or for three days and Carl drives to Toronto or somewhere, I guess. I have nowhere to go, so I spend the next three days in the Empire Hotel trying to stay out of trouble.

Table Scrap

Carl told me afterwards that when he and Dieter reached the station wagon that day, they cracked a bottle of rum and spent an hour toasting each other, and we all know how a few shots can clear your mind and allow you to make good decisions. They decided that rather than take two station wagon loads to the Austin Airways hanger, Dieter would follow Carl by skidoo on the five-mile trip on a snow packed road. I guess the rum befuddled the technocrat because he dumped the sled right in the middle of the road. Carl said the gravity meter fell off the sleigh and rolled 200 feet down the road until it veered off and hit a tree. I bet that quartz spring thingy is still bouncing!

Side bar: It turns out that we had found a mine on that claim block in Langmuir Township! Within a year or two a follow up drill program would prove up the ore body and the McWatters Mine would come into production. The deal, as I understood it, was that the claims had been staked on an Inco airborne anomaly and an arrangement had been made with Noranda (Mining Corp) to work it. If a mine was found then a

259

company would be formed to develop it. If it turned out to be nickel, it would be shared 60/40 with Inco being the 60 - if copper, then Noranda would get the 60.

Table Scrap

During one of my tea breaks with Jack at his boat on Nighthawk, he mentioned that he had run into a guy working at an adjoining property on the eastern shore of the lake. Jack brought my name up and it turned out the guy knew me, so one evening I took our Ski-Doo across Nighthawk and found the guy in a cabin he had rented.

It turned out to be Merv, the stump sitter of Bisset in 1962. He was still full of BS, but what the heck. I had a cup of tea and listened to him brag on how much money he was making. He was on contract for a promoter, and was getting paid by the mile for a mag survey. Things did not add up – either he was getting paid a phenomenal rate, or he was traveling at 20 miles an hour. I kept picking at him. How could he produce so much work?

"It's easy." said Merv, "I read every third line and fill in the rest. It's just moose pasture anyway."

I was disgusted. Once a stump reader, always a stump reader, I guess.

The problem as I see it is this; it might be moose pasture all right, but a shoddy survey does no one any favours. Mines have been found in moose pasture. If Merv's crappy job writes off a property, it may be years before interest can be rekindled. I know that our survey and that of Jack and Garth's are legit, but in times like these, others may be doing the same thing as Merv. Even today, new discoveries are being developed in the Timmins camp. How many of these are in so-called moose pasture, I wonder?

Sidebar: Moose pasture is a derogatory term to describe claims staked on ground of limited potential. In every stalking rush, claims on the periphery of the strike will be extolled by some promoter to be, "On strike with such and such," or "In an area of significant interest." These promoters are scorned by smart investors, but guess what? Mines can be and are found in moose pasture, and dumb investors suddenly become smart. In Merv's case, the property that he was jerking around was in fact down-strike from two future mines.

The man in the field has no business making a judgment call. Do the work, buddy. Let things sort themselves out.

A Three Day Break

I pick up the Olds and check into the Empire Hotel for three days. A nice hot shower gets rid of two months buildup of crud, then another nice hot shower leaves me clean and refreshed. I will find a laundromat before I head back into the bush but right now its time for a little R&R. I head for the bar.

Timmins is a hot spot this winter. The bars are busy and Stompin' Tom Connors is holding down the stage at the Maple Leaf Hotel just around the corner. It is impossible to get a seat in the Maple Leaf, but the radio is playing his songs. "I Got a New One in Rouyn" and another tells about life "Way Down in the Hollinger Mine." The Texas Gulf rush made Stompin' Tom, proving that there is more than one way to hit paydirt.

The rush created as many problems as it solved, as I learn. One of the first people I meet in the bar is Luke, our chopper pilot at the end of the winter season in '62 at Chibougamau. Luke is no longer his happy self – In fact he looks absolutely terrible, and tells me a sad tale of woe.

It seems that he had been under contract to Texas Gulf, flying men and gear into the drill sites and flying core boxes out. One day the head geologist loaded some core boxes onto the cargo racks, climbed into the chopper and told Luke to buy every bit of Texas Gulf stock he could manage. The core boxes, it turned out, held a humongous intersection of massive sulphides with phenomenal ore grades. Luke sank everything into Texas Gulf, made a huge pile of money, ($40,000) a huge pile of so-called friends, and partied hugely!

Now, six months later, he is flat broke and unemployed. He asks me to lend him a 100 bucks for old time's sake, which I do. Luke invites me to join him and his friends for drinks, which I also do. We go to some seedy bar with hard looking women and seedy types dressed up in mafia good-will suits. Luke is proud to introduce me. I am not quite so enthused, and neither is the gang. They soon write me off as a bit player and it is obvious that Luke has outlived his usefulness. I soon remember a previous commitment and make my escape.

(That hundred bucks was going to be sent home to Mom. I mail her 50 and explain that the other half has been lent to a friend in need. Mom writes back that it is OK, you should always help friends. I never did tell her that I may just as well have lit a match to that hundred bucks. I never saw Luke again, but I hope things changed for the better in his life.)

I have a beer in the Empire the next afternoon and it's pretty quiet. I strike up a conversation with a gent at the next table and we sit together for an hour or two. It turns out he is a geologist for Phelps-Dodge and I'm darned if I can remember his name. He seems like a pretty good guy. He has been there, done that, and I pick his brain a little. I have seen and learned enough to make me feel that there are possibilities for me to go contracting on my own and I ask his opinion. He tells me to avoid the lure of the role of entrepreneur – in other words, do the work yourself – don't trust hired men to do it for you. Six years later I will remember that conversation in the Empire Bar. Too bad I didn't listen better.

I also ask him about the old fellow I see in the Empire lobby from time to time. The gent is in his late 60`s, I estimate, and very dapperly dressed, with a floral waist coat/vest and an impressive gold pocket watch and chain.

"Does he wear a tie with a diamond stick-pin? The Phelps-Dodge guy asks. I confirm that.

"Why that's Diamond Jim! He says, "He's a local legend!"

So here is the story – truth or legend? You be the judge.

Table Scrap: Diamond Jim.

It seemed that Jim had been an insurance salesman and for sure a good one. He blew into Timmins in the late 40`s and sold the first group insurance policy to a Timmins mine, probably the Hollinger. Others, such as the Paymaster followed suit, and no doubt at least some of them were still Jim's policy holders. Jim soon retired on his residuals,

261

bought his stick-pin and became Diamond Jim. He spent his days at the local stock broker's and hobnobbed with the elite and not-so-elite in the lobby of the Empire. I had chatted with him a bit, and I thought he was a pretty good guy.

Anyway, it came to pass that sometime in the 50`s a prospector buttonholed Jim in the lobby and told him of a good looking strike he had found, and asked Jim who he should go to with his samples.

"Why here is just the man to talk to," says Jim, and introduced him to a mining big-wig who happened to be walking past.

As it turned out the prospector had a mine in his pocket. Sometime later with the mine nearing production status, Jim ran into the prospector.

"You owe me some money!" says Jim.

"Like hell I do," says the prospector.

Jim took him to court, was awarded a 10% finder's fee, and a precedent was set.

Table Scrap

I also run into an Inco airborne guy in Timmins. He tells me they are flying from an ice strip near Foleyet, a small lumber mill town 40 miles west, and the man in charge is Bill Hurley. I know Bill and we get along well, so that evening I drive to Foleyet, find the tourist camp they have rented and share a few jars with Bill.

Bill tells me that he is a taking a year off (with Inco`s blessing) and joining two geologists/professional engineers who have secured a contract in Mexico. They will set up an exploration department for the Mexican government, with an airborne division and ground crew follow up. Bill will head the airborne and they need good men on the ground to train Mexicans in the art of anomaly chasing. Bill says the wages are unreal and since he has considerable stroke in the operation, he would like me to come down to Mexico with him. I ask Bill who the two principals involved are. One of them is (identity withheld), an ex-Inco area supervisor. I decline the offer. I had worked for that man in Shebandowan/Marathon in '61 and I would not go near him with a ten-foot sombrero.

We are headed into our last Timmins grid, after the most interesting three days off I have ever experienced. I am happy to be going back to work – I need the rest.

We are going in by Skidoo. There are lots and lots of good winter trails in this area of flat swampy spruce bush, and André and I take in the first load. It's about sixteen miles from road's end west of town, and when we arrive on the claim block we find an abandoned cabin right in the middle of our grid. We scope it out.

Now, you have to put yourselves in our snowshoes. After two months of living cheek to cheek in a tent, any kind of shack looks like the Royal York. This one even has windows and a door! Inside it's not too bad at all. It's twice the size of our tent, with a big old barrel stove with intact pipes, a good solid floor, kitchen counter, shelves, and a decent table with a couple of chairs. Utopia! (In 1965, Skidoos were used for work, rather than recreation, and the remoteness of the cabin has saved it from vandalism.) I send André back to Carl with a note: "Hold the tent and bring in the groceries, we have a motel room!"

It needs a nail or two for some floor boards, but by supper time with the floor swept and a nice hot fire going, we are snug as bugs in a rug. There is plenty of dry spruce nearby for firewood, and we have a wood shed. This is living, man! We can even get a weather check without going outside. On a clear night we can see the stars through the holes in the tar-paper roof.

The weather is great. We are into the last two weeks of March, and every day is sunny with crispy cold and clear nights. We hit the bush at 6 am every day, and in ten days we are finished, dang it anyhow. One of the best winter camps I have ever been in, and we are done already! Such is life.

We pull out just before the end of March. Carl will drop Ti-Loup and André at Val D'Or, and continue on to spend some time with his future wife. I will overnight in Timmins and drive to Mattagami, Quebec the next day, where I will join Gerry and Raymond to finish out the season.

(When I had picked up the freshly serviced Olds for my three-day break, the service station guy told me it needed a $440 valve job. What a bummer. I have neither the money nor the time, so Olds will just have to suck it up.)

When I leave Timmins it is running pretty good, with just an occasional miss, and still cruises pretty good at 70 miles per. I cross into La Belle Province well before lunch time, and continue on another 40 miles past Rouyn to Highway 109, where I hang a left to head for Amos and thence on to Mattagami.

East of Rouyn, I get to expand my French/English dictionary. All the road signs are in French – no problem. It's a lovely morning, traffic is light and I'm really cruising. A sign says, "Plage 500 pieds." I pass a little lake with a small beach. OK, "plage" is beach, and I know "500 pieds" is 500 feet.

The next sign says, "Cahot 200 pieds" in great BIG letters. What is a "cahot" I wonder?

Two hundred pieds later, I sure as hell find out. I hit a huge frost heave at 70, get to see some country-side from a higher point of view, and when the Olds returns to earth, she bottoms out and sparks fly out from under both sides. Everything seems to be in place, so I give her a pat on the dash, apologise and continue on to Mattagami – a slower, wiser and more bilingual man.

Mattagami is 80 miles north of Amos, and exists because of the Mattagami Lake Mine, another subsidiary of Noranda Mines. Most of the road is a straight shot, but I resist the urge to highball it. As usual, there has been tons of snow in Northern Quebec, and the banks are high on each side of the road. It is just after 12 noon, and the sun is warm enough to melt the snow, but the meltwater has no place to run off so I'm driving through ponds deep enough to give the car a good soaking. I arrive at the Mattagami Inn, force the driver's door open, and find the Olds encased in ice from the beltline down. The old girl is squatting like an LA low-rider, and if I wanted to lift the hood (which I don't) I would need a blowtorch – the whole front end is solid ice. I know there are headlights and a grille under there somewhere, if they didn't fall off back there at the cahot.

No problem. The sun will soon melt the ice, and since the trunk opens ok, I haul out my packsack and go to check in. Nope, it's Sund and there is no one at the front desk.

263

Again, no problem. Gerry and Raymond are staying here and I'll just have to wait for them to come out of the bush. I see a door leading somewhere, find it is unlocked, and enter the bar.

"Good old Quebec," I think. "Maybe I can find a bite to eat."

It's a nice looking bar, but practically deserted. Gee whiz – Is everyone in church this afternoon? There is only one other customer, a middle-aged guy, and he is either drunk as a skunk or goofy, it's hard to tell which. He is having a conversation with himself, and it sounds like his ying is mad at his yang. I pick a table as far away as possible. I later learn that he is an underground miner, and a good one, but he spends his days off in a semi-stupor. What a life!

The waitress has pretty good English, thank goodness, but to "manger" is "pas possible." The dining room doesn't open until 5 pm, so I order a beer and sit back to wait for the guys.

It's only 2 pm, and man, am I hungry – breakfast at Timmins was toast and coffee, and that was seven hours ago. I look around the bar for a peanut machine or something. No such luck – but every table, including mine, has a saucer in the middle with a cheese sandwich cut into quarters, along with a little dill pickle. The cheese is kind of curled up where it sticks out past the bread, and the pickle is a little bit shrivelled up, but what the hell – I'll eat anything.

So I take a bite, and it's just like rubber between two slices of asbestos. I soldier on. I chomp on another quarter and order another beer. The girl brings it, looks at the plate and stands back, both hands on her hips, and fire in her eyes. Holy cow! I'm in trouble again! "You hate my sandwich!" She cries, "That sandwich not for heat!"

She grabs the plate and stomps off to the kitchen and brings back another. "Don't you heat that!" she says, and I promise not to. I don't get any more beer, either.

If not for heat, why the sandwich? I find out that local laws allow for Sunday opening as long as food is available. The bar is acting to the letter, if not the spirit of the law. I assume they make up the sandwiches every New Year's Day and probably shellac them. They no doubt will turn green by St Patrick's Day, and are replaced quarterly.

The rest of my stay in Mattagami is uneventful. As usual we go out early to catch the frozen snow. Then we get a cold snap and finish the job by April 15th, and it's time-off time again: Winter is over.

The End of Book One

The doggone book just got to be <u>too</u> long – we had to cut it in half! Stay tuned for Book Two, subtitled,

How to Find Your Way Out of the Bush Without a Compass

Heading off for Book Two

Glossary
(For more information, go google.)

Ambroid: A tube of material which hardens when exposed to air: used to repair cedar-strip, canvas-covered canoes.

Azimuth: Compass direction, usually in terms of degrees of the compass, i.e.: 30 degrees east of true north (our Contwoyto Lake declination.)

Bombardier: Bombardier Auto-Neige, a Quebec company, manufactured two enclosed types of snow vehicles. They made two sizes, narrow track and wide track. A Chrysler flat head six-cylinder motor, located in the rear of the unit, powered them. A three-speed transmission coupled to a high-numbered (low ratio) differential delivered power to sprockets at the rear of a rubber belted track on each side. These rubber belts had cross-cleats, which rolled on solid rubber tires. At the front were two skis on knee action struts to steer the thing with, although the turning circle was anything but tight. When running on a previously broken bush trail the skis followed the track fairly well, but if you were breaking a fresh track the sharper corners required a lot of back-and-forth. Two bucket seats up front held the driver and one passenger, the rest of us sat on two canvass upholstered benches on each side over the track tunnels, with another upholstered bench across the rear in front of the enclosed engine. It would seat eight men comfortably. On cold days we could open a hatch on the engine compartment to get more heat. The centre floor had lots of room for gear or cargo. Top speed was 30 mph, but bush road conditions seldom allowed more than 15 to 20 mph. It was pretty darn comfortable, really, unless you found it necessary to use a road frequented by Muskeg tractors – these roads were more of the roller coaster variety.

Canadian Nickel had seven of these units. Numbers 2 to 6 were used regularly, #7 was a standby unit. Number 1 was Hector's personal machine. Hector was the road commissioner/maintenance chief. I won't say much about him, but the fact that he felt he was entitled to his own personal machine says it all.

Bore Hole: Diamond drill hole

Bushwhacker: Often used as a pejorative, for all those who make their living slogging through muskeg while fighting off black flies and mosquitoes.

Canadian Nickel: Inco's Manitoba exploration wing.

Candled Ice: In the spring the melting ice will get vertical holes in it, looking like closely packed candles: also known as rotten ice. Stay Off!

Chalcopyrite: A copper/iron sulfide mineral.

Churchill Line: CN built this spur line to the newly created port of Churchill, Manitoba in the '50's, later owned by Omnitrax.

266

Cookee: The cook's helper and all around good guy, in charge of washing dishes, carrying water, cleaning counters, sweeping floors and carrying garbage.

Contact: The joint-plane where two different formations meet. Ore bodies frequently occur along contacts.

CNR/CN/Canadian National Railways: Once Canada's National rail line, now owned primarily by Bill Gates.

Dike: Any mass of igneous rock, which, in a state of fusion, has entered a fissure in other rocks and has chilled and solidified. Dikes are usually vertical and cross-strike.

DOT: Federal Department of Transport: among other duties, the DOT maintains national regulations and standards for the flying part of Canadian life.

Float: Loose scattered rocks, not easily identified as to source: If a prospector can establish glacial flow direction, he may be able to find the source – the smoother the float, the farther it has traveled. When Grandpa was picking stones in the barley field, he was harvesting float.

Fold/folding: A roll or bending in rock structure.

Gossan: The iron-bearing deposit filling the upper parts of veins or covering masses of pyrite. Oxidization of the iron component of the deposit lends it a rusty colour.

Muskeg Tractor: This was also built by Bombardier. It had wider tracks than the snow machine (generally referred to in this text as the Bombardier), with two rows of bogeys on each side, and was steered by hand clutches like a caterpillar tractor. A cargo deck on each side sat over the track tunnels. Inco used the J-5: The driver sat low in the centre, between the two cargo decks and in front of the motor, and a canopy swung back to access the "cockpit." It was geared lower than the Bombardier – top speed 15 mph. It was used extensively by Midwest and other drilling companies winter and summer. It had good flotation on muskeg, hence the name.

These machines – snow and muskeg – were the mainstay of the exploration camps. The snow machine is still widely used by commercial fishermen on the lake ice and you'll also find them in the tourist industry. The Skidoo was still being refined and I would not see them in camp until the winter of '62 in Northern Quebec.

NFB: National Film Board of Canada – funded by tax dollars.

Packsack Diamond Drill: Developed and manufactured by Reg Minogue in North Bay, Ontario, it was meant for shallow drilling to test anomalies or surface showings. It was basically a 7 hp chain saw motor with the chain sprocket replaced by a heat-treated coupling to drive the drill string – five-foot sections of 1 1/4 inch hollow steel rod. A five-foot core barrel, reaming shell and diamond set drill bit, preceded the rods down

267

the hole. A swivel tee between the motor and drill string allowed for water to be pumped down the hollow rods to the drill bit, to flush away rock cuttings and bring them up the outside of the hole to the surface, as well as providing cooling and lubrication for the process. The diamond faced hollow drill bit and reaming shell kept the hole constant and made a rock core, which would fill the five-foot core barrel. When the core barrel was full, the crew would pull the drill string by hand, remove the 7/8" rock core and store those from the target area in core boxes. (This is basically the same principal used by the larger drills, the only real difference being the larger drill string is pulled by a winch.)

A "Johnson Bar" was used to apply pressure to the bit while drilling. It was simply a steel bar five feet long with a yoke on the drill, attached by chain to a hook fixed to a convenient stump or rock bolt near the hole collar. One man held the drill motor steady while another applied downward pressure on the long end of the Johnson Bar. As the hole progressed, more sections of drill rod were added.

Casing, steel tubing slightly larger than the drill string, also with a diamond encrusted "shoe," would first be driven through any overburden to bedrock to seal the hole, keeping gavel, dirt and small rocks from falling in.

Supposedly good for 200 feet, it was seldom used past 100. The steel rods were heavy to pull, and lack of power to overcome downhole friction limited the depth. In extremely hard rock such as quartzite, it was difficult to keep enough down pressure to avoid polishing the diamonds: Then the string would have to be pulled, and the bit replaced with a fresh one, the old one being sent back to the manufacturer to have the diamonds reset with fresh, sharp points protruding.

Deep overburden and boulders could also negate the Packsack drill's effectiveness, once again because of lack of power.

The drill was hard on the motorman, a position that was often traded off during the workday. Your hands would continue to tingle long after the supper dishes were done.

By 1964 the Winkie drill would replace the Packsack drill. But The Packsack drill did its job well within limits and was a good prospecting tool. Some are still to be found in use by low-budget prospectors.

Spotting the Borehole: A picket would be planted at the hole location. On it one would write the azimuth and the angle of the hole to be drilled.

Strike: General direction of the rock formation in an area.

Step: Aircraft floats are made with a break in the bottom surface about half-way down, with the rear bottom surface about two inches higher than the front. As the plane picks up speed and some lift is generated, the higher rear surface lifts out of the water, reducing the amount of drag and surface tension so the plane can accelerate to flying speed.

Swash Plate: The plate fastened to the rotor hub on a helicopter that controls the angle of attack of the blades, increasing it on the backward sweep and decreasing it on the forward sweep of the blades' travel during forward motion. This gives uniform lift on both sides of the craft. It is a major wear area, and is also very critical to the safe

268

operation of the craft. In cold weather the wear and stress can easily cause failure, hence rotorcraft such as the Bell G series had lower limits on the temperature they could be flown in.

Test Tube Etching: A mild acid/water solution is inserted in a glass tube which is lowered into the drill hole and left for a short period; the acid etches the glass, and when inspected shows the angle of the hole from vertical.

Wobble/wobbling: an unscheduled, unplanned and usually illegal work stoppage, most often of a unionized work force.

****

This Book can be purchased from Amazon.com

****

44451351R00149

Made in the USA
Lexington, KY
10 July 2019